1994

Handbook of

Techniques for Aquatic Sediments Sampling

Second Edition

Handbook of

Techniques for Aquatic Sediments Sampling

Second Edition

Edited by
Alena Mudroch, M.Sc.
Research Scientist
Sediment Remediation
Aquatic Ecosystem Restoration Branch
National Water Research Institute
Canada Centre for Inland Waters
Burlington, Ontario, Canada

Scott D. MacKnight, Ph.D.
President
Land & Sea Environmental Consultants Ltd.
Dartmouth, Nova Scotia, Canada

Library of Congress Cataloging-in-Publication Data

Handbook of techniques for aquatic sediments sampling / editors,
Alena Mudroch, Scott D. MacKnight. — 2nd ed.
p. cm.
Rev. ed. of: CRC handbook of techniques for aquatic sediments sampling. 1991.
Includes bibliographical references and index.
ISBN 1-56670-027-2
1. Marine sediments—Sampling. 2. Lake sediments—Sampling.
I. Mudroch, Alena. II. MacKnight, Scott, 1949—
III. CRC handbook of techniques for aquatic sediments sampling.
GC380.2.S28H36 1994
551.46´083—dc20 94-14255
CIP

International Standard Book Number 1-56670-027-2
Library of Congress Card Number 94-14255
Printed in the United States of America 1 2 3 4 5 6 7 8 9 0
Printed on acid-free paper

DISCLAIMER

The information on the equipment, mention of trade names or commercial products is provided as a guide and should not be construed as an endorsement of any particular device.

PREFACE

Recently, contaminated sediments in rivers, lakes, and oceans have become a worldwide issue. It has been shown that sediment-associated contaminants can be transported by resuspension of sediment particles, may accumulate in the food chain or affect the health of biota and water quality in aquatic environments. Therefore, different techniques have been developed for remediation of contaminated sediments in many countries. In North America alone the amount of sediments dredged annually to maintain commercial access to the ports and harbors is in the order of 600 million cubic meters at an associated cost of approximately five billion dollars.

Most of the dredging projects and assessment of impact of contaminated sediments on aquatic environments require collection of sediment samples to adequately define the physical and chemical characteristics of the sediments, transport of sediment-associated contaminants by resuspension of sediment particles or migration through sediment pore water, and test the effects of contaminated sediments on biota. At present, there is no comprehensive monograph on sampling of bottom and suspended sediments and sediment pore water, and on handling of recovered samples prior to physico-chemical analyses and other tests. This book was written to provide the essential background information on these subjects to those interested in defining the physical and chemical characteristics of aquatic sediments and evaluating effects of contaminated sediments on aquatic ecosystems.

Alena Mudroch
Scott D. MacKnight

EDITORS

Alena Mudroch, M.Sc., is a Research Scientist with the National Water Research Institute, Environment Canada, Burlington, Ontario.

Mrs. Mudroch graduated with a diploma from the Chemistry Department, State College, Prague, Czech Republic, and obtained her M.Sc. degree in 1974 from the Department of Geology, McMaster University, Hamilton, Ontario.

She is a member of the Canadian Association on Water Pollution Research and Control and of the International Association for Great Lakes Research (IAGLR). She served as 1989/90 President of the IAGLR. She is a member of several national and international committees, and the Chief, Sediment Remediation Project, Aquatic Ecosystem Restoration Branch, National Water Research Institute, Department of Environment, Canada.

Mrs. Mudroch has published over 100 scientific papers and reports and presented over 50 papers at national and international conferences and workshops. Her current major research interests include characterization of aquatic sediments and defining their role in the pathways, fate, and effects of sediment-associated contaminants in aquatic ecosystems.

Scott D. MacKnight, Ph.D., is President of Land & Sea Environmental Consultants Ltd., Dartmouth, Nova Scotia, Canada.

Dr. MacKnight graduated with a B.Sc.(Hon.) and an M.Sc. in Chemistry and a Ph.D. in Oceanography from Dalhousie University, Halifax, Nova Scotia, Canada. Specializing in the environmental aspects of port and harbor development over the past seventeen years, he has undertaken studies in over three hundred harbors in Canada, China, Indonesia, and St. Lucia. His project work has included assessment of dredging/dredged material disposal, monitoring of waterfront development/landfill projects, assessment of sediment quality and trace contaminant fluxes across the sediment/water interface.

He is a member of the American Chemical Society, Air and Waste Management Association, and the Western Dredging Association. He is a former member of the Natural Sciences and Engineering Research Council Strategic Grants Committee for Environmental Quality.

CONTRIBUTORS

Donald D. Adams
Professor
Center for Earth and Environmental Sciences
State University of New York
Plattsburgh, New York

José M. Azcue
Research Scientist
Sediment Remediation Project
Aquatic Ecosystem Restoration Branch
National Water Research Institute
Canada Centre for Inland Waters
Burlington, Ontario, Canada

Jürg Bloesch
Research Scientist
Institute of Aquatic Sciences (EAWAG)
Swiss Federal Institute of Technology (ETH)
Dübendorf, Switzerland

Richard A. Bourbonniere
Research Scientist
Aquatic Ecosystem Protection
National Water Research Institute
Canada Centre for Inland Waters
Burlington, Ontario, Canada

Jean R.D. Guimarães
Research Scientist
Coordenação de Proteção Radiológica Ambiental
Instituto de Radioproteçãoe Dosimetria, CNEN
Rio de Janeiro, Brazil

Scott D. MacKnight
President
Land & Sea Environmental Consultants Ltd.
Dartmouth, Nova Scotia, Canada

Olaf Malm
Research Scientist
Laboratório de Radioisótopos
Instituto de Biofisica, UFRJ
Rio de Janeiro, Brazil

Alena Mudroch
Research Scientist
Sediment Remediation Project
Aquatic Ecosystem Restoration Branch
National Water Research Institute
Canada Centre for Inland Waters
Burlington, Ontario, Canada

Paul Mudroch
Physical Scientist
Protection and Prevention Branch
Environment Canada
Nepean, Ontario, Canada

David E. Rathke
Research Scientist
U.S. EPA Region VIII
Water Management Division
State Programs Branch
Denver, Colorado

Fernando Rosa
Physical Scientist
Sediment Remediation Project
Aquatic Ecosystem Restoration Branch
National Water Research Institute
Canada Centre for Inland Waters
Burlington, Ontario, Canada

TABLE OF CONTENTS

Chapter 1
INTRODUCTION

Alena Mudroch and Scott D. MacKnight

Bottom sediments consist of particles of different size, shape, and chemical composition that have been transported by water, air, or ice from the sites of their origin in a terrestrial environment and have been deposited on the river, lake, or ocean floor. In addition to these particles, bottom sediments contain materials precipitated from chemical and biological processes in river, lake, and ocean waters. The relative proportions of the terrigenous and precipitated particles in sediments can vary widely. Strong biological productivity results in deposition of nearly pure calcareous and siliceous oozes on the ocean floor and in some lakes. Nearly pure terrigenous sediments accumulate on the section of the ocean floor that lies on the path of the supply of the major river and wind systems.[1] Particles transported in the water become sorted and deposited according to their textural properties in different areas of the lakes and oceans. Generally, coarse material, such as sand and pebbles, settles in the nearshore zone and fine-grained particles, such as silt and clay, become deposited in deep waters with restricted currents. The particles accumulate on the bottom at different rates varying between tenths of millimeters and several centimeters per year.

Different concepts were used for classification of freshwater and marine sediments. For example, sediments were classified by their geographical and geological origin, geochemistry, or the physico-chemical properties, such as color, texture, grain size, structure, organic matter content, etc. The classification of marine sediments appears to be more advanced than that of freshwater sediments.[2] Mineralogy of many sediment particles in large lakes is similar to that in marine sediments, and the distribution of the sediments in large lakes depends on factors similar to those in the oceans. On the other hand, the physico-chemical properties of sediments in small lakes vary widely, as they are affected by different geological, hydrological, and biological conditions. In addition, the composition of lake sediments is affected by many characteristics of the lake, such as trophic level, genetic type, climatic conditions, etc.

Natural processes responsible for the formation of bottom sediments are altered by human activities. The erosion of soil is accelerated by the construction of buildings and roadways. Many man-made compounds in gaseous, liquid, and solid form, with complex chemical composition and physico-chemical properties, have entered streams, lakes, and oceans through atmospheric deposition, runoff from the drainage basin, or direct discharge into the water. Most hydrophobic organic contaminants, metal compounds, and nutrients entering water bodies become associated with particulate matter. This particulate matter is carried by currents into quiet areas in the rivers, lakes, and oceans where it settles and accumulates in bottom sediments. Under certain conditions the contaminants in bottom sediments may become released into water or accumulate in the food chain. Consequently, bottom sediments are a sink as well as a source of contaminants in an aquatic environment. A particular distribution equilibrium of contaminants is established among suspended and

1-56670-027-2/94/$0.00+$.50

bottom sediments, sediment pore water, overlying water column, and biota. This equilibrium is affected by the physico-chemical regime which controls the kinetics of different reactions taking place within the system.

Sediments have been studied for many years to characterize their nature and properties for different purposes. Recently, contaminated sediments have become one of the most important environmental issues. They appear to have the potential to become a significant regulatory issue with important scientific implications. The contaminants in bottom sediments are often identified as "in-place toxics" that can pose a high risk to the environment, and a serious and costly environmental issue whose management may require a special approach including sampling and analyses of the sediments, interpretation of the results, establishment of guidelines, and remedial action plans. However, it is difficult to quantify the extent and severity of sediment contamination when in many countries criteria for distinguishing between clean and contaminated sediments are either nonexisting or are still being developed.

Generally, sediments in freshwater and marine harbors contain great concentrations of contaminants which originate from local urban and industrial sources, and shipping activities. Sometimes tributaries, runoffs, and effluent discharges contribute a load of contaminated sediments into the harbor. The sediments in the harbors need to be removed by dredging to maintain navigable waterways. The evaluation of toxicity of dredged contaminated sediments is necessary for their proper disposal and treatment. At freshwater and marine sites, such as streams, lakes, estuaries, and nearshore areas in the oceans, which do not require dredging, contaminated sediments have to be evaluated to assess their effects on the ecosystem and develop proper plans for remedial action.

Extensive survey, monitoring, and research activities, generally very expensive, are required to assess the extent and severity of sediment contamination, to evaluate the effects of contaminated sediments on freshwater and marine environment, and to prepare a plan for proper remedial action. The research on behavior of contaminants in an aquatic ecosystem includes laboratory experiments, for example, different bioassays, testing sediment properties, and identification of chemical forms of sediment contaminants. In field studies, monitoring is carried out of spatial and temporal changes of concentrations of different contaminants in water, suspended sediments, sediment pore water, and biota. These activities require sampling of all compartments of the aquatic system. However, a detailed description of sampling techniques is neglected in many reports on investigation involving sediments. To compare the results of the studies carried out at sites with different environmental conditions and contaminated with different compounds, identical sampling techniques must be used. To process a nonrepresentative or incorrectly collected or stored sample may lead to waste of money and effort and to erroneous conclusions.

The objective of this book is to provide sufficient information on sampling techniques for suspended and bottom sediments and sediment pore water to those involved in characterization of sediments for different purposes, and in studies of sediment/water interaction processes. The information is divided into the following seven chapters of this book. The information necessary for preparation of a sediment sampling program is outlined in Chapter 2. Three different projects dealing with environmental pollution issues are included in the chapter to illustrate the necessary steps in the preparation of sediment sampling programs. Chapter 3 provides a guide in sediment sampling strategy and preparation of sediment sampling design. Included examples show different approaches to the selection of positions and the number of sampling stations. A review of different equipment for bottom sediment sampling and their use under different sampling conditions is outlined in Chapter 4. The definition, origin and fate of suspended particulate matter, its sampling strategy, the state of the art of different sampling devices and sample processing prior to analysis are described in Chapter 5. Details of preservation, processing and storage of collected sediment samples are discussed in Chapter 6. Chapter 7 contains a review of different sampling techniques for sediment pore water and sediment gases, and an introduction of recently developed methods for sampling pore water gases. A general outline of the objectives, methods, and cost of sediment sampling programs is given in three case studies in Chapter 8. We hope that the text of the book with included figures and tables will help the reader to understand the importance of choosing proper techniques in sediment sampling programs, and provide a guide in preparing and executing such programs.

REFERENCES

1. **Lerman, A.**, *Geochemical Processes: Water and Sediment Environments*, John Wiley & Sons, New York, 1979, 260.
2. **Håkanson, L. and Jansson, M.**, *Principles of Lake Sedimentology*, Springer-Verlag, Berlin, 1983, 17.

Chapter 2
PROJECT AND DATA REVIEW

Alena Mudroch and Scott D. MacKnight

I. INTRODUCTION

Sediment characterization plays an important role in many projects. These projects have been carried out for a wide variety of reasons, such as testing of scientific hypotheses, survey of environmental conditions, evaluation of fish habitats, or construction involving sediment removal or displacement. Adequate and representative characterization is a function of both sample collection and analyses. No matter how much care is taken in laboratory analyses, such factors as improperly located sampling sites, collection of inadequate number or quality of samples, and inappropriate sample handling can generate false information on sediment properties.

To avoid such mistakes, detailed information is necessary about the outline of the whole project prior to selecting sediment sampling techniques and proper methods for handling and analyses of the collected samples. This information is achieved by the project review which should include:

- A definition of the problem(s) to be assessed
- A statement of the hypotheses to be tested or objectives to be achieved
- The collection and review of all available data
- The preparation of the project plan to meet the objective(s) including the design of sample collection and analyses
- The design of quality control/assurance program
- An estimation of cost and degree of effort to undertake project activities

The project review should provide sufficient information on the requirements for sediment sampling and determination of sediment properties to assist in the preparation of a detailed plan for sampling strategy and logistics; selection of sampling stations and equipment for sediment collection; sample handling, preservation, storage, and transport; and methods for sediment analyses.

II. REVIEW PROCESS

A. PROBLEM DEFINITION

The problem to be solved by conducting the project needs to be clearly defined. The impact of contaminants on the aquatic environment is the most common problem in pollution studies. The lack of specific information on the distribution and geotechnical properties of sediments can be a problem at a pipeline route area. In many research-oriented projects the problem to be solved is a lack of information for validation of a hypothesis. A project to determine the interactions between physical, chemical, and biological processes in water and their effects on marine or freshwater sediments, or the

1-56670-027-2/94/$0.00+$.50

influence of bottom fauna on physico-chemical properties of sediments, can provide a complete data set to confirm the results of preliminary studies or to verify modeling of the processes. The problem does not need to be defined in studies which are initiated by a specific environmental regulation or regulatory permit process. However, in such cases, study objectives are often narrowly defined and many of the processes or parameters which should be measured or tested are considered mandatory in the project plan.

B. STATEMENT OF THE OBJECTIVES

Project objective(s) or the reason for conducting the project should be formulated in such a way that the execution of the project will contribute to the solving of the problem. A proper statement of the objective(s) is critical to ensure proper development of the project plan, particularly for the design of the sediment sampling program and processing collected samples. For example, the objective of a project dealing with sediment characterization may be of scientific interest in the nature of sea, lake, or river floor in an area for which no information exists, such as the distribution of different types of sediments, sedimentation rates, relationship between sediment geochemistry and physico-chemical and biological processes in the water column, etc. On the other hand, the objective of a project carried out in oil, gas, or mining exploration or laying cables on the sea, lake, or river floor could be the assessment of geotechnical properties of sediments in the project area. The objective(s) of a project involving environmental pollution could be the determination of the effects of contaminants in the project area and/or a proposal of a proper remedial action for the cleanup of the area.

C. DATA COLLECTION AND REVIEW

Depending on the nature of the project and the site to be investigated, there may be a considerable body of historical information and data relevant to the project objective(s). The gathering of historical data with a comprehensive review of literature, reports, and all available previously published data generated by surveys and studies, including the characterization of the sediments, should be completed before the preparation of the project plan.

Historical data can be obtained from a variety of sources. For example, if construction in a particular harbor is taken as an example, then there are several important sources of information:

- Data specific to that harbor
- Data of interest for the harbor
- Data for the region in which the harbor is located
- Data for the watershed draining into the harbor of interest

Data specific to the area of the harbor could include that derived from geotechnical investigations for the construction of port infrastructure, sediment analyses for dredging permits, benthic investigations in conjunction with ecological studies, or environmental impact studies.

Data from the harbor at large could include the above data from other areas of the harbor as well as data related to:

- The types of industry and businesses
- Data from effluent monitoring

- The occurrence of sewage treatment plants
- Watershed activities
- Harbor activities (commercial traffic)
- Storm drainage
- Residential development and extent

Data from regional reconnaissance surveys could provide information on a broad scale, such as concentrations predicted from the known geology and mineralogy of the area; geochemistry of the harbor sediments; general "background" concentrations or concentrations of different chemicals in soil which, through weathering or erosion within the watershed, would contribute material to the harbor sediments. Material could enter from the watershed in either dissolved form or associated with eroded soil materials and could include, for example, pesticides or fertilizers from agricultural practices, mining wastes or excavated materials, or industrial/mining processing by-products/effluents.

The important factor to consider is that even very old or incomplete data can be used to provide a first estimate of the concentration of a parameter or the likelihood of sedimetary processes, or provide sufficient information to warrant additional sampling at the area. In some cases, even simple commentary from local citizens about the industry for which there is little documentation can prove to be valuable.

There are particular pieces of data which are relevant to the project planning. These include:

- General information on the watershed, including quantity and quality of runoff, climatic conditions, general or specific land use, types of industries, effluent, and urban runoff
- Distribution, thickness, and types of sediments, particularly fine-grained sediments (this will assist in assessing the physical extent of sediment accumulation, zones of deposition and erosion, and sediment transport)
- Quantity, particle size, geochemistry, and mineralogy of suspended sediments discharged by tributaries, stormwater runoffs or originating from shoreline erosion (knowledge of the nature and quantity of dissolved and particulate materials entering the area is necessary for the calculation of contaminant and nutrient loading)
- Horizontal and vertical profiles of physical (e.g., porosity, geotechnical properties, water content, bulk density, grain size) and chemical (e.g., organic matter content, concentrations of nutrients, metals, and organic contaminants) characteristics of bottom sediments
- Biological community structure, composition and diversity, bioaccumulation of contaminants, or bioassay results

1. Data Collation

The data and information collected must be carefully reviewed as to their:

- Relevancy (to the overall objective of the project)
- Completeness (taking into account that parameters or processes of interest may not have been measured in previous studies, and the objective for previous study was different)
- Quality of data (based on reported limits of detection and precision compared to precision now required)

An important aspect which is often overlooked is a site inspection. The visit to the project site permits an assessment of the completeness of the collected information and identifies any significant changes at the project site.

D. PROJECT PLANNING

From the review of collected data the gaps in the information should be identified and the sampling program designed to fill these gaps and to achieve the overall objective(s) of the project. The project plan should describe in detail which objectives will be selected and how these objective(s) will be achieved within a given time frame and budget for the project. Further, the project plan should contain a detailed work plan, manual of standard operating procedures, and the rationale for the proposed techniques which will be used to achieve the objectives. The project plan should include an identification of individuals responsible for each operation, the equipment, and special requirements to perform the work. The plan may be outlined as a flow-chart and, in cases where the project has to be carried out in a given time frame, the timetable schedule can also be outlined.

The selection of the number and locations of sediment sampling stations, and description of methods for sediment sampling, handling, and analyses are a key part of the project plan. Sampling locations affect the quality and usefulness of data in environmental studies. Selection of the sampling locations should mainly be based on the project objective(s).

E. QUALITY CONTROL/ASSURANCE PROGRAM

The project plan should contain an adequate quality assurance (QA) program for sediment sampling and sediment analyses. For example, under the established U.S. Environmental Protection Agency Mandatory Quality Assurance Program, data quality acceptance criteria and QA project plans are prepared for all data collection projects.[1] These project plans clearly describe which operations will be performed at each stage of data collection (i.e., sediment sampling site selection, techniques for sediment collection, sample handling and analyses, and data handling and analyses), and include instructions or standard operating procedures for each field and laboratory activity. Properly described and standardized methods for sediment sampling, analyses, and data processing allow for the comparison of results from different studies, and support the confidence in study conclusions and, in projects dealing with pollution, in selections of proper remedial measures.

The character and quantity of data which need to be collected in order to draw conclusions should be accurately identified in the project plans. In projects dealing with assessments of the effects of contaminated sediments, different actions which have to be applied as a result of project conclusions should be outlined together with the action level (for example, criteria or guidelines for disposal of dredged sediments).

Consider the example where approximately 100,000 m^3 of sediment needs to be dredged from a waterway. A part of this sediment has been reported to be "contaminated" with cadmium (containing greater concentrations of cadmium than those given in the criteria for clean sediments), but the spatial extent of sediment contamination is not known. According to local regulations, dredged sediments containing a greater concentration of cadmium than 1 μg/g have to be disposed upland, and those with concentrations less than 1 μg/g, into open water. Further, sediments with cadmium concentrations less than 2 μg/g can be disposed on adjacent agricultural land, and those with cadmium concentration greater than 2 μg/g have to be totally confined. The construction of a confined disposal facility with 100,000 m^3 sediment storage capacity would cost $5 million. Consequently, the action level, in this case 1 and 2 μg/g of cadmium in the sediment, has to be considered in the preparation of project plans including sediment sampling and analyses, to determine the portion and location of the dredged sediment requiring open water, unconfined land, and confined land disposal, which, in turn, would affect the cost of disposal.

F. ESTIMATION OF COST

In most cases, organizations funding a project require a detailed project proposal including a budget from those entrusted with performing the work. Usually, most project funding organizations are concerned with the economic feasibility of the project and most likely will not support a project which appears to be too expensive for justification of expected results. Therefore, the project plan has to be prepared in such a way that the execution of the project will meet the stated objectives within given cost constraints. The project budget should closely follow the project plan by estimating the cost of each step of the work plan. The estimated cost of individual steps should be obtained from the personnel responsible for a particular task. The responsibility of the project manager is to make the sampling program accurate and most cost-effective.

The cost of sediment sampling and processing includes all expenditures for collecting bottom and suspended sediments into proper containers, measurements and sample handling in the field, transportation and storage of the samples, sample preparation, different sediment tests and analyses, quality control procedures, data processing, and preparation of reports.

The cost of sediment sampling depends, in a decreasing order, first, on the number of sampling locations; second, on the selected, and in many cases expensive, sampling procedures; third, on the number of samples to be collected and/or on the amount of sediment material required for analyses and different experiments; and fourth, on sample handling and preservation procedures. The number of sampling locations and the selected sampling procedure greatly affect the cost of sample collection and are factors which, with regards to available funding, will contribute to the decision about the feasibility of the project. Obviously, there is a difference between the cost for collection of a sediment sample in a small shallow lake from a small vessel operated by two workers, and the cost for collection of a sediment sample in a large, deep lake or sea that requires a vessel of appropriate size with experienced crew.

Although it should be expected that the cost of sediment sampling will vary from one project to another, cost items commonly encountered in sediment sampling programs are

- Renting/leasing/purchasing and operation of vessels required for the sampling
- Renting/leasing and operation of cars and/or trucks, and snowmobiles in winter for sediment sampling from ice, required for transporting sampling equipment and technical personnel to the sampling location
- Cost of shipping of sampling equipment to and from the sampling location
- Purchasing or renting essential equipment, particularly sediment grab samplers and corers, sediment traps and centrifuges, selected spare parts,

instruments and tools, containers for handling and storage of sediments, freezers or refrigerators, etc.
- Cost of quality assurance and quality control programs considered essential for the project.
- Cost of accommodation when the sampling program will be carried out for more than 1 day
- Salaries, wages, and travel expenses for scientists and technical workers
- Hiring experienced personnel, for example, divers
- Hiring local labor
- Overhead (allowance to cover miscellaneous service and maintenance costs, such as tools used, depreciation, power, etc.)

Costs considered expenses and those considered assets owned at the end of the project should be separated in the project budget. Expenses are the cost of goods and services consumed during project duration. For example, a small boat or a sampler purchased for a project and retained after finishing the work is an asset. However, the same equipment rented is an expense.

It should be recognized that even well-planned activities can go astray, and that various problems can arise, such as sickness of the workers, lost sample containers, and poor quality samples requiring additional sampling. Contingencies for such incidents as well as additional funds should be considered during the planning stage. One factor which should always be considered in all cost estimates is the weather and its caprices. Poor weather, for example, storms, can interfere in sediment sampling, causing delays and unexpected expenses.

The cost of chemical, physical, and geotechnical analyses, bioassays, and other tests outlined in the project plan has to be estimated as accurately as possible and added to the estimated cost of sediment sampling. The costs of laboratory measurements can be obtained from private or governmental laboratories involved in the type of analyses required in the project.

If the project budget exceeds available funds, alternatives have to be considered, for example, canceling or postponing the project or revision and changes of project objectives and working plan.

III. EXAMPLES

Three examples of projects dealing with environmental pollution issues are given below. Project title, problem statement, and objectives for the investigation of sediment properties are outlined for each example together with a list of information relevant to the preparation of the project plan, in particular to the selection of the sediment sampling stations and determination of which tests need to be run on collected sediments. A sketch of each project area is shown in Figures 1 to 3.

A. EXAMPLE 1

1. Project Title

Evaluation of environmental effects of offshore hydrocarbon exploration on the nearshore zone of an arctic marine system.

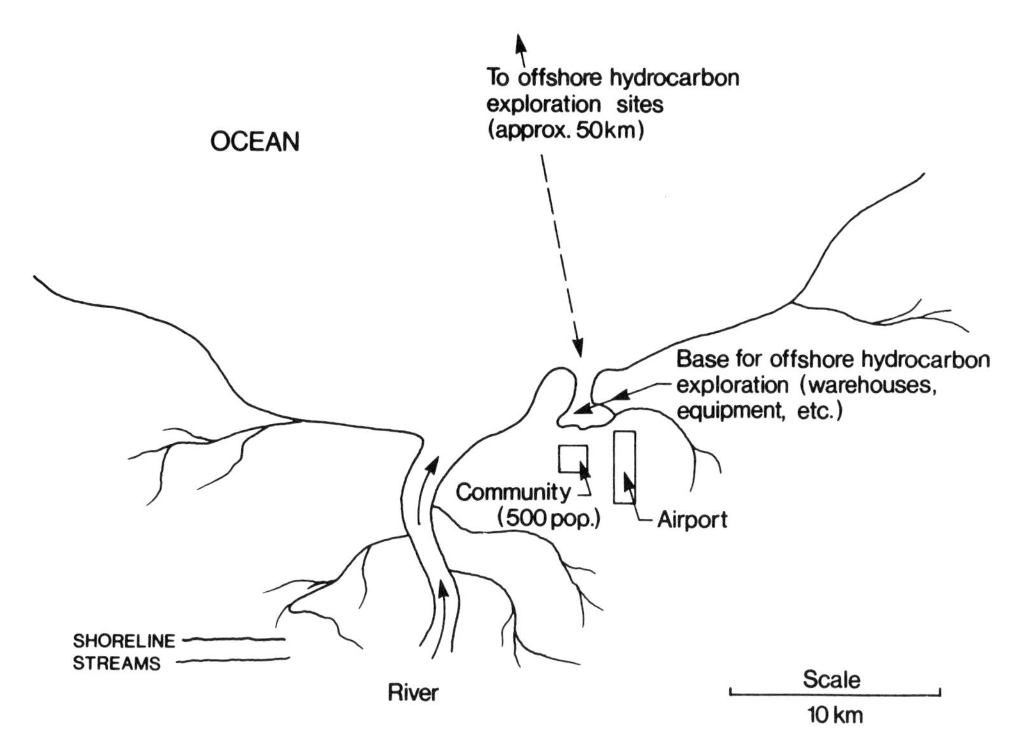

FIGURE 1. Example 1.

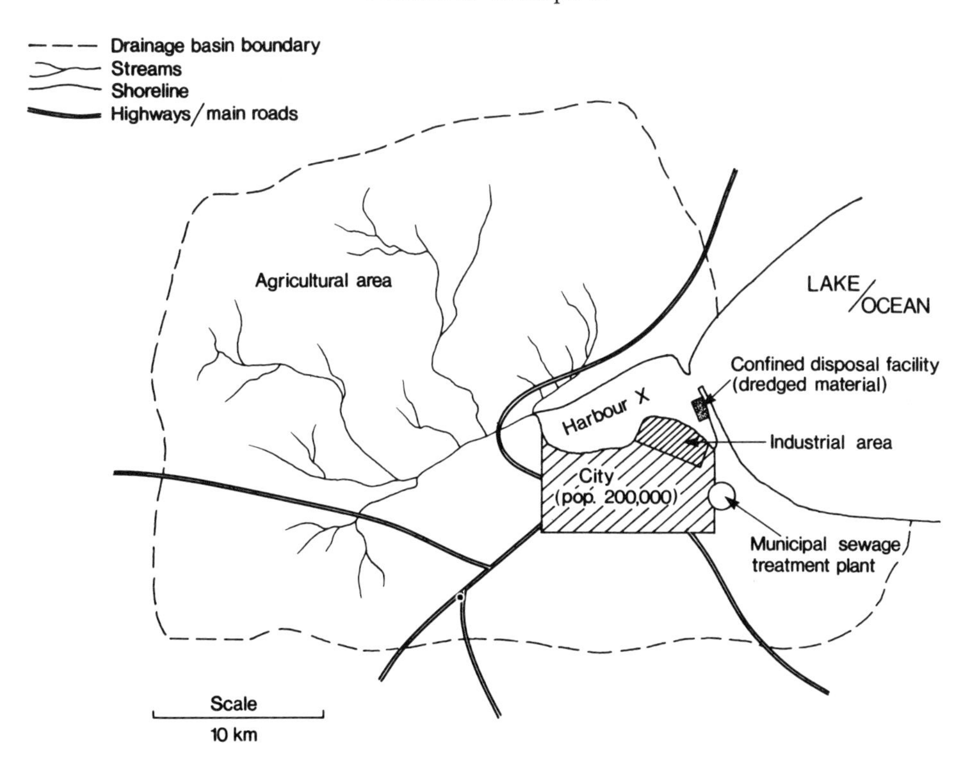

FIGURE 2. Example 2.

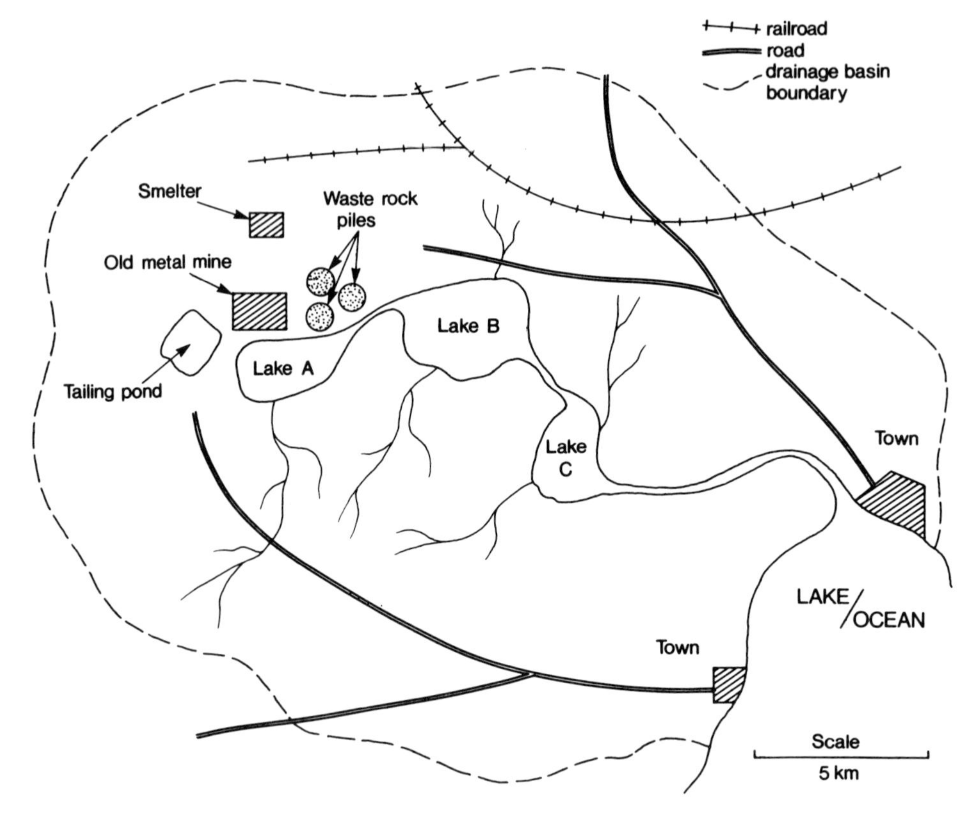

FIGURE 3. Example 3.

2. Problem Statement

Offshore hydrocarbon exploration produces materials which, when discharged to the arctic marine ecosystem, can generate an adverse effect on the environmentally sensitive nearshore zone near the area of these activities.

3. Objectives of the Investigation and Sediment Sampling

To determine the presence, distribution, and effects of wastes from the offshore hydrocarbon exploration activities in the nearshore zone at the area of these activities.

Most of the contaminants from discharged wastes become associated with particulate matter in the sea, and some waste is already discharged in solid form. Consequently, concentrations of contaminants are higher in suspended matter and bottom sediments than in the water, and determination of the contaminants in the sediments allows for a better evaluation of the extent of contamination from hydrocarbon exploration. In addition to the determination of concentrations of contaminants in the sediments, a survey of benthic community structure and uptake of contaminants by different benthic species will be carried out to evaluate the effects of contaminants in the sediment on benthic fauna.

4. Relevant Information

Information on other possible sources of different pollutants is necessary to differentiate between the effects of contaminants generated by the offshore exploration activities and other sources which would affect the investigated area.

a. Natural Setting: Land-Based Information

- Geology of the area adjacent to the investigated nearshore zone
- Soil erosion and transport of eroded particles into the sea
- Bedrock and soil geochemistry
- Hydrology, e.g., discharge of tributaries, quality of water and suspended sediments in the tributaries
- Climatic conditions, e.g., permafrost line, snow and ice cover, rainfall

b. Land Use

- Industrial activities, e.g., products, by-products, production technology; waste disposal, and treatment technology
- Population density, e.g., municipalities, sewage treatment technology
- Landfills and waste disposal sites (runoff, management practices)
- Activities associated with offshore hydrocarbon exploration, e.g., shipping; storage and handling of materials, such as drilling fluids, gasoline, oil, diesel fuel, etc.
- Historical changes in land use

c. Input from Land

- Quantity and quality of industrial effluent discharges into tributaries or nearshore zone
- Quantity and quality from landfills
- Runoff and discharges of solid wastes from the area used for land-based preparation for offshore exploration activities
- Municipal sewage-related inputs

d. Natural Setting: Sea-Based Information

- Outline of area of concern and control sites
- Distribution, particle size, and geochemistry of bottom sediments
- Occurrence of benthic invertebrates, e.g., community structure
- Water depth and water column density
- Frequency and duration of high energy events
- Quantity and quality of suspended sediments in the sea water
- Bottom sediment resuspension and transport

e. Activities Associated with Offshore Drilling

- Number and location of drilling sites
- Activities for the preparation of drilling (dredging, construction of artificial islands and ice-formation enhancement, any construction or cable-laying on the sea floor)
- Quantity and physico-chemical properties of material used for drilling, e.g., drilling fluids, chemicals for ice enhancement
- Disposal of wastes from drilling activities, e.g., packing material, drilling fluids, oil, human wastes

B. EXAMPLE 2

1. Project Title

Rehabilitation of the aquatic environment of a harbor.

2. Problem Statement

The aquatic environment of a harbor X deteriorated by eutrophication and the presence of different contaminants in the harbor such that fishing, swimming, and other recreational activities are impossible. Moreover, dredging of the harbor to keep shipping for a local industry was required.

3. Objectives of the Investigation and Sediment Sampling

The objectives are to evaluate the role of contaminated sediments on the degradation of the ecosystem of the harbor, and propose proper treatment of contaminated sediments; to prepare a long-term management plan for the disposal of sediments dredged from the harbor and shipping channel.

4. Relevant Information

In this project, it will be necessary to gather information to determine if the sediment is the only source of pollutants, or if there are other significant sources which have to be considered before sediment treatment. In addition, any historical information on the extent and type of sediment contamination will help in the preparation of a sediment sampling program for the assessment of the quality of sediments in the harbor, and to prepare a long-term management plan for dredging and disposal of contaminated sediments from the harbor and shipping channel.

a. Drainage Basin

- General information: geology; soil type and chemistry; soil erosion; climate (e.g., rain- and snowfall); hydrology, e.g., runoff period; water levels in streams
- Current land use: urban (e.g., population, waste and sewage treatment practices), industrial (e.g., type of industries, products and by-products, processes used in the production), agricultural (e.g., agricultural practices; use of fertilizers, pesticides, and herbicides), landfill sites (e.g., size, type of disposed material, management practices, runoff), historical changes in land use (e.g., changes in population, type of industry, agricultural practices)

b. Inputs to the Harbor

- Industrial facilities, e.g., location; quantity and quality of effluents and discharged solid wastes; historical changes in discharges
- Municipal sewage treatment plants, e.g., location, treatment technology, quantity and quality of effluent, historical changes, variance in flow due to storm waters, discharge of industrial effluents via municipal sewage
- Sewers, e.g., construction and location, sewers combined with other effluent discharges, overflow during storm events, runoff from roads, quantity and quality of effluents
- Tributaries, e.g., quantity and quality of water, quantity and quality of suspended solids
- Landfill runoff, e.g., quantity and quality
- Navigation-related inputs, e.g., dredging and disposal practices, quantity and quality of disposed sediments, management of disposal sites, spills
- Atmospheric input, e.g., from local industries, significance of long-range transport to the harbor.
- Occassional spills of contaminants or nutrients in the drainage basin

c. Environmental Conditions in the Harbor

- Bottom sediments, e.g., physico-chemical properties: particle size distribution, geochemistry, concentration of contaminants and nutrients

- Dredging and disposal of sediments, e.g., quality and quantity of disposed material, disposal sites in the harbor or adjacent area
- Water depth and quality, e.g., concentrations of nutrients, contaminants, dissolved oxygen
- Water circulation, e.g., exchange with adjacent waters, residence time of water in the harbor, water column stratification
- Fish, e.g., species found in the harbor, bioaccumulation of contaminants, incidence of tumors/lesions/deformities
- Benthic plants and animals, e.g., community structure, species diversity, bioaccumulation of contaminants

C. EXAMPLE 3

1. Project Title

Survey of contaminated sediments originating from past mining activities.

2. Problem Statement

Contaminants from past metal mining activities, in particular arsenic, copper, lead, nickel, and zinc, were transported through a chain of lakes, and impaired the lakes' ecosystem. Mining waste rock, containing specific contaminants, was used for the construction of roads and the railroad in the area. Some contaminants leached into the groundwater and contaminated local wells. The population growth in the lake drainage basin resulted in the construction of new buildings, an increase in soil erosion, and a promotion of leaching of contaminated soil and old mining wastes.

3. Objectives of the Investigation and Sediment Sampling

The objectives are to determine the leaching and transport of contaminants at old metal mining sites into the lakes; to evaluate the effects of contaminated sediments on the quality of lake water and biota; to prepare a plan for the remediation of mining wastes and consider remedial action for the cleanup of the lakes.

4. Relevant Information

Information indicating how much and from where the contaminants entered the lakes will help in the preparation of the sediment sampling plan.

a. Drainage Basin

- General information: geology (e.g., information on concentrations of trace elements associated with metal ore, bedrock, and soils; chemistry of the metal ore), soil erosion, hydrology (e.g., runoff period, water table, discharge of tributaries)
- Current land use, e.g., agricultural practices or industrial activities which could release different contaminants into the lakes in addition to those from the past mining activities
- Historical land use: mining operations, e.g., technology used in metal extraction, water intake for the operation, waste disposal practices; quantity and quality of disposed solid wastes, erosion and quantity and quality of runoff from disposal sites
- Construction practices, e.g., material used in the construction of roads and railroads, quality and quantity of runoff from roads and railroads
- Construction of new buildings and housing

b. Input to the Lakes

- Seepage and runoff from waste rock piles and tailings, e.g., quantity and quality
- Tributaries, e.g., discharge quantity, water and suspended sediment quality
- Runoff from roads and railroads, e.g., quantity and quality
- Effluents from other industries in the drainage basin
- Submerged tailings (location, quantity, and quality)

c. Environmental Conditions in the Lakes and Connecting Streams

- Water depth and quality, e.g., concentration of contaminants, nutrients, and dissolved oxygen
- Water circulation, e.g., residence time of water in the lakes, lake stratification, winter conditions — ice cover
- Bottom sediments: physico-chemical properties, e.g., accumulation areas of fine-grained sediments, geochemistry, concentration of contaminants in question

REFERENCES

1. **U. S. Environmental Protection Agency**, Sediment sampling quality assurance user's guide, Report EPA/600/S4-85/048, Center for Environmental Research Information, Cincinnati, OH, 1985.

Chapter 3
SELECTION OF BOTTOM SEDIMENT SAMPLING STATIONS

Scott D. MacKnight

I. INTRODUCTION

Funds spent on sample analyses by the most sophisticated techniques are wasted on samples collected at inappropriate locations or where an insufficient number of samples are taken to represent the project area. Consequently, the selection of the number and positions of sampling stations needs to be carefully designed. There is no one formula for design of a sediment sampling pattern which would be applicable to all sediment sampling programs. This chapter was prepared to provide a guide to those involved in sediment sampling strategy and preparation of a sediment sampling design, by using examples to illustrate different approaches to the selection of positions and numbers of sediment sampling stations.

II. FACTORS AFFECTING THE SELECTION AND NUMBER OF SAMPLING LOCATIONS

When defining the positions and number of sediment sampling stations, the following factors should be considered:

- Purpose of sampling
- Study objectives
- Historical data and other available information
- Bottom dynamics at the sampling area
- Size of the sampling area
- Available funds vs. estimated (real) cost of the project

A. PURPOSE OF BOTTOM SEDIMENT SAMPLING

Generally, the reasons for bottom sediment sampling can be divided into the following categories:

- Geochemical survey
- Environmental assessment of contaminants in the sediments
- Evaluation of sediment for a dredging/disposal permit
- Research of sedimentary processes

Although the strategy and goal of sediment sampling in each type are different, the sampling techniques are similar, and the method for selection of space and number of sampling stations for one purpose may be applicable to the others. The selection and number of sampling stations depends on the project objective(s), and must be modified for peculiarities of each project.

1-56670-027-2/94/$0.00+$.50

B. STUDY OBJECTIVES

Careful definition of the project objectives is highly critical to the successful completion of the sediment sampling program. Generally, samples will be collected from the study area to investigate the distribution of parameters of interest at the project site. In studies of distribution of contaminants, sediment samples featuring the most suitable grain size for analyses and scheduled experiments, such as bioassays, are preferred. The objectives of a research scientist studying sedimentary processes at an estuary are naturally different from the objectives of a project proponent applying for an open-water disposal permit for sediment to be dredged from a channel within the same estuary. Although both workers will collect samples to characterize the sediment, their sampling strategy will often be distinctly different.

C. USE OF AVAILABLE HISTORICAL DATA AND OTHER INFORMATION

It is widely recognized that sediment sampling operates in an environment of uncertainty. Most decisions are based on information that is often limited both in quality and quantity.

Before carrying out an expensive and time-consuming sampling program, it is recommended that the literature and/or the files of local governmental agencies be reviewed for relevant historical data. The sampling area, such as a harbor or a contaminated site in a lake, river, or marine coastal area, may have been surveyed before, and thus much work can be saved. The process of reviewing historical data and other information relevant to the designing of a sediment sampling program is discussed in detail in Chapter 2.

Most harbors have up-to-date bathymetric maps. If such maps or background data do not exist, then a two-tiered approach is essential: an initial reconnaissance survey and a systematic mapping survey. The selection of spacing and number of sediment sampling stations can proceed without further detailed mapping if bottom sediments consist of a uniformly distributed fine-grained material (e.g., silt/clay) over the entire sampling area. This does not often occur. Another measure of sediment homogeneity at an area was suggested by Håkanson.[1,2] He showed that for direct analysis of the sediments for cadmium plus lead plus copper to give mean values with a 10% error in the value, it would require 789 samples to be collected at the river mouth, 106 in the river, and only 6 in the receiving lake. Only by further specifying subsampling and sample processing procedures could the number of samples required be reduced. This is also an excellent example where poorly stated objectives for a project may not take into account the natural variability of a contaminant of interest in a natural system. Håkanson's data clearly show that there is considerable variability in the sediments at the mouth of the river compared to the more homogeneous sediments in the lake.

In most cases, the results of the first survey will illustrate a variability in sediment grain size, and a comprehensive sediment mapping program will be needed. Such a mapping program typically consists of a bathymetric survey and sampling of sediment units composed of different material, such as sand, gravel, silt/clay, etc., to collect sufficient data for sediment mapping. It is well known that fine-grained sediments tend to contain the greatest concentration of contaminants.[3] In environmental studies, samples of fine-grained sediments are usually selected to determine the extent of contamination, unless there is an indication that certain contaminants are associated with coarse-grained material at the project site.

The position of the sampling station should allow for a reliable, rapid repetition of sampling in the future without difficulty. It is imperative that each sampling station be properly referenced to a survey grid on a map and properly labeled.

D. BOTTOM DYNAMICS AT THE SAMPLING AREA

The distribution of the sediment on the lake, river, or ocean floor is affected by energy-controlled processes. Sand, gravel, and boulders are the sediment units on the bottom of a fast flowing river. Fine-grained sediments (i.e., silt and clay) may accumulate in areas of low energy zones, such as bays or the outer side of the main channel of a meandering river. Sediment deposits in large lakes, although strongly influenced by the characteristics of source material, reflect the changes of various energy-controlled processes, such as wave action, current circulation, etc.[4]

Håkanson[5] suggested three processes characterizing the bottom dynamics in a lake: erosion, transportation, and accumulation. Areas of erosion are characterized by exposed bedrock, gravel, sand, or hard glacial clays and tills. Fine-grained sediments (grain size typically smaller than 63 μm) exist in accumulation areas and typically contain the greatest concentrations of organic matter and different contaminants. In most lakes, an accumulation area is found at the deepest point of the lake. Fine-grained sediments may accumulate over a short period at areas of transportation. Principal sediment transport mechanisms are resuspension by wave- or ship-induced turbulence and movement by local currents. At such areas sediments become resuspended and further transported into accumulation areas. Unless there is a specific interest in the investigation of sediments at erosion or transportation areas, sediment sampling sites and stations should be located at areas of fine-grained sediment accumulation.

Scientists involved in the selection of sediment sampling stations should have at a least a basic knowledge of bottom dynamics at the project area. Ideally, sediment particle size distribution should be mapped prior to the selection of sediment sampling sites.

A survey of sediment deposits and geochemistry in a lake can be a project. In such a case, sediment mapping will be carried out as a part of the project, and sampling stations will be selected to provide sufficient information for sediment mapping. On the other hand, the selection of sampling sites in projects dealing, for example, with the evaluation of sediment contamination in an area, requires a knowledge of sediment distribution to locate the stations of fine-grained sediment accumulation.

Maps of the sediments on the sea, lake, and river floor should be prepared with special attention to areas of erosion, transportation, and accumulation. One of the basic tasks of planners is the proper selection of locations considered suitable for sample collection. The goal is to maximize the probability of detecting the areas with the greatest concentrations of pollutants, or conversely, to minimize the cost of collecting improper samples or the loss of collecting no samples.

E. SIZE OF THE SAMPLING AREA

The number and spacing of sediment sampling stations also depends on the physical size of the project area, and how large an area each sample has to represent. In addition, as already mentioned above, the density of sampling stations required for the characterization of sediments is determined by the variability or gradients in the processes which control the distribution of the investigated sediment parameter or property. When the distribution of sediment parameters is relatively homogeneous, stations can be widely spaced. If the distribution of the parameters is heterogeneous, a more dense sampling grid will be required. For example, Moore and Heath[6] described examples of spacing sampling stations in several different studies in the oceans. These included a successful mapping of the influence of the Gulf Stream and the North Atlantic Drift using foraminiferal assemblages of 161 surface sediment samples from the North Atlantic, with the location of one sample per 300,000 km^2, and a study of the sediment dispersal pattern in the Panama Basin, using a sample density of approximately one sample per 17,000 km^2, in definition of large topographic features and bottom flow on the distribution of different particle size material, clay minerals, and biogenic debris. Thomas et al.[7,8] used a sampling grid with one sampling station at intersections of a 10-km grid to determine the physico-chemical properties of bottom sediments in the Laurentian Great Lakes, and to map concentrations of different contaminants in the sediments.[9] In projects dealing with environmental pollution of small areas, sediment sampling stations need to be located usually much closer, in particular at areas with heterogeneous distribution of different sediment units and many contaminant sources.

F. AVAILABLE FUNDS/ESTIMATED COST OF THE PROJECT

When totaled, individual costs of sample collection, sample handling, chemical analyses and bioassays, data processing and interpretation, and preparation of a report represent the overall cost of one sample. Depending on many factors, as discussed in Chapters 2 and 4, the cost of one processed sediment sample can vary over a wide range.

For example, in projects dealing with contaminated sediments, available funding usually impacts the design of spacing sampling stations, and the number of samples that can be analyzed to fit the budget. The cost limitations require all workers involved to work as efficiently as possible and to keep the cost of all operations as low as possible. Collected samples must meet required quality standards.

To characterize the sediments for dredging projects, the sampling stations must be established in such a way that the whole project area is investigated. Although the grid system for the selection of sampling stations is generally applied, the number and spacing of sampling stations are controversial parameters. The guidelines are vague and consensus is rarely achieved. However, it should be considered that the cost of sediment sampling and analyses represents only a portion of the total cost for sediment dredging and disposal, typically between 1 and 25%. If financially and technically possible, replicate samples should be taken even if only one sample is scheduled for analyses.

III. SAMPLING DESIGNS

A. DESIGN OF POSITIONS AND NUMBER OF SAMPLING STATIONS

Numerous reviews have been published on the statistics of sampling,[10] with more recent publications illustrating the application of statistical sampling to environmental studies including both traditional studies[11,12] and the identification of hazardous wastes.[13-15]

The traditional approach for sediment sampling is the haphazard pattern, with samples taken from areas easiest to access or easiest to sample. This type of pattern is somewhat upgraded when conducted by a sampling team with an understanding of the project site, thereby placing a certain amount of judgment on the location of the haphazard samples. Drawbacks of this approach include missing areas which should be sampled (i.e., an inadequate characterization) or requiring a good knowledge of a project area, information, or expertise which may not always be available. The result is considerable difficulty in applying a statistical treatment to the data and often an inability to resample at the same sites.

A statistical sampling pattern permits subsequent manipulation of the data to determine trends or locations with high concentrations of parameters of interest, or to assist in further sampling within a particular area. Random statistical sampling may appear haphazard, but, in fact, takes into account a pattern, or assumption of a pattern, of distributions of chemical compounds. Typically, an area is divided into a series of blocks or triangles and sampling sites identified either in the center of each unit or at the intersection of each unit. Depending on the size of the project and the predicted distribution of the parameter or constituent of interest, the intensity of sampling can be selected.

For example, Cahill[16] described an extensive sediment sampling to determine the sediment distribution and geochemistry in Lake Michigan. Two grids were prepared, the first one with the dimensions of 12 × 12 km for the general survey, the other one with the dimensions of 7 × 7 km for the areas with expected greater concentrations of contaminants. A Decca 416 radar unit with a variable range marker was used for navigation. Fixes were obtained at arrival and departure from each sampling station and at 15-min intervals between stations. Continuous profiles were made by an echosounder operating at 14.25 kHz. The locations of the sampling stations were plotted on the navigation chart and recorded as longitude and latitude. The recorded positions were accurate within approximately 500 m in the center of the lake. Grab samples were collected at the intersection of the grids.

Thomas et al.[17] investigated the geochemical properties of surficial sediments in Lakes Ontario, Erie, Huron, and Superior. Samples were collected at the intersections of a 10-km Universal Transverse Mercator (UTM) grid. In addition, at some areas the samples were collected at alternative intersections of a 5-km UTM grid to obtain a close sampling density. Positioning of the sampling stations was carried out by radar using a Decca 416 with variable range marker. The accuracy of the positioning was estimated as being better than ±500 m in mid-lake positions.

Sly[18] designed a special sampling grid for sedimentological, geochemical, and biological studies at specific areas of the Laurentian Great Lakes. The objectives of the study were to compare the characteristics of the sediment within and between the selected areas. The sampling grid, which provides for comparison of samples spaced from 30 to 3000 m apart, was devised as a cell structure, with each unit controlled by four corner samples. The smallest unit of each cell has an area of about 930 m^2, and each succeedingly larger unit within each cell increased by a multiple of 10. The complete grid was composed of a number of separate cells forming a large square. The spacing of the sampling stations was sufficiently close to that previously used in the Great

Lakes which allowed comparison to results of other studies. A high order of precision in positioning was used to place each sample within 50 ft of its plotted position.

On smaller lakes, a grid system can start with one sample at the point selected according to the morphology of the lake or with a grid with 10 × 10-m grid blocks covering the whole area. In projects dealing with environmental pollution of small areas, sediment sampling stations usually need to be located very close, in particular in areas with many contaminant sources.

Håkanson and Jasson[19] proposed a pilot sample formula to provide an even cover of sampling stations over the whole lake. The formula is based on two morphometric standard parameters: the lake area (i.e., more samples should be taken in larger rather than in smaller lakes), and the shore development, which is used as an indirect measure of the bottom roughness.

In sediment sampling in the marine environment, one of the main objectives is to cover the largest possible area in a minimum amount of time and, consequently, at the minimum cost.

B. STRATIFIED RANDOM SAMPLING

Stratification permits the division of a heterogeneous population into more homogeneous subpopulations that are less variable than the original population and a sampling independent of different portions of the population at different rates when this appears advisable. This sampling technique seems to be suitable for project areas that have already been mapped, are slated for dredging and sediment disposal, and consist of well-defined zones with different sediment types, such as gravel, sand, silt, and clay. Locations with fine-grained sediments and high organic matter concentrations which have a greater affinity to contaminants should be sampled in greater detail, for example, at 100%, than locations with sand and gravel, sampled at 50 and 10%, respectively. Once the stratification principle is established, the sampling of the subsets is then undertaken on a random basis within each subset.

C. BOTTOM SEDIMENT SAMPLING IN A RIVER

River surveying is complicated, requires sustained supervision, and all operations involved must be conducted with meticulous care to obtain samples truly from the selected sampling points. In areas where there are rapids or falls, it is useless to take samples because erosion is such that fine-grained sediments do not remain long and are quickly transported downstream. Consequently, sampling upstream of the rapids, before reaching the eddies, is the recommended approach.

In a reconnaissance survey of stream sediments sampling is carried out on a grid with a density of one to several samples per square kilometer.[20] This density provides a satisfactory picture of the local geological and geochemical background. The density of sampling as well as station spacing, usually 1 km but sometimes 750 or 500 m apart, is determined before starting the survey in accordance with the geology of the region. If it is decided that a sector should be examined in detail, a grid with shorter spacing should be used. The survey region is covered by a grid consisting of the succession of rectilinear blocks. The baseline is the axis of the river. From this baseline parallel lines (rows) are drawn perpendicular to each side. The distance between specific lines is decided either on the map or directly on the spot. As a rule, the density of the rectilinear blocks varies according to the local conditions and the accuracy desired. In such cases, two sediment samples can be taken 10 to 20 m apart where there is fine-grained sediment. In contrast, gravel sites are ignored but when gravel is sampled, it is poured into a 5-mm sieve placed above a bucket and screened until a sufficient quantity of material containing sand and fine-grained material is obtained (from 100 g to several kilograms). The oversize is discarded.

In an environmental survey of stream sediments the goal is to check the extent of pollution, and the presence and abundance of specific pollutants. Sediment taken within a grid block is representative of the pollution for the area surrounding the sampling point.

D. SEDIMENT SAMPLING NEAR A POINT SOURCE

Where the effect of an outfall or other point source is to be considered in terms of sediment contamination, the sampling pattern is typically based on the decrease in impact with the distance from the source of contamination. However, in natural systems the distribution is also strongly affected by currents and other physical factors affecting the dispersion of materials from the point source.

The typical grid from a point source would begin with a fixed distance, "x", and then sampling stations at points of "2x", "4x", "8x", etc., out to a distance designated as being equivalent to background. If the point source is within a water body (e.g., dispersion of drilling mud residues from an off-shore drilling platform or dispersion of disposed dredged materials from a barge), then the sampling pattern is designated with sites at the intersection of each distance line and each major point of the compass (i.e., 90° separation) or more frequently (e.g., every 45°). Figure 1 illustrates this pattern.

IV. EXAMPLES OF SAMPLING PROGRAMS

A. INTERIM SAMPLING GUIDELINES (CANADA)

Canada, as a signatory to the London Dumping Convention (LDC), regulates the disposal of wastes in the sea under Part IV of the Canadian Environmental Protection Act (CEPA). The CEPA regulations require the dredging project proponent to provide information on the contaminants in the material to be disposed. Approximately 90% of the ocean disposal permits issued regulate the disposal of dredged sediments from the Canadian marine environment.

Sediment samples from the dredging project site need to represent both the vertical and horizontal distribution of contaminants of interest. An inadequate number of samples or incomplete description of the vertical distribution of contaminants has most often been the reason for refusing a dredging permit or requiring additional sampling and information on sediment quality. Consequently, Environment Canada developed a set of guidelines for the collection of sediment samples with the following objectives:

- To provide a pattern of sediment sampling stations sufficient for a statistically significant description of the distribution of contaminants of interest at the project site
- To consider the cost of sample collection (and analyses)
- To provide guidelines which can be followed by personnel with no experience in sediment samples collection

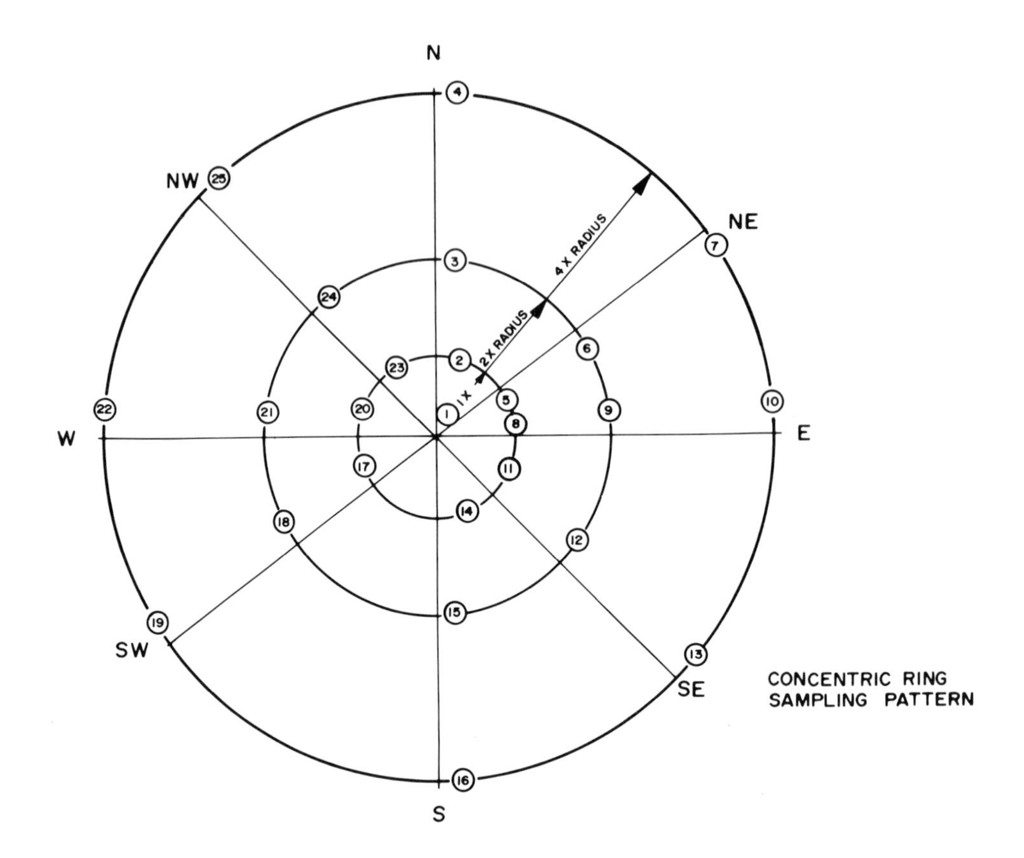

FIGURE 1. Sampling grid at a point source.

The guidelines use the concept of a management unit of 1000 m^3 to describe the division of the project area into a number of blocks to be sampled. The selected size of a unit represents a compromise between a typical volume of a disposal barge and the size of an area which can be defined by electronic survey equipment. The pattern of sampling stations is similar to that used in geochemical reconnaissance studies.[21] As in geochemical programs, it is assumed that any location of sediments with contaminant concentrations in excess of a regulated limit is randomly distributed within the area of the proposed dredging project. The sampling station is defined as the center point of each grid block.

The main steps to follow in the guidelines are

1. Define the dredging area (in m^2 or km^2) and volume of material (in m^3) to be dredged.
2. Calculate the dimensions of a sampling block using 1000 m^3 as a management unit, and the thickness of sediment planned for the removal. For example, if the required thickness of the sediment to be removed is 1 m and the management unit is 1000 m^3, then the area of the sampling block is 1000 m^2, which corresponds to an approximately 32×32-m square. On the other hand, if the thickness of the sediment to be removed is 0.2 m, the area of the sampling block is 5000 m^2, which corresponds to an approximately 71×71-m square.
3. Calculate the number of the sampling blocks by dividing the entire project area by the dimensions of one sampling block.

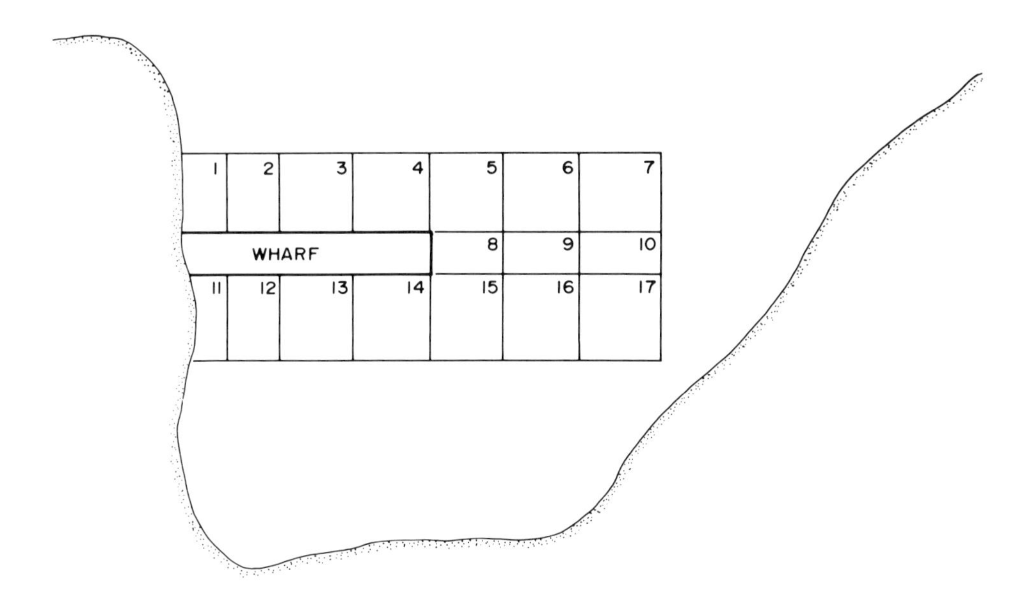

FIGURE 2. Random selection of sampling stations with no data available on the project.

4. If no historical data are available for either sediment particle size or contaminant distribution at the project area, 60% of the blocks are sampled on the basis of truly random selection. Sampling sites are designated at the center of each block. An example of this case is outlined in Figure 2. According to the bathymetry, a uniform removal of 1 m of sediment is required. Ten management units are defined for sediment sampling.
5. If a significant outfall or other point source of contamination is known, the block into which the effluent enters must be sampled. This would be in addition to sampling 60% of all blocks. An example of such a case is shown in Figure 3. The sampling pattern is adapted to the need to sample the area influenced by the defined point source.
6. If the data on sediment particle size are available, then the blocks are to be designated as containing gravel, sand, or fine-grained sediment on the basis of median particle size. Again, 60% of the total number of blocks is to be sampled, but the same proportions of gravel:sand:fine-grained sediment must be maintained as for the overall project. These sediment types have to be sampled at a minimum of one sampling station. An example of this case is given in Figure 4.
7. Samples are to be taken at each designated sampling station as a core or borehole to ensure a complete vertical description of the material to be dredged.
8. Each core or borehole sample is to be subdivided into horizontal sections of a specific size (for example, into 5-, 10-, or 15-cm sections). To constrain costs, it is recommended that the topmost section of each core or borehole be analyzed first and the concentrations compared for the contaminants of

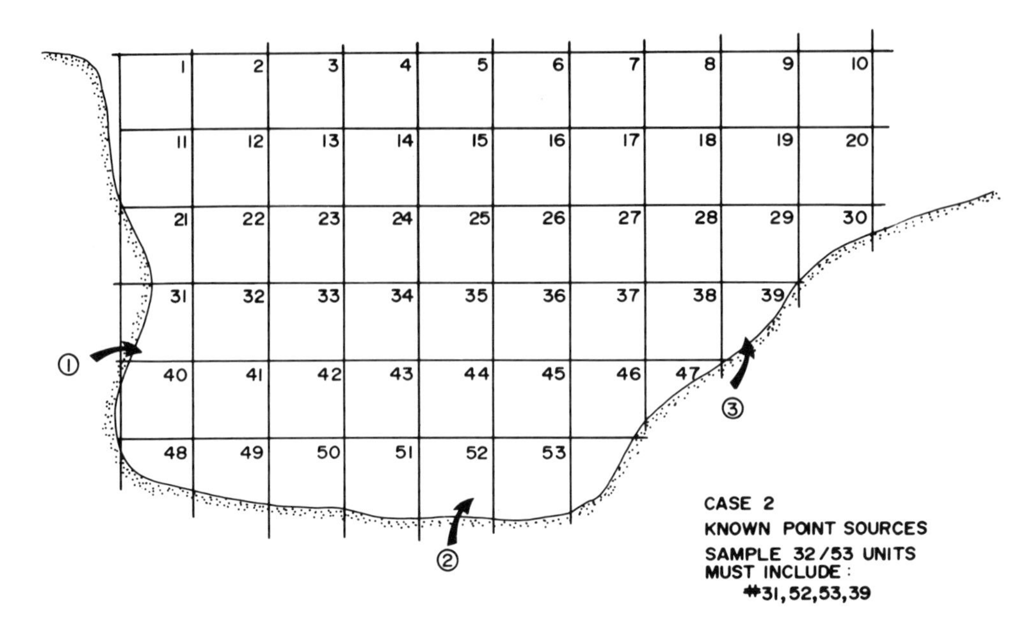

FIGURE 3. Selection of sampling stations with available information on point sources.

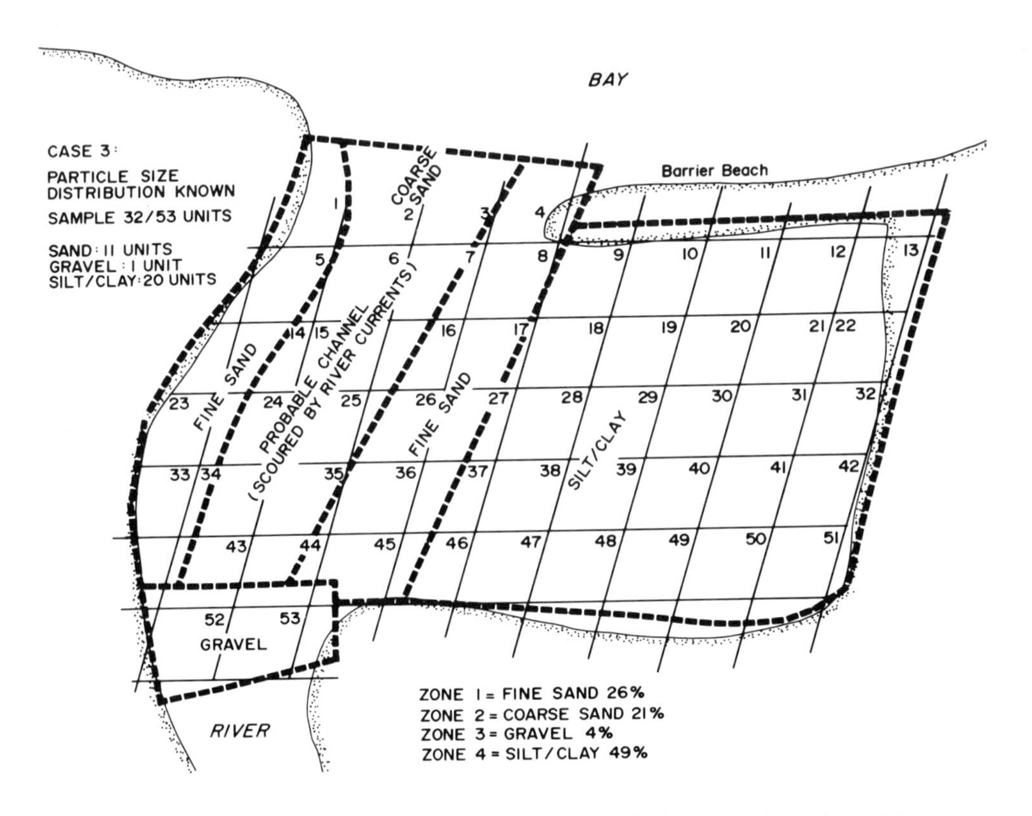

FIGURE 4. Selection of sampling stations with available data on sediment particle size.

interest to the regulated concentrations. If concentrations of any one contaminant exceed the regulated concentrations, a deeper section of the core (or borehole) is to be analyzed until a horizon is reached in which concentrations are below the regulated limits. Those materials containing contaminants in excess of the guidelines are to be disposed of in a confined disposal facility; the remainder is acceptable for open-water disposal.

9. If the results of sediment analyses indicate blocks with a high concentration of contaminants, then either the sediment from the remaining 40% of the originally designated blocks is to be sampled and analyzed or the block(s) with contaminated sediment are to be subdivided into smaller blocks generating a new pattern of blocks (i.e., a secondary pattern). Sediment samples are collected from these blocks by the same manner as for the first pattern. The purpose is to define, as closely as is reasonable, those areas requiring special methods for disposal of the contaminated sediment.

The guidelines emphasize the use of historical data to assist in the planning of the sampling design. Incomplete or outdated data should be used carefully, but can still prove useful in the design of the sampling stations.

B. PUGET SOUND, U.S.

The guidelines for sampling in Puget Sound, U.S.,[22] are less elaborate than those developed by Environment Canada, yet there are many similarities. The project area is divided into management units of 2900 m^3 (4000 yd^3) size, representative of a typical disposal barge volume. At each designated site, a 1.2-m (4-ft) vibra-core or borehole is taken and the whole core (or borehole) is composited to create one sample. This reflects a characterization of each load being disposed as a representation of the dredged materials. In areas where there is a low “reason to believe” there is serious contamination, several cores may also be composited; primarily to reduce analytical costs.

Environment Canada’s procedure of dividing the cores (boreholes) into smaller sections (usually 10, 15, or 20 cm) reflects the smaller dredges and disposal barges used in many Canadian harbors.

The major differences arise from the number of management units to be sampled. The Puget Sound program uses 2900-m^3 units with each unit sampled; the Environment Canada program uses 1000-m^3 units with initially only 60% of the designated units being sampled.

The Puget Sound guidelines also recommend the review of earlier or historical data to assist in planning, and they recommend electronic survey positioning to facilitate proper sample site placement and to permit resampling of a site.

REFERENCES

1. **Håkanson, L.,** Determination of characteristic values for physical and chemical lake sediment parameters, *Water Resour. Res.*, 17, 1625, 1981.
2. **Håkanson, L.,** Sediment sampling in different aquatic environments: statistical aspects, *Water Res. Resour.*, 20, 41, 1984.
3. **De Groot, A.J., Salomons, W., and Allersma, E.,** Processes affecting heavy metals in estuarine sediments, in *Estuarine Chemistry*, Burton, J.D. and Liss, P.S., Eds., Academic Press, New York, 1976, 131.

4. **Sly, P.G.,** The significance of sediment deposits in large lakes and their energy relationships, in *Proc. Symp. Hydrology of Lakes*, I.A.H.S.-A.I.S.H. publ. 109, Helsinki, 1973, 383.
5. **Håkanson, L.,** The influence of wind, fetch, and water depth on the distribution of sediments in Lake Vanern, Sweden, *Can. J. Earth Sci.*, 14, 397, 1977.
6. **Moore, T.C., Jr. and Heath, G.R.,** Sea-floor sampling techniques, in *Chemical Oceanography*, Vol. 7, Riley, J.P. and Chester, R., Eds., Academic Press, London, 1978, 75.
7. **Thomas, R.L., Kemp, A.L.W., and Lewis, C.F.M.,** Distribution, composition and characteristics of the surficial sediments of Lake Ontario, *J. Sediment Petrol.*, 42, 66, 1972.
8. **Thomas, R.L., Kemp, A.L.W., and Lewis, C.F.M.,** The surficial sediments of Lake Huron, *Can. J. Earth Sci.*, 10, 226, 1973.
9. **Thomas, R.L. and Mudroch, A.,** Small Craft Harbours — sediment survey, Lakes Ontario, Erie and St. Clair 1978: dredging summary and protocol, Report to Small Craft Harbours, Ontario Region, from the Great Lakes Biolimnology Laboratory, Dept.of Fisheries and Oceans, 1979, 149.
10. **Cochran, W.G.,** *Sampling Techniques*, 2nd ed., Wiley Interscience, New York, 1963.
11. **Green, R.H.,** *Sampling Design and Statistical Methods for Environmental Biologists*, Wiley Interscience, New York, 1979.
12. **Howarth, R.J. and Thornton, I.,** Regional geochemical mapping and its application to environmental studies, in *Applied Environmental Geochemistry*, Thornton, I., Ed., Academic Press, New York, 1983, 41.
13. **Provost, L.P.,** Statistical methods in environmental sampling, in *Environmental Sampling for Hazardous Wastes*, Schweitzer, E.G., and Santolucito, J.A., Eds., ACS Symp. Ser. 267, American Chemical Society, Washington, D. C., 1984, 79.
14. **Parkhurst, D.F.,** Optimal sampling geometry for hazardous waste sites, *Environ. Sci. Technol.*, 18, 521, 1984.
15. **Gilbert, R.O.,** *Statistical Methods for Environmental Pollution Monitoring*, Van Nostrand Reinhold, New York, 1987.
16. **Cahill, R.A.,** Geochemistry of recent Lake Michigan sediments, Illinois State Geological Survey, Circular 517, Champaign, IL,1981, 94.
17. **Thomas, R.L., Jaquet, J.-M., Kemp. A.L.W., and Lewis, C.F.M.**, Surficial sediments of Lake Erie, *J. Fish. Res. Board Can.*, 33, 385, 1976.
18. **Sly, P.G.,** Sedimentological studies in the Niagara area of Lake Ontario, and in the area immediately north of the Bruce Peninsula in Georgian Bay, in *Proc. 12th Conf. on Great Lakes Research*, International Association of Great Lakes Research, Ann Arbor, MI, 1969, 341.
19. **Håkanson, L. and Jasson, M.,** *Principles of Lake Sedimentology*, Springer-Verlag, Berlin, 1983, 39.
20. **Chaussier, J.-B.,** *Mineral Prospecting Manual*, North Oxford Academic Publishers, London, 1987 (English transl.).
21. **Singer, D.A.,** Relative efficiencies of square and triangular grids in the search for elliptically-shaped resource targets, *J. Res. U.S. Geol. Surv.*, 3, 163, 1975.
22. **Kendall, D.R.,** Dredged material sampling, testing, and disposal guidelines application report. Puget Sound Dredged Disposal Analysis Report to U.S. Army Corps Engineers (Seattle Dist.), 1989.

Chapter 4
BOTTOM SEDIMENT SAMPLING

Alena Mudroch and Scott D. MacKnight

I. INTRODUCTION

The purpose of sampling is to collect a representative, undisturbed sample of the sediment to be investigated. There are many factors which need to be considered in the selection of suitable equipment for bottom sediment sampling. These factors include the sampling plan, the type of available sampling platform (vessel, ice, etc.), location and access to the sampling site, physical character of the sediments, the number of sites to be sampled, weather, number and experience of personnel who will carry out the sampling, and the budget. Because of these many factors, the standardization of the sampling techniques is difficult. Generally, the selected sampling equipment should recover an undisturbed sediment sample.

Several excellent comprehensive reviews are available on bottom sediment sampling devices. These reviews have tended to discuss the limitations of equipment for particular purposes: for example, grab samplers and corers suitable for sampling of benthic organisms;[1] a wide variety of bottom sediment samplers designed for biological and geological work mainly in the marine environment, but which may be used in pollution studies in the marine or freshwater environment;[2] sediment sampling techniques in studies of sedimentary structure;[3] description and evaluation of performance of commercially available and custom-made bottom sediment samplers;[4] review and testing of bottom sediment samplers used in the lacustrine environment;[5] description of different instruments, including bottom sediment samplers, and their use in oceanographic research;[6] efficiency of grab samplers and corers in benthic organism sampling;[7] description of bottom sediment samplers suitable for studies of sediment microbiology;[8] bibliography of samplers for benthic invertebrates;[9,10] bottom sediment sampling strategies and sampling devices in marine studies;[11] sampling devices, including bottom sediment samplers, for studies of marine pollution;[12] description and evaluation of different bottom samplers for studies of geology of the continental shelf;[13-15] equipment currently used and under development to investigate the ocean bottom for surveying offshore construction sites;[16] theoretical and practical aspects and advantages and disadvantages of various types of sediment samplers.[17]

Many samplers described in the literature are only variations of a few early models modified to overcome observed deficiencies or to be used for specific objectives and for different operating conditions. The many different names applied to sediment samplers are usually confusing to those who need to choose a suitable one for a project.

In addition to the above reviews, commercially available or custom-made sediment samplers have been described in studies involving bottom sediments in marine and freshwater systems: for example, the investigation of benthic fauna, the collection of samples at sites with different water depths and

1-56670-027-2/94/$0.00+$.50

sediment textures for geological interpretation, or simply the comparison of the performance of newly designed samplers to fulfill a specific task.

II. SEDIMENT SAMPLERS

The purpose of collecting the sample is to obtain an accurate representation of the nature of the bottom in the study area. Therefore, the retained sample should resemble the original material as closely as possible without loss of a particular size or geochemical fraction. Disturbance or sample alteration can occur through sediment compaction, mixing, or fractional loss. Major sources of these disturbances are maneuvering of the sampling vessel in the shallow water prior or during sampling, the pressure wave in advance of the lowered sampler, frictional resistance during sediment penetration by the sampler, tilting or skewed penetration of the sampler, and washout or other loss during retrieval to the sampling platform.

Regardless of the equipment chosen for the sampling, it is useful to know the water depth at each station before starting the sampling. If water depth information is unavailable, it is recommended that the water depth should be first measured. Measurement equipment can range from a weighted chain to a highly accurate bathymeter. The purpose is to ensure adequate cable/rope length for operation of the correct equipment and to control the speed of entry of the sampler into the sediment. The speed of deployment of the sampler can be critical to good operation and sample recovery. Too rapid deployment generates and increases the shock wave advancing in front of the equipment. This shock wave can displace the soft unconsolidated surface sediments. Too fast deployment may also cause equipment malfunction, such as the trigger arm of a piston corer being activated before achieving correct positioning.

It is also useful to have some understanding of the currents at the sampling site. Strong near-bottom currents can lead to poor equipment deployment, deflect a grab sampler, or require a long cable/wire to be deployed. Care should be taken to ensure that the weight of the sampler is adequate for working at the particular current conditions and the sampler collects sediment at or very near the desired sampling site.

Generally, there are two types of samplers (commercially available) used for collecting bottom sediments: grab samplers for collecting surface sediments, thereby providing material for the determination of horizontal distribution of parameters; and corers for collecting a cross section of sediments, thereby providing material for determination of vertical distribution of parameters. The depth of the sediment collected by different surface sediment samplers and corers is given in Table 1.

The dimensions of individual grab samplers and corers described in this chapter are, in most cases, those of commercially available samplers. The actual dimensions of a particular sampler produced by different manufacturers may vary slightly.

A. GRAB SAMPLERS

Simplified drawings of grab samplers with their essential components are shown in Figure 1. Grab samplers consist either of a set of jaws which shut when lowered to the surface of the bottom sediment or they contain a bucket which rotates into the sediment upon reaching the bottom. A large, vented top

TABLE 1
Sediment Depth Collected by Different Samplers Under Optimal Conditions (About 2 m of Fine-Grained Sediment)

Sediment depth sampled	Sampling equipment
0—10 cm	Lightweight, small-volume grabs (for example, Birge-Ekman, Ponar and mini-Ponar, mini-Shipek)
0—30 cm	Heavy, large-volume grabs (for example, Van Veen, Smith-McIntyre, Petersen)
0—50 cm	Single gravity corers (for example, Kajak-Brinkhurst and Phleger corers)
	Box corers
	Multiple corers
0—2 m	Single gravity corer (for example, Benthos and Alpine corers)
Deeper than 2 m	Piston corers

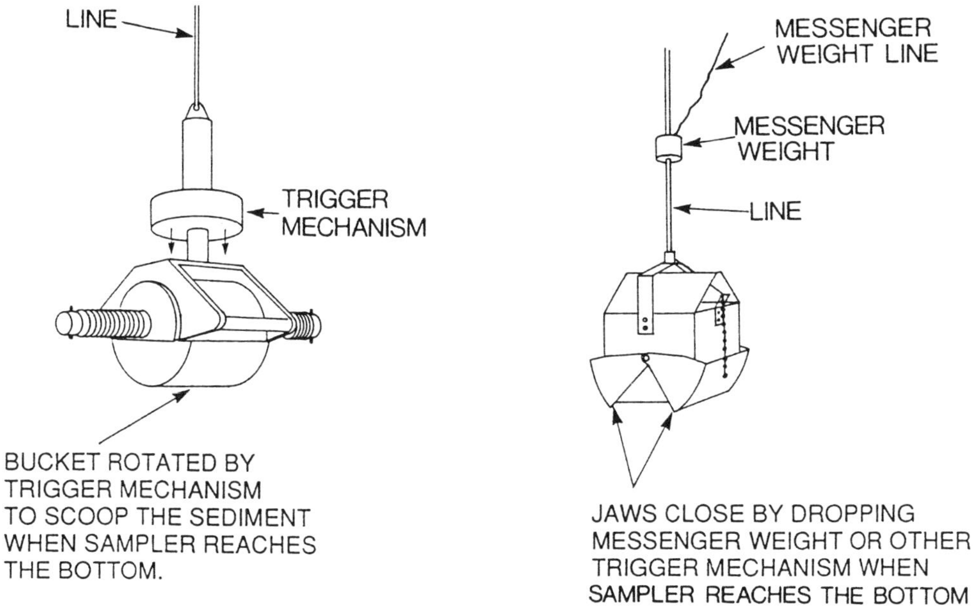

FIGURE 1. Grab samplers with their essential parts.

(usually a screen), or an opening in the back of the sampling bucket greatly reduces sediment disturbance caused by the shock wave in front of the descending sampler. For example, Wigley[18] tested the effects of shock wave and found a relation between the disturbance of the sediment surface and the area of the vented openings in the bucket of the grab sampler.

The properties of grab samplers which have to be considered for general operational suitability are their stability and protection of sample from washout. From the grab samplers described below, the Shipek and Ponar grabs have proved to be excellent general purpose samplers capable of collecting most types of surface sediments. Both samplers maintain a near-perfect vertical descent and stable stance on the bottom in most waters with relatively weak currents, such as harbors and lakes. They are less suitable in fast-flowing rivers or the marine environment with strong currents. The use of small or lightweight samplers, such as the Birge-Ekman grab sampler, is advantageous because of easy handling, particularly from a small vessel, but becomes a disadvantage in areas with strong currents or during poor weather conditions with high waves and intensive vessel's motion. Under poor weather conditions, lightweight grabs are continuously lifted and dropped or are dragged along the bottom during sampling. The lightweight samplers are also less stable during sediment penetration and tend to fall to one side as a result of inadequate or incomplete penetration.

A surface layer of 2 to 3 cm of fine-grained, soft sediments can be lost due to washout. The loss of fine-grained sediments from a Shipek grab sampler during retrieval was investigated at the National Water Research Institute, Environment Canada, Burlington, Ontario, by comparison of concentrations of metals in surficial sediments collected by the Shipek grab sampler and a box corer in a depositional basin in Lake Ontario. The results indicated a loss of the topmost 2 to 3 cm of very fine, unconsolidated soft sediments from the Shipek grab sampler.

Samples collected by some grabs in firm, cohesive sediments, such as glacial till or glacio-lacustrine clays, are often disturbed. For example, when the bucket of the Shipek grab sampler rotates into firm sediment, it cuts only a small sample filling about one third to one half of the bucket. The cohesive sediment has sufficient space in the bucket to turn as one piece upside down during the ascent of the sampler from the bottom, making the recognition of the sample's surface difficult upon retrieval.

Sample volume also needs to be considered when choosing a proper surface sediment sampler. Grabs that can collect more material are favored for sampling in biological studies requiring a large sample volume, for example, the Petersen grab sampler.

The following is a description of the basic, most commonly used and available surface sediment samplers. The choice usually depends on the sampled sediment depth and volume, handling suitability under given sampling conditions, and, very often, on personal preference. For example, three surface sediment samplers, the Shipek and Ponar grabs and Birge-Ekman sampler, are most commonly used by the personnel at the National Water Research Institute, Environment Canada, Burlington, Ontario, for collecting surface sediments from Canadian inland waters. The larger, heavier Van Veen or Ponar grab samplers are more commonly used for the collection of surface sediments in the Canadian marine environment.

1. Birge-Ekman Sampler — Petite (Standard Size in Brackets)

Sampled area: 15 × 15 cm (23 × 23 cm)
Cutting height: 15 cm (23 cm)
Sample volume: 3400 cm^3 (13,300 cm^3)
Weight: empty about 5 to 10 kg — depending on the material (13 kg)
Weight with the sediment: 10 to 15 kg (40 kg)

The Birge-Ekman sampler (Figure 2) is available in several sizes, with sample chambers ranging from 3500 to 28,320 cm^3 and with carrying cases. A tall version is also available, with optional weights for deeper penetration into sediment and a 1.5-m-long operating handle for shallow water applications. The standard Ekman sampler can be operated manually. For the larger models a winch or crane hoist is recommended. All models are a stainless steel or brass box with a pair of jaws and free-moving hinged flaps. The spring-tensioned,

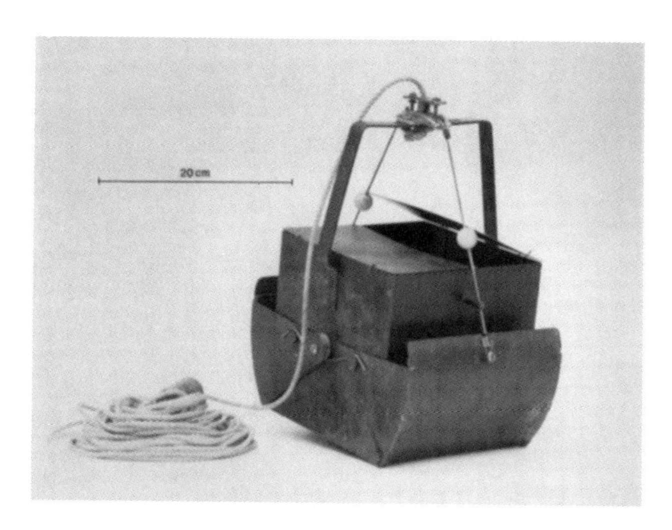

FIGURE 2. Birge-Ekman grab sampler.

scoop-like jaws are mounted on pivot points on opposite sides of the box. The jaws are held open by stainless steel wires which lead to an externally mounted trigger assembly, activated by a messenger. After closure the jaws meet tightly along the seams to prevent washout during retrieval. A weighted messenger is secured to the sampler's line so it can freely move along the line and not become lost during sampling. It is highly advisable to have a spare messenger weight on board of the sampling vessel. When the sampler has reached the sediment the messenger is sent down the rope or wire to activate the jaw closure mechanism. During descent through the water, the flaps are forced open by the pressure of water passing through the open jawed box. The flaps cover the surface of the box during retrieval of the sample preventing disturbance of the collected sediment. The Birge-Ekman sampler is suitable for sampling fine-grained, soft sediments and a mixture of silt and sand. Larger objects, such as gravel, shells, or pieces of wood, trapped between the jaws will prevent jaw closure and result in sample loss. Due to its weight and the need to use an activating messenger, the sampler has to be used under low current conditions and penetrate perpendicular to the sediment. In very soft sediments with a high water content the sampler tends to penetrate too deep due to its weight. This can be prevented by dropping the messenger weight immediately after the sampler reaches the bottom. With careful deployment, a small but relatively undisturbed sample can be obtained. Upon retrieval, the sediment can be divided into several subsamples through the flaps at the top of the sampler. However, to empty the sampler quickly, sediment has to be removed through the bottom by opening the jaws over a container. In the latter case, the sample has to be treated as a bulk surface sediment sample.

Reliability of soft sediment sampling has been improved by the modification of the standard Ekman grab sampler.[19]

2. Ponar Grab Sampler — Standard

Sampled area: 23 × 23 cm
Weight: about 23 kg
Maximum sample volume: 7250 cm^3
Required lifting capacity: 100 kg

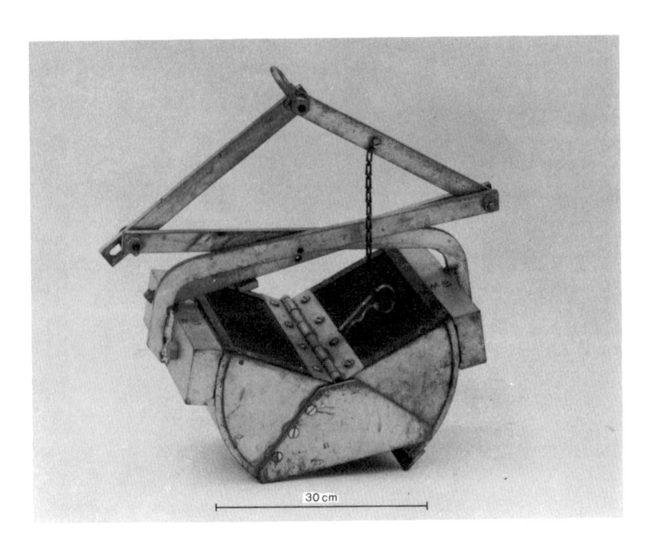

FIGURE 3. Ponar grab sampler.

The Ponar grab sampler (Figure 3) is used with a winch or crane hoist. It consists of a pair of weighted, tapered jaws held open by a catch bar across the top of the sampler. On touching the bottom, the tension on the bar is released, allowing the jaws to move and close. A special mechanism of the Ponar grab sampler prevents accidental closing during handling or transport. The device is activated by the release of the cable/rope tension on the lifting mechanism when the sampler reaches the sediment. During retrieval the tension on the cable keeps the jaws closed. The sampler has to be lowered slowly under the water surface to avoid premature triggering on impact with the water surface. When in water, the sampler can be lowered at any speed until approximately 5 m from the bottom, then it must be lowered slowly. A steady, slow winch speed should be maintained to lift the sampler from the bottom after its penetration of the sediment. The jaws of the sampler overlap to minimize sample washout during ascent of the equipment. The upper portion of the jaws is covered with a mesh screen and rubber flap, allowing water to pass through the sampler during descent, reducing disturbance at the sediment-water interface by a shock wave. Upon recovery, the mesh screen can be removed, providing easy access to the recovered sediment for subsampling. Where a bulk sediment sample is required, or the entire sediment sample needs to be sieved for benthic organisms, it is easy to empty the sampler by opening its jaws over a sufficiently large container. The Ponar grab sampler has a pair of metal side plates which prevents the sampler from falling over after the jaws' closure, reducing washout, and helping to preserve a very good sediment sample. This sampler is suitable for sampling most sediment types from soft, fine-grained to firm, sandy material, with the exception of hard clay, in both freshwater and marine environments with little or no disturbance.

Commercially available, the petite Ponar grab sampler (weight about 10 kg, sampled area 15 × 15 cm, sample volume 1000 cm^3) is designed for hand line operation, but its construction and operation are similar to the standard Ponar grab sampler. Similar to the standard Ponar grab sampler, the petite Ponar grab sampler is very good for sampling coarse and firm bottom sediments.

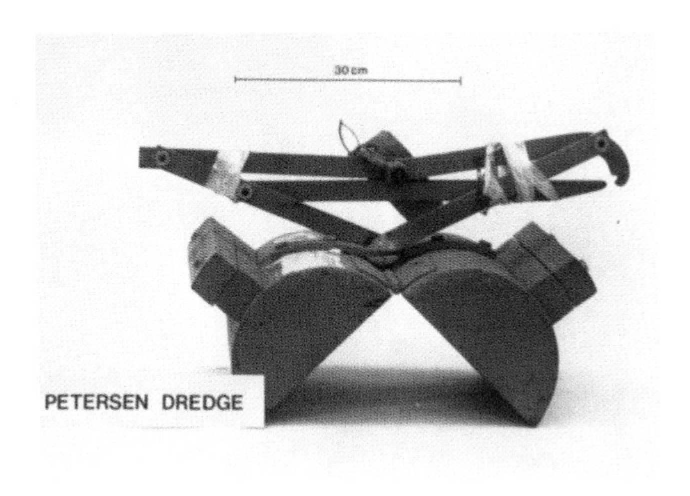

FIGURE 4. Petersen grab sampler.

3. Petersen Grab Sampler

Sampled area: 30 × 30 cm
Weight: about 35 kg
Sample volume: 9450 cm^3
Required lifting capacity: 150 to 250 kg

The Petersen grab sampler (Figure 4) is suitable for obtaining bulk samples of hard bottom material, such as sand, marl, gravel, and firm clay. The sampler consists of a pair of weighted semicylindrical jaws which are held open by a catch bar. Vent holes, located in the jaws, reduce the frontal shock waves. On touching the bottom, the tension on the bar is released, allowing the jaws to move and close. The sampler maintains a near-vertical descent under most conditions. Extra jaw weights (about 9 kg) can be added for better penetration of hard material. When the sampler is used in very coarse or shelly sediment, large sediment grains and pebbles may be trapped between the jaws, preventing their closure and causing severe sample loss. In the usual design, there is no access to the retrieved sample from the top of the sampler. The retrieved sediment has to be transferred from the sampler into a container by opening the sampler jaws.

4. Van Veen Grab Sampler — Standard (Large Size in Brackets)

Sampled area: 35 × 70 cm (50 × 100 cm)
Weight: 30 kg (65 kg), with weights 40 kg (85 kg)
Sample volume: 18 l (up to 75 l)
Required lifting capacity: 150 to 400 kg

The Van Veen grab sampler (Figure 5) is suitable for obtaining bulk samples ranging from soft, fine-grained to sandy material for biological, hydrological, and environmental studies in deep water and strong currents in the marine environment. The Van Veen grab sampler is manufactured in several sizes from hot-galvanized steel or stainless steel which is particularly suitable in pollution studies. The weighted jaws, the chain suspension, and doors and screens allow flow-through during lowering to the bottom and assure vertical descent where strong underwater currents exist. The relatively large surface area and the

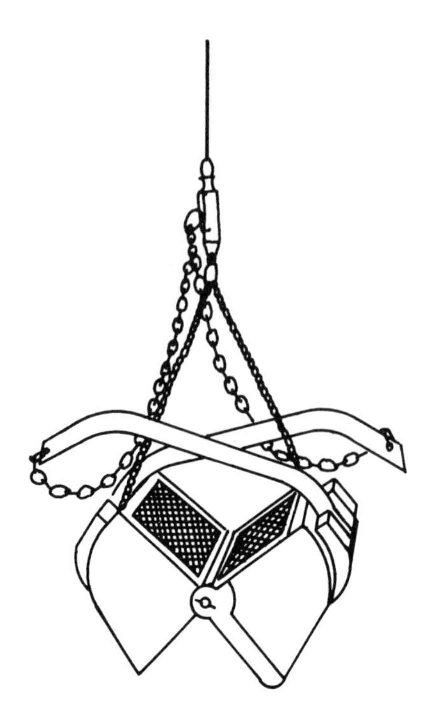

FIGURE 5. Van Veen grab sampler.

strong closing mechanism allow the jaws to excavate relatively undisturbed sediments. A shock wave is not created when the sampler settles on the bottom. As the lowering wire slackens, the hook on the release device rotates and the short suspension chains fall free. When the wire is slowly made taut, the chains attached at the top of the release exert great tension on the long arms extending beyond the jaws, causing them to lift, dip deeper into the sediment, and trap material as they close tightly. The stainless steel door screens have flexible rubber flaps which, during lowering, are lifted. When the grab settles on the bottom, the flaps fall back and cover the screens completely, preventing any loss of sediment during retrieval. Catches on the jaws are provided to lock the doors. When the grab is on the sampling platform, the sediment sample may be dumped into a box or container for further handling. Alternatively, the doors can be unlocked and the samples collected before dumping. The Van Veen grab sampler is similar in operation to the Petersen or Ponar grab samplers, but tends to be larger and heavier, has top access doors for subsampling, and has no internal parts to contaminate the sample, as for example, the chain in the Petersen grab sampler. Because of its typical weight and size, the Van Veen grab sampler is more commonly used in the marine environment, where there is deeper water and strong currents.

The Petersen and Van Veen-type grab samplers (all sampling 0.2 m^2 of sediments) were tested for suitability in studies of benthic organisms by Birkett.[20] In testing the performance of Petersen, Van Veen, and Smith-McIntyre samplers, Gallardo[21] found that the sampling performance of the first two grabs was relatively similar in both soft and hard substrates, most likely due to their particular closing mechanism. However, he found that the Smith-McIntyre grab performance was affected by the consistency of the substratum. Lassig[22] described the tendency of premature closing of the van Veen grab sampler during sediment sampling under rough sea conditions. However, the addition of an improved release mechanism to the sampler made feasible sediment sampling in unsheltered waters. Lie and Pamatmat[23] found the coefficient of variation from 7.4 to 20% in measuring the volume of sediment collected with a 0.1-m^2 Van Veen grab sampler, negligible effect of the shock wave introduced by the sampler for studied infauna, and in only 8 of 37 cases, a significant difference in sample counts for the most abundant species obtained by hand digging and the Van Veen grab sampler. Wigley[18] found that a strong shock wave introduced by the Van Veen sampler forced aside unattached benthic animals up to 8 cm long, and that the Smith-McIntyre sampler created only a weak oscillatory shock wave. Subsequent modification of the Van Veen grab sampler considerably increased the screened opening area at the top of the sampling bucket to reduce the shock wave, and the addition of rubber flaps over the screened opening to prevent the loss of sediment during ascent. An addition of lead weights (teflon-coated in environmental studies) to the upper edges of the Petersen and Van Veen grab sampler jaws improved their penetration of firm sediments.[24] A method for subsampling and measuring volume of sediment collected with a 0.2-m^2 Van Veen grab sampler in benthic survey was described by Kennedy.[25]

5. Smith-McIntyre Grab Sampler

Sampled area: 31 × 31 cm
Weight: about 90 kg (lead weights may be added for deeper penetration)
Sample volume: 10 to 20 l
Required lifting capacity: 200 to 300 kg

The Smith-McIntyre grab sampler was designed for the collection of bulk sediments ranging from soft, fine-grained to sandy material and microbenthos sampling in the marine environment. The sampler shown in Figure 6 is a modified version of the original design of the Smith-McIntyre grab sampler. It is mounted on a sturdy, weighted, steel frame suspended from the lowering wire, with springs to force the two-jaw bucket into the sediment, when released, achieving good penetration. When the sampler comes to rest squarely on the bottom, the springs are released and drive the bucket into the sediment. Two tripping pads, positioned below the square-based frame on which the sampling bucket is suspended, make contact with the bottom first and are pushed upward to release two latches holding the spring-loaded bucket jaws. A free-fall from about 10 m above the lake or ocean floor is usually sufficient to collect a sample even from firm sediments. Externally mounted side and bottom plates on the jaws push stones, gravel, etc. away to prevent jamming or improper closure. A removable frame, fitted with a 2.5-m-aperture brass screen, is attached at the top of each jaw. During the lowering operation sturdy rubber flaps fastened to the screen frames lift to allow water to flow freely through the screens and eliminate a shock wave which might disturb the surface layer of the sediment. The rubber flaps drop to completely cover the brass screens during the retrieval operation and prevent entrance of water which might wash out any of the trapped material. A long-handled bar is furnished to provide easy cocking of the strong bucket springs. When released, the springs exert a force to insure good penetration of the open-mouthed bucket into hard sediments. Safety pull-pins are provided to prevent any premature or accidental release of the cocked assembly.

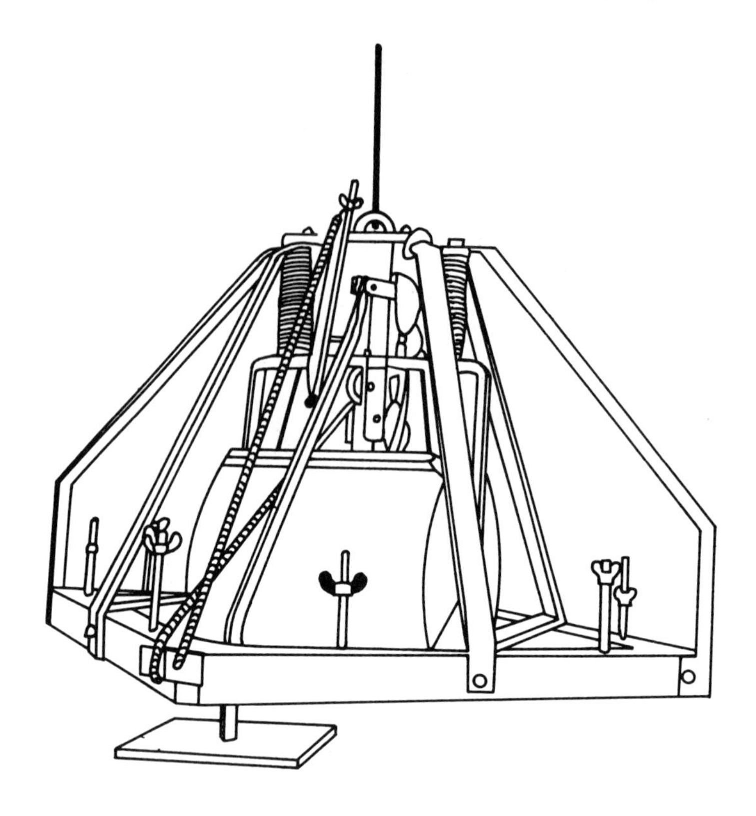

FIGURE 6. Smith-McIntyre grab sampler.

6. Shipek Grab Sampler
Sampled area: 20 × 20 cm
Weight: 50 kg
Sample volume: 3000 cm^3
Required lifting capacity: 200 to 300 kg

The Shipek grab sampler (Figures 7 to 9) was designed to obtain relatively undisturbed samples of sediments ranging from soft, fine-grained to sandy material, even from sloping bottoms, from any depth. The sampler operates by spring-driven rotation of a bucket upon contact with the bottom encircling the sediment to about 10 cm depth. It consists basically of a steel shank, weight, and bucket. Of the two concentric half cylinders, the inner half cylinder is rotated at high torque by two helically wound external springs. The sampler is lowered with the bucket in an inverted position until it contacts the sediment surface. An internal weight triggers a release mechanism. The bucket is then forced to rotate on its axes at high speed by two helical springs. It cuts through the sediment and at the conclusion of a rotation of 180° is stopped and held in an upward closed concave position. Cast into each end of the sampler frame are large stabilizing handles.

The Shipek grab sampler should be lowered slowly until the trigger weight is submerged. The lowering velocity should not exceed the terminal velocity of approximately 100 m/min. The sampler must be raised cautiously from the bottom. Retrieval rates up to 200 m/min can be used. On the sampling platform the bucket is released by pulling outward on the pivot pins supporting the bottom of the bucket with both hands during the lifting of the holding

FIGURE 7. Shipek grab sampler.

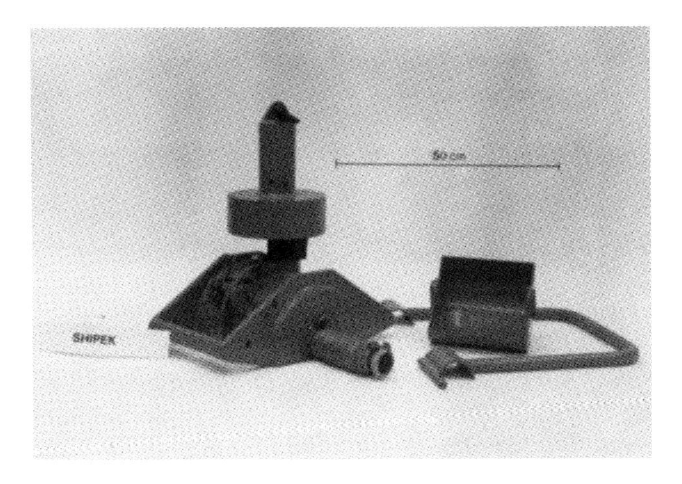

FIGURE 8. Shipek grab sampler with the sampling bucket and cocking lever.

system. Washout can occur during the ascent of the sampler from the bottom, particularly if the uppermost material is soft and fine-grained. Sampling of hard sediments, such as firmly packed sand and glacio-lacustrine clay or till, can be unsuccessful and the samples can be tilted or disturbed. There is easy access to the sample when the bucket is removed from the holding system. This enables visual observation and description of the individual sediment layers, and horizontal or vertical subsampling from the bucket. Larger objects, such as gravel, pieces of wood, shells, etc., trapped between the edge of the bucket and the body of the sampler can cause disturbance and washout of the sample. They should be removed only by rotating the bucket into an open position, but they should never be pulled out of the sampler by hand. The triggering mechanism is very sensitive, and extreme caution is necessary when the bucket is rotated inside the sampler and is ready for lowering to the bottom.

A double-Shipek grab sampler consists of two single Shipek grab samplers joined by a metal bar. It is suitable for obtaining duplicate samples at the sampling site. However, the weight and sufficient space at the sampling platform for handling the double Shipek have to be considered.

A mini-Shipek grab sampler (weight about 5 kg, sampled area 10×15 cm, sample volume up to 500 cm^3), constructed at Environment Canada, Canada

FIGURE 9. Shipek grab sampler prepared for sampling.

Centre for Inland Water, Burlington, Ontario, operates on the same principle as the standard Shipek grab sampler. The sampler is most suitable for collecting fine-grained, soft sediments. In sand and other firm material, such as glacio-lacustrine clay or till, up to 3 cm of surficial sediment only is recovered due to the light weight and small size of the sampler. In addition, the sample from a hard bottom may be tilted and will not represent an undisturbed surface sediment. To obtain a good quality sample, the mini-Shipek grab sampler has to penetrate the sediment surface vertically. A consistent, slow lowering speed of the sampler is necessary to prevent the sampler from triggering before reaching the bottom. Washout of very fine material from the surface of the sample is very likely to occur during the retrieval, similar to the standard Shipek grab sampler. Bacause of its weight, the mini-Shipek grab sampler is suitable for hand-line operation from different sampling platforms.

7. Published Information on Grab Samplers

Although the design of grab samplers has not changed considerably, several useful modifications have been made to the standard samplers to increase their efficiency and reliability. Relevant published information on testing and the design of different grab samplers was selected and is presented together with a brief outline of the studies in which the samplers were used.

LeRoy[26] described details of the operation of a 10-kg pressure disk sediment sampler used from a small boat to collect fine-grained sediments, sand, and coral debris in 30 to 90 m water depth off the west coast of Java. Kutty and

Desai[27] tested the efficiency of the Van Veen and Foerst-Petersen grab samplers in sampling bottom fauna, finding that in sandy sediments the heavier Van Veen grab sampler appeared to be more efficient than the lighter Petersen grab sampler. Sixteen grab samplers were tested for sampling the macrofauna[28] with no instrument recommended for general use. The performance of the samplers will always depend on many factors such as the size of the vessel and available hoisting gear, type of sampled sediment, water depth, and sampling in sheltered areas or open sea. Smith and Howard[29] compared the sampling efficiency of the Smith-McIntyre grab sampler and a spade corer using microfaunal abundance, biomass, and size as indicators, finding that the spade corer was more efficient than the grab sampler due to the greater depth penetration. Christie[30] found an exponential relationship between the grab sample volume and sediment texture, as well as a linear relationship between the grab sample volume and species abundance in samples obtained from water depths between 280 and 440 m. A design of a new benthic grab sampler was based on previous experience. Details of the design and relative merits of the new sampler were described.[31] The efficiency of the Day and Smith-McIntyre grab samplers were compared in sampling benthos organisms.[32] No significant differences were found between the grabs in the sampling of the shallow burrowing infauna. However, the Day grab sampler was more efficient in sampling deep-burrowing species. Bascom[33] described 15 types of devices suitable for the characterization of water, biota, and bottom sediments in measuring pollution in the sea. He emphasized the necessity of careful operation of the devices by experienced technicians and the importance of a well prepared plan of investigation and analyses of samples. Dall[34] described a hand-operated grab sampler suitable for collecting replicate unit-area samples from the shallow stony literal zone in studies of zoobenthos. The performance of seven grab samplers weighing less than 25 kg (Van Veen, weighted and unweighted Ponar, Friedinger version of the Petersen grab sampler, Dietz-La Fond mudsnapper, pole-operated Birge-Ekman, and pole-operated Allan grab sampler) used in an operation from a small vessel was described.[35] The samplers were tested using four types of sediments ranging from fine-grained material to gravel. Jensen[36] compared the efficiency of the Van Veen grab sampler and HAPS corer in sampling benthos, and found that the corer gave the highest abundance and diversity with the least use of working resources. Hulle and Jestin[37] proposed a special device for sampling benthos communities in testing the hypothesis of homogeneity underlying the classical sampling method. Statistical tests showed significant differences in meiofaunal abundance in testing the efficiency of the Van Veen grab sampler, Phleger corer, and scuba divers in sampling silty sediments.[38] Six features were designed for a quantitative grab sampler for sampling benthos to make it outperform other grabs. The sampler was used in sampling sediments ranging from very soft silty-clay to very firm sand.[39] A hydraulic lift sampler was modified and its efficiency compared with the Petersen grab sampler using a hydrobiid snail as a test organism.[40] Both samplers were equally efficient, but the hydraulic lift sampler had a number of advantages. After evaluating the performance of different Ekman grab samplers and other sediment sampling equipment, Blomqvist[41] suggested that a properly operated Ekman grab of an adequate design remains a suitable instrument for collection of certain types of samples, such as benthic macrofauna, and infauna in particular.

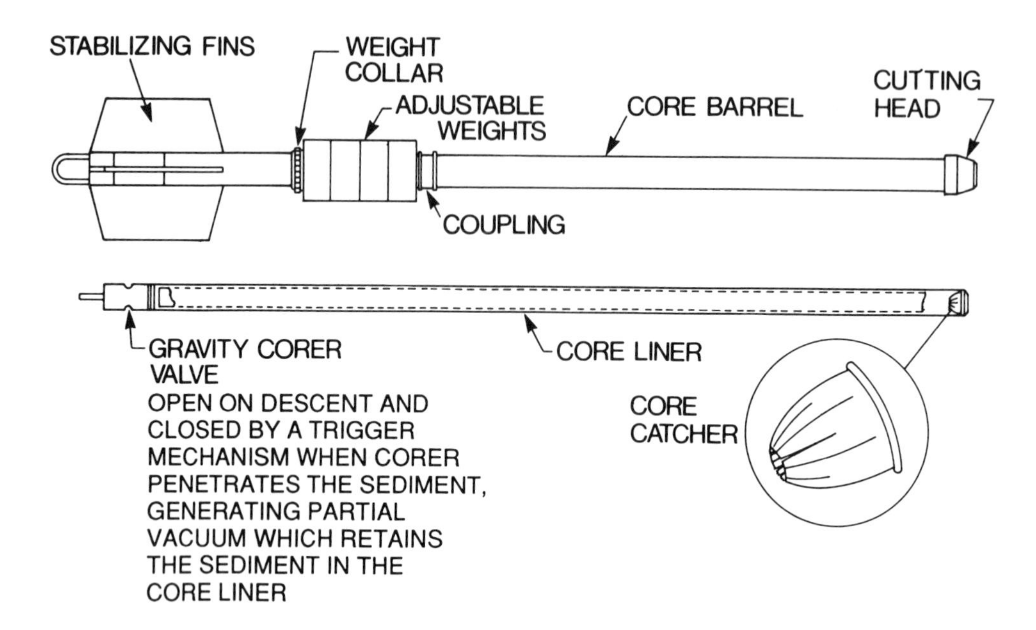

FIGURE 10. Typical parts of a sediment corer.

B. CORERS

Corers are fundamental tools for obtaining sediment samples for geological and geotechnical survey and, recently, for the investigation of historical inputs of contaminants to aquatic systems.

Generally, corers consist of a hollow metal or, rarely, a plastic pipe, the core barrel, varying in length and diameter; easily removable plastic liners or core tubes which fit into the core barrel and retain the sediment sample; a valve or piston mounted on the top of the core barrel which is open and allows water to flow through the barrel during descent, but shuts upon penetration of the corer into the sediment thereby preventing the sediment from sliding from the corer during the ascent; a core catcher to retain the sediment sample; a core cutter for better penetration of the sampler; removable metal (usually lead) weights to increase penetration of the corer into the sediment; and stabilizing fins to assure vertical descent of the corer. Typical parts of a corer are shown in Figure 10.

The cutter is mounted on the end of the core barrel to achieve better and deeper penetration into the sediment. Commercially available cutters are typically made of stainless steel or steel, and have a screw, bayonet, or setscrew type mount. Brass or plastic cutters have also been used. The cutting edge should be easy to sharpen when it becomes dull or damaged by gravel or other materials in the sediment. A core catcher is inserted inside the cutting head of the corer to prevent loss of the sediment during retrieval. The sample is retained in the liner by a series of spring-loaded metal fingers which allows the sediment to enter, but not to fall out of, the liner. A core catcher is less effective in soft sediments with a high water content than in consolidated ones. It also "rakes" the sample during entry disturbing sections of the core. There is also a possibility of contamination from lubrication of the spring which is a part of some core catchers. Plastic core catchers can be used but have a poor ability to rebound to their former shape. Different types of core catchers were reviewed by Bouma.[3]

TABLE 2
Estimated Weight of Dried Uppermost 1 cm Sediment Layer Subsampled from Core Liners of Different Diameters

I.D. tube (cm)	Approximate weight of dry material in uppermost 1 cm sediment layer[a]	
	Soft fine-grained sediment[b] (g)	Firmer silty clay[c] (g)
3.5	0.7—1.4	1.1—2.2
5.08	1.5—3.0	2.3—4.6
6.6	2.6—5.2	3.2—7.9
10.0	5.9—11.8	8.8—17.7

[a] Based on 90—95% sediment water content.
[b] Based on specific gravity of sediment 1.5 (high organic matter content).
[c] Based on specific gravity 2.3 (low organic matter content).

1. Size of the Core Tube

Different dimensions of core tubes (metal tubes or plastic liners) used with the corers affect the quantity of the recovered sediment sample. The sample size is important, particularly when more parameters need to be determined in the sediment sample. An important part of the core, the uppermost 1- to 3-cm layer, often consists of soft material with a high water content. Estimated weights of an uppermost 1-cm layer of sediment subsampled from core liners of different I.D. are given in Table 2.

The quantity of the sediment in a core tube below the topmost 1-cm layer gradually increases due to the lower water content and sediment compaction. In fine-grained material the water content is about 80% at 10 cm sediment depth, 70% at about 20 cm sediment depth, and about 50 to 60% at 30 to 40 cm sediment depth. Below 50 cm the sediment usually becomes more compacted and there is little change in the water content. An exception can be sediment with high organic matter content from small lakes with restricted water circulation. Water content in this type of sediment can be 90 to 95% at 50 to 100 cm sediment depth, and sampling will require a special coring device to retain the sediment in the core tube during retrieval.

Several specific coring devices have been developed for sediment sampling:

- Single gravity corers featuring a core barrel penetrating the sediment by gravity and collecting up to 2 m of sediment
- Multiple gravity corers, featuring 2- to 4-core barrels
- Box corers for collecting a rectangular sample from the upper 50-cm sediment layer
- Piston corers featuring a core barrel with a liner and piston for collecting cores 20 m (and longer) at deep water
- Boomerang corer for taking samples from the seafloor
- Vibracorers featuring a vibrating device and a stationary piston for collecting samples from hard clays, shales, and recent calcareous sandstones[42]

Commercially available coring equipment ranges from a small, hand-operated corer that can be used in shallow water, to a heavy, oceanographic core sampler.

The shock wave produced during the free-fall of most corers can disturb the surface sediment. Open core barrels and core valves with unrestricted water flow allow water to pass freely through during the lowering of the equipment to the bottom and limit the shock wave. The degree of disturbance appears to be a function of the speed of impact and the surface area of the core. With the lowering of the speed of entry of the corer into the sediment, the sediment disturbance becomes smaller, but there is also less penetration and, consequently, a shorter core is recovered. The effects of the speed of entry of coring equipment on the retention of the surface sediment characteristics were investigated by Baxter et al.[43] The standard free-fall corer created a greater disturbance of the sample than a soft-landing corer with a slow velocity entry creating a greater loss of very fine surface material.

During the sediment penetration by a corer, frictional resistance results in sediment deformation and compaction. Thin, smooth walls and sharpening the lower end of the core tube to a small angle reduce frictional resistance.

A corer may enter at an angle not perpendicular to the bottom or, in the worst case, the corer may plunge sideways into the sediment. This usually occurs when the vessel drifts during sampling or the corer is lowered too quickly. If the upper sections of the coring device emerge upon sample retrieval covered with sediment, there is a strong likelihood that the corer tilted in the sediment because the rate of entry was too fast or it encountered a firm underlying material, such as compacted sand or rock. A sampler covered with sediment may also be a sign of over-penetration by high speed entry into fine-grained sediments. The sloping surface of the sediment recovered in a core tube indicates the corer penetrated at an angle.

The degree of disturbance of a sediment core that can be tolerated varies with the intended use of the results obtained. Studies of sediment contaminants are primarily concerned with the profile of the concentrations of recently deposited contaminants in the sediments. Assuming an annual rate of sediment deposition of 0.01 to 1.00 cm, the concentration of contaminants in the uppermost 1-cm sediment layer indicates the input of contaminants during the past 10 years for an area with a sedimentation rate of 0.1 cm/year. The loss of a few millimeters of the surface sediment will underestimate the contaminant loadings to the site.

Sediment cores with a minimum of disturbance are required for continuous long-term monitoring of concentrations of contaminants in sediments carried out to assess the efficacy of remedial actions implemented in the drainage basins of rivers and lakes, and to determine the rate of natural burial of contaminants in sediments by deposition of clean material. For example, a new waste treatment technology was implemented at a plant discharging its effluent into a stream entering a lake in a remote area. The efficacy of the new technology and the burial of contaminated sediments may be assessed by sampling and analyzing sediments collected at depositional areas in the lake over next 15 to 20 years. The changes in the concentrations of contaminants in the sediment column will reflect the changes of contaminant concentrations in the effluent and indicate the rate of burial of contaminated sediments. Consequently, sampling an undisturbed sediment column will be required during the 15 to 20 years of the monitoring.

a. Hand Corers

Most hand corers are suitable for collecting soft or semicompacted sediment samples by hand in marshes, tidal flats, rivers, and other shallow water areas, or in deep water by a diver. Different models, commercially manufactured, usually consist of a metal or plastic core tube 3.5 to 7.5 cm I.D. and extension handles on the top end for driving the corer into the sediment. Colored plastic caps should be available for sealing and identifying the top and the bottom of the core tube or the liner. Commercially supplied hand corers can have extra handles of various length (about 1 to 5 m) which, attached to the hand corer, allow the collection of samples from depths equal to the length of the handle with the corer. The core tubes and liners are usually 50 to 120 cm long, threaded on both ends and tapered on the bottom for easier penetration of the sediment. A nose piece can be attached at the bottom and, if found advantageous, a core catcher can also be installed. The weight of hand corers varies from 5 to 17 kg; the extension handles add another 4 to 12 kg.

b. Single Gravity Corers

i. Manually Deployed Corers

There are few corers which can be operated without a mechanical winch. Phleger and Kajak-Brinkhurst corers and their modifications are the representatives of this group. Surface sediment sampling from a small vessel can be a single-person operation; however, at least two operators are required to stabilize the vessel at the sampling station, to lower the corer over the board, and cap the bottom of the retrieved core.

Phleger Corer
Weight: about 8 kg (without lead weights, additional 7 kg per each added lead weight)
Core tube size: 3.5 cm I.D.

The Phleger corer (Figure 11) is suitable for sampling different types of sediment ranging from soft to sandy, semicompacted material, as well as peat and vegetation roots in shallow lakes or marshes. The length of the obtained core is up to 50 cm. A relatively narrow 3.5-cm-I.D. core liner recovers a small quantity of material, which is a disadvantage, particularly when the core needs to be subsampled into small sections. The core barrel has a bayonet fitting nose cutter. The upper part of the core barrel screws into a further section of tubing on which ring weights are mounted. The upper tubing supports the weight rings and provides excellent vertical stability during core descent. The upper tube is capped with a valve assembly consisting of a neoprene bung mounted on a metal pin which slides in two locations. The bung is slightly tapered and fits into a similarly shaped metal seating. The water pressure within the tube forces the bung up and clear of its seat on lowering and penetration. On withdrawal, the pressure is maintained within the tube by the bung as it slides back into its seat, sealing perfectly, thereby retaining the sediment sample.

Kajak-Brinkhurst (K-B) Corer and its different modifications
Weight: about 9 kg (standard size), lead weight 7 kg
Core tube size: 5 cm I.D.
Core tube length: 50 cm, 75 cm

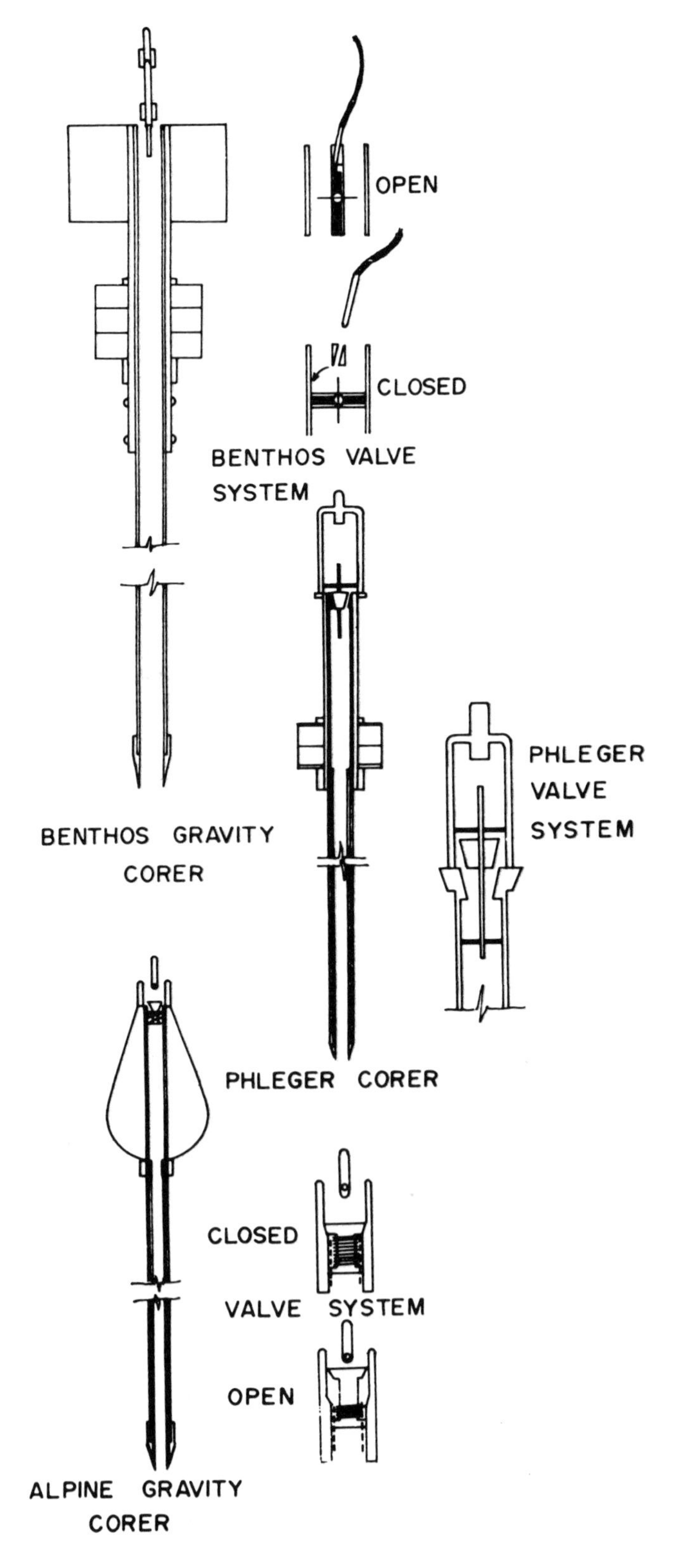

FIGURE 11. Phleger, Benthos, and Alpine gravity corers with their valves.

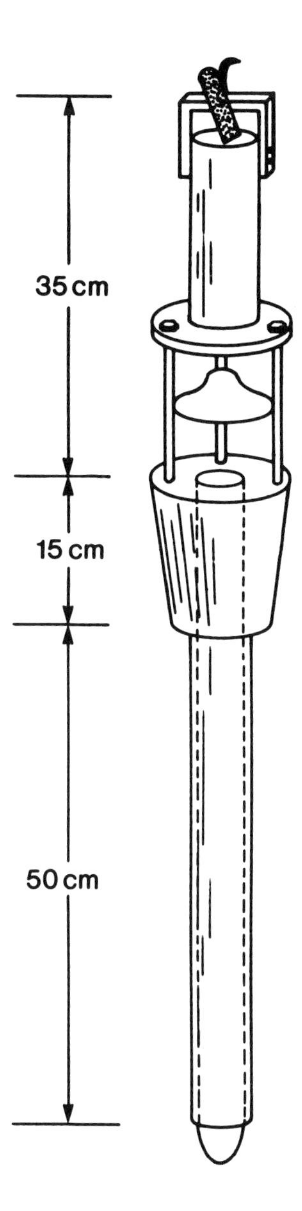

FIGURE 12. Kajak-Brinkhurst corer.

The corer (Figure 12) is suitable for sampling soft, fine-grained sediments and recovering up to about 70-cm-long cores. The standard K-B corer is a messenger-operated sampler with unrestricted water flow during its descent. The corer is suitable for sampling soft, fine-grained sediments. The closure of the messenger-operated valve allows the operator to choose the closing time of the valve when he feels that the sampler has sufficiently penetrated the bottom sediment. Closing the valve by the messenger creates a partial vacuum inside the core tube during the ascent of the sampler from the bottom, and assists in the retention of the sediment in the tube. The K-B corer was improved by the addition of an automatic trigger mechanism to replace the messenger. The major problem in using the messenger was that after the corer vertically penetrated the sediment, the line was often "streaming", e.g., was not vertical from the water surface to the corer, in which case the messenger would not

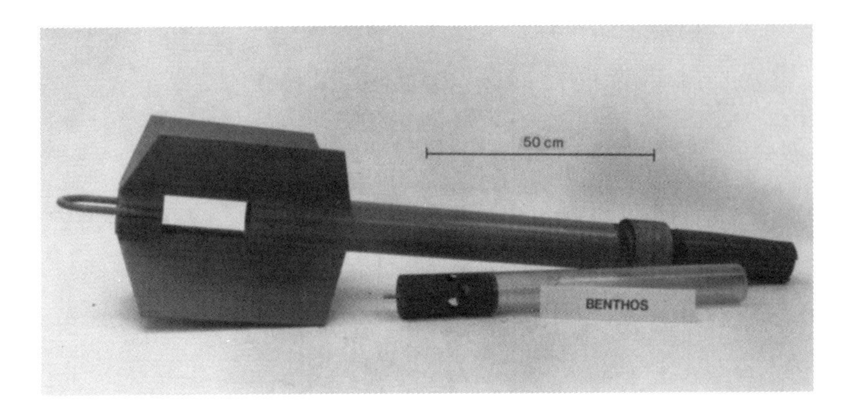

FIGURE 13. Benthos gravity corer.

properly activate the valve. There are various types of this corer commercially available with various accessories, such as stabilizing fins, extra weights, and various core tubes (PVC, Lexan, brass). The standard K-B corer can be operated manually, but a winch is recommended for a "heavy" K-B corer with accessories and 75-cm-long core tubes. A 5-cm-I.D. core tube used with this corer recovers a greater quantity of sediment than the 3.5-cm-I.D. core tube used with the Phleger corer.

ii. Winch- or Crane-Deployed Corers

Benthos Gravity Corer

Weight: 25 kg and up to 6 × 20 kg of additional lead weights
Core tube size: 6.6 cm I.D., 7.1 cm O.D.
Required lifting capacity: 350 to 500 kg

The Benthos gravity corer (Figure 13) was designed to recover up to 3-m long cores from soft, fine-grained sediments. On the recent model stabilizing fins on the upper part of the corer promote vertical penetration into the sediment. To enhance penetration up to 6 × 20 kg weights can be mounted externally on the upper part of the metal barrel. A valve system at the top of the liner prevents loss of the sample from the tube. The valve is fitted to the top of the core liner which is then inserted into the core barrel. The valve is a critical part of the corer and the success of sampling depends on its proper operation. This led to various designs of the valve by the corer manufacturing company. For example, the valve presently used at the National Water Research Institute, Burlington, Ontario, is an auto-valve which is held open by the water flow during descent and penetration. Upon retrieval, the suction of the sediment attempting to slide out of the core tube and the force of the spring push the plunger into a machined seat. The created vacuum holds the sediment in the tube. The valve should always be carefully sealed in the liner and its operation regularly inspected.

Alpine Gravity Corer (model 211)

Weight: 110 kg, lead weight about 45 kg
Plastic liners: 3.5 cm I.D., 3.8 cm O.D.
Required lifting capacity: 500 kg

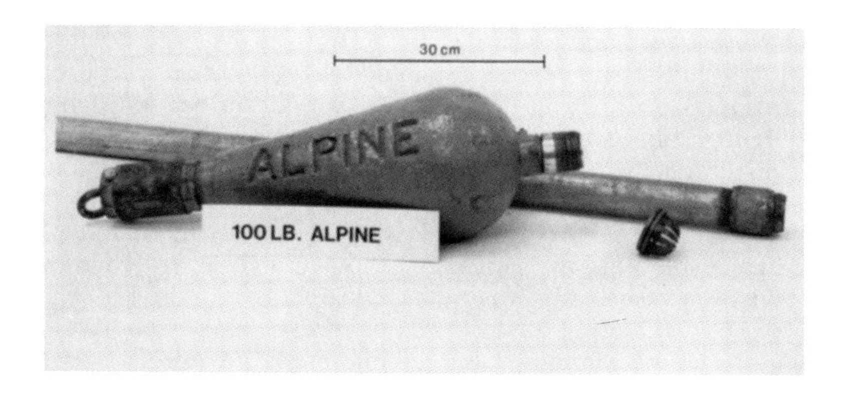

FIGURE 14. Alpine gravity corer.

The Alpine gravity corer (Figure 14) is finless and has an interchangeable steel barrel (4.1 cm I.D., 4.8 cm O.D.) in lengths of 0.6, 1.2, and 1.8 m. A lead weight is mounted on the corer above the barrel. Attached to the top of this is a combination attachment point/valve assembly. The valve system uses a light compression spring to retain a plastic and rubber leg and cap assembly against a beveled, circular seat. During penetration, the increased pressure in the barrel causes the cap assembly to lift off its seat and allow the necessary displacement of water from the barrel. Penetration and pressurized displacement cease simultaneously. The compression ring then forces the cap valve to retreat and seal, prior to withdrawal.

Sly[5] tested an Alpine gravity corer with extremely variable results. The most successful cores were obtained by allowing free-fall for a distance up to twice the barrel length used. However, due to the lack of fins, vertical penetration of the corer was not obtained in many cores. The worst entry observed was approximately 25° from the vertical. Under good working conditions the angle reduced to 5°, and the corer embedded itself deeply in the mud. Sheared laminae and disturbed surfaces were observed on radiographs of hundreds of cores.

In addition to these corers there are several other corers available from different commercial suppliers of aquatic sampling equipment.

iii. Published Information on Single Gravity Corers

There is extensive published information on design, operation, and evaluation of different gravity corers used in the past. Pettersson and Kullenberg[44] described a corer modified for sampling deep-sea sediments. A gravity corer was redesigned to permit proper interpretation of core samples.[45] Piggot[46] tested sediment coring, particularly the penetration of the coring instrument and the length of the recovered core. Hvorslev and Stetson[47] described the efficiency of a free-fall coring tube in sediment sampling, particularly in coring firm sediments. A gravity corer equipped with a simple core catcher was used in studies of foraminifera.[48] Heezen[49] discussed different problems observed during sediment coring using various instruments, such as the estimation of the time when the corer reaches the bottom and proper tripping of the instrument. A portable corer was described suitable for a two-man operation from a rowboat to collect 6-m long undisturbed sediment cores at a water depth up to about 250 m.[50] An underwater camera was used to monitor bottom coring operations with a Stetson corer.[51] A plastic-barrel sediment corer, operating with or without a piston, and its operation were described by Richards and Keller.[52] Photographic studies, mechanical recording, detailed stratigraphy, radiocarbon dating, and temperature measurement of the sediment indicated that sediment cores obtained by open barrel gravity corers were shorter than the depth of penetration.[53] A coring device was described which could penetrate up to 1.5 m of shallow marine sands. The corer could be operated from a small launch or fishing vessel, and was automatically converted upon withdrawal from the sea bed, retaining the loose sands in the core barrel.[54] Willermoes[55] used a sampler suitable for quantitative sampling of the unicellular organisms such as foraminifera, nematodes, and harpacticoids. This sampler obtained a core several centimeters long with sediment from an area of 3 cm^2 from the sea floor. A 15-cm-diameter coring device was developed and successfully tested with barrels up to 6 m long.[56] An undisturbed sediment core obtained by this device was suitable for sediment analyses requiring large volumes of sediment. Tests of behavior of free-fall small, lightweight gravity corers indicated an optimum free-fall setting of 2 to 3 m, with general caution on the minimum setting and specific caution on the maximum one for corers not fitted with stabilizing fins.[57] A wide-diameter corer with a watertight core catcher was described together with an electric release system which prevented accidental triggering by sudden shocks.[58,59] Disturbance was negligible or not observed in cores obtained with this device. Developments and tests at sea led to the design of modified and improved coring devices for taking large, well-preserved, oriented cores in deep water.[42] Burke[60] described a coring device with core barrels of 21 cm diameter, equipped with a valve at the top of the barrel and a nose cone with a core cutter and core catcher at the leading edge of the core barrel. The corer was suitable for the recovery of 1-m-long cores at the water depth up to 600 m. Inderbitzen[61] used shear strength, water content, and unit quantity of sediments in evaluating performance of seven different corers, and found that each tested corer appeared to yield good results for at least one of the measured properties but yielded the worst results for another property. A simple large coring device weighing approximately 140 kg with a 7-m-long tube, 13 cm I.D., equipped with a novel piston and pullout relief mechanism, was constructed and described.[62] Menzies and Rowe[63] described a large sampler for quantitative sampling of soft bottom sediments in studies of benthic organisms. A stabilizing framework was designed to fit around the Knudsen sampler to prevent the instrument from falling on its side before sediment penetration.[64] The design of a hand-operated short corer for sediment sampling was described by Mackereth.[65] A simple corer made of brass with a plastic lining tube was successfully used for the sampling of soft estuarine muds.[66] A corer with a top valve and a core catcher, supported by a frame, was successfully used for sampling soft sediments and sands. The sampler was considered suitable for *in situ* studies of bottom invertebrates.[67] Keegan and Konnecker[68] described a corer designed for the collection of 50-cm-long cores with minimum disturbance. The device was suitable for *in situ* quantitative sampling of benthos organisms. A simple lightweight gravity corer using an acrylic glass tube was described by Meischner and Rumohr.[69] A hydraulically operated device for obtaining cores of 30.5 cm diameter and up to 46 cm long was described by Thayer et al.[70] This corer was able to penetrate the firm bottom and could be operated in water depths of up to 4 m. Axelsson and Håkanson[71] discussed the general principles of coring with open-barreled gravity corers. They described a gravity corer with a new, simple, and efficient valve system and rectangular coring tubes specially designed for scanning the cores by X-radiography before extrusion. A 12-m version of the Mackereth corer with a magnetic orientation system was described by Barton and Burden.[72] Hongve and Erlandsen[73] tested light-weight open-barrel gravity corers and found that shortening of the sediment cores depended on the diameter of the corer and its velocity during sediment penetration. Experiments showed that soft sediment layers were more reduced in thickness than the stiffer ones. A lightweight corer was developed for the collection of 10-cm-diameter cores up to 1 m long from unconsolidated, fine-grained fluvial and lacustrine sediments including organic-rich deposits.[74] This corer was hand-operated in sediment sampling from float planes and small vessels. Avilov and Trotsyuk[75] described a device for recovery of undisturbed sediment samples. This equipment permitted hermetic sampling of bottom deposits and their vacuum degassing on board for measuring gases in sediments. The performance of a conventional small-diameter gravity corer (6 cm I.D.) and the Craib corer, designed to soft-land on the seabed and retain intact, the light, superficial sediment layer, were compared.[43] The results of the study showed that the small-diameter gravity coring, and possibly other techniques using heavy and high-velocity sampling devices in unconsolidated surface sediments, can cause an extensive loss of material and distort the vertical distribution of sedimentary parameters of interest. Lebel et al.[76] compared the performance of gravity and box corers in studies of profiles of pore water alkalinity and dissolved iron, manganese, and phosphates. Increased alkalinity gradients in pore water indicated significant shortening of sediment cores collected by the gravity corer. A relative shortening of the cores was also observed in the concentration profiles of the other parameters. It was suggested that a concurrent box core or other independent measure of the *in situ* gradients be obtained to correct for the shortening of a core recovered by a gravity corer. A corer for surface sediment sampling in shallow (<6 m) water was developed, described, and evaluated.[77] To core unconsolidated sediments, particularly clay and sand, in an intertidal environment with the minimum disturbance, a portable hand-coring device was developed with a telescopic tube and a stationary piston mounted on a tripod frame.[78] Satake[79] described a small, light corer suitable for sampling recent sediment deposits at up to 300 m water depth for chemical and microbiological analyses of sediments. McCoy and Selwyn[80] described coupling of a hydrostatic motor to a corer for the recovery of deep-sea sediments. The equipment enabled driving a gravity (open-barreled) corer into stiff marls where the penetration by a conventional gravity corer was unsuccessful. A lightweight (7 kg) coring device was described[81] suitable for sampling soft and hard substrates in shallow lakes (<1 to 4 m water depth) in studies of sediment chemistry, biology, and microbiology. Using concentration profiles of Zn and Pb as indicators, Evans and Lasenby[82] compared the performance of a modified K-B corer lowered slowly into soft, recent sediments from a vessel with facilities for hand-coring sediments by a scuba diver, and obtained comparable results for a core tube of 3 cm diameter. Twinch and Ashton[83] used a gravity corer with a continuous-flow adaptor in their studies of sediment/water chemical gradients to evaluate potential exchange rates. Blomqvist[84] studied the shortening of sediment cores, and proposed an accurate correction factor for the degree of core shortening. A suction corer was evaluated in sampling benthic fauna.[85]

Large-volume, undisturbed sediment cores, approximately 3 m long, were obtained using a modified gravity corer with a square cross section 12.7 × 12.7 cm.[86] The corer was designed to obtain and subsample undisturbed fine-grained sediment cores, particularly at the sediment-water interface. Pedersen et al.[87] evaluated a lightweight, portable gravity corer suitable for use with small (less than 5-m-long) vessels for recovery of up to 3-m-long, undisturbed sediment cores.

c. Multiple Corers

Multiple corers typically consist of several core barrels mounted on a single fin and weight system. They have been developed for multiple sampling at one site, comparative studies, evaluation of sediment sampling precision, and determination of sediment heterogeneity over a small area.

For example, a Triple Benthos corer, built at the Canada Centre for Inland Waters, is based on the same operational principles as the single Benthos gravity corer. It has an outer ring which houses three evenly spaced core barrels welded to a fourth nonfunctional barrel. The valves and trigger mechanism are original Benthos products. The core tubes are 50 cm in length and have an outside diameter of 7.1 cm.

Brinkhurst et al.[88] described a multiple unit of the K-B corer and its application in studies of sediment biota. A lightweight multiple corer (approximately 7.9 kg) was used for sampling the benthos of profundal sediments.[89] The sampler collected simultaneously four cores, and tests indicated that in soft sediments these samples were generally superior to those collected with most of the tested conventional samplers. A triple corer was designed based on the modification of the Benthos gravity corer.[90] The equipment collected satisfatory quality cores from fine-grained sediments at water depths up to 250 m. Jones and Watson-Russell[91] described a diver-operated, lightweight multiple coring system for collecting undisturbed sediment samples.

d. Box Corers

Box corers are gravity corers that were developed in late the 1950s, and later modified and refined to improve their operation.[3,92-95] They were designed for collecting large rectangular sediment cores in biological and geological studies at various water depths, variable penetration rates, and different sediment types. There are two basic designs to the bottom mechanism of the box corer: (1) the "Ekman" design in which there are two bottom flaps which can be triggered and closed much like the Ekman grab sampler, and, (2) the "Reinecke" design in which a large shovel-like device is activated and slides across the bottom of the box corer. There are several box corers of different design and size commercially available. Recognition of the excellent quality of the undisturbed sediment samples collected by box corers, particularly in studies of sediment-water interface, initiated design and construction of different custom-made box corers for special requirements.

Generally, a box corer consists of a stainless steel box of a variable size. Most box corers are equipped with a frame which ensures vertical penetration also on low slopes, and stabilizes the sampler on the bottom. Due to its large weight, up to 800 kg, size (up to 2 × 2 m), and required lifting capacity on the order of 2000 to 3000 kg, the box corer can be operated only from a vessel with a large lifting capacity and sufficient deck space. Unless specially designed, the core retainer of a box corer can be damaged by penetration into a thin layer (less

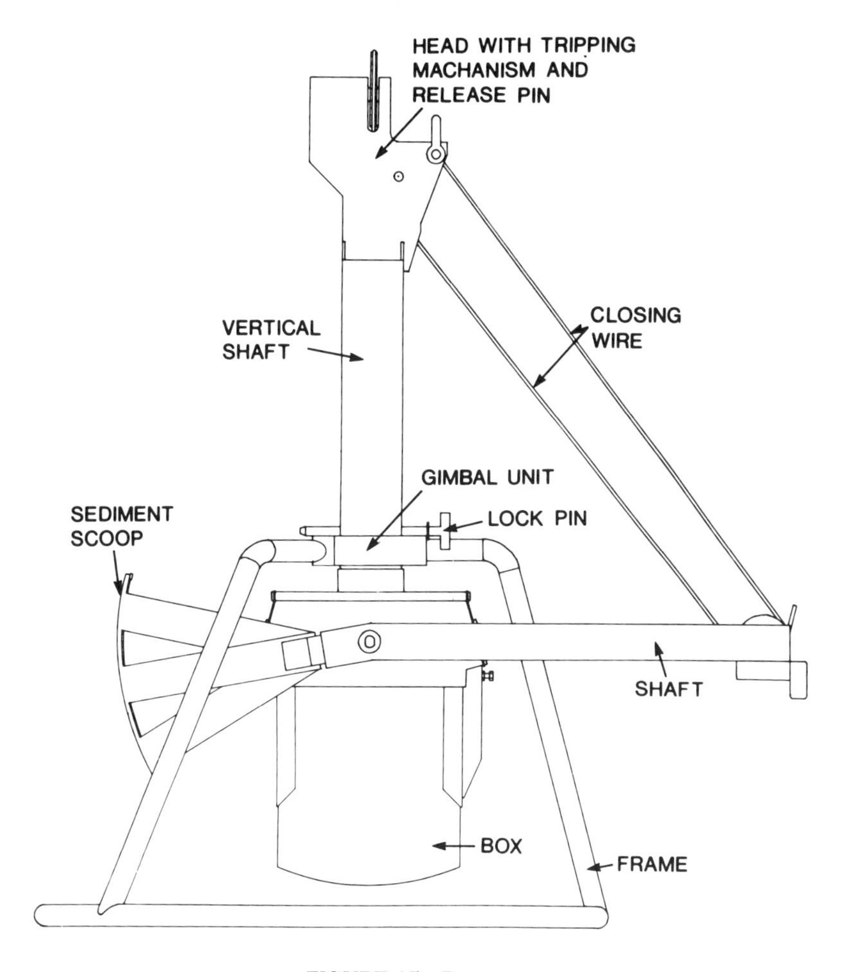

FIGURE 15. Box corer.

than 30 cm) of soft, fine-grained sediment underlain by firm material containing gravel or boulders. Therefore, such box corers should be used only in areas with a minimum of 1-m layer of soft, fine-grained sediments. Box corers are triggered mechanically when they reach the bottom. However, the actual sediment coring is carried out after the device is on the bottom.

The box corer described below and shown in Figures 15 to 19 was redesigned and custom-made using the design of the box corer described by Bouma[3] and is used at the National Water Research Institute, Burlington, Ontario. The box corer consists of a number of basic parts: gimballed frame, control stem with two box holders, closing mechanism, tripping mechanism, and sampling box. The central frame slides through the gimballed top of the frame in such a way that samples are always taken vertically. The closing mechanism consists of a blade at the end of a double arm which pivots about the box holder. A tripping mechanism on top of the central stem makes it possible to use only one wire for lowering, sampling, closing, and returning to the surface. Very little free-fall is possible which allows penetration into the sediment primarily based on gravity. Weights can also be added to the central stem by removal of a plate on the side to have deeper penetration of the sediment. The hollow pipe frame ensures vertical penetration at slopes up to 18° and prevents the sampler from falling over on the bottom.

FIGURE 16. Removal of the safety pin prior to the lowering of the box corer to the lake bottom.

FIGURE 17. Retrieval of the box corer.

The sediment inside the box corer can be subsampled by inserting core tubes into the sediment (Figure 20). The top of the core tubes has to be sealed by a core cap to prevent the sediment from falling out during recovery of the cores from the box corer. Since all subsamples come from the same small area of the

FIGURE 18. Lowering the box corer with sediment to the ship deck.

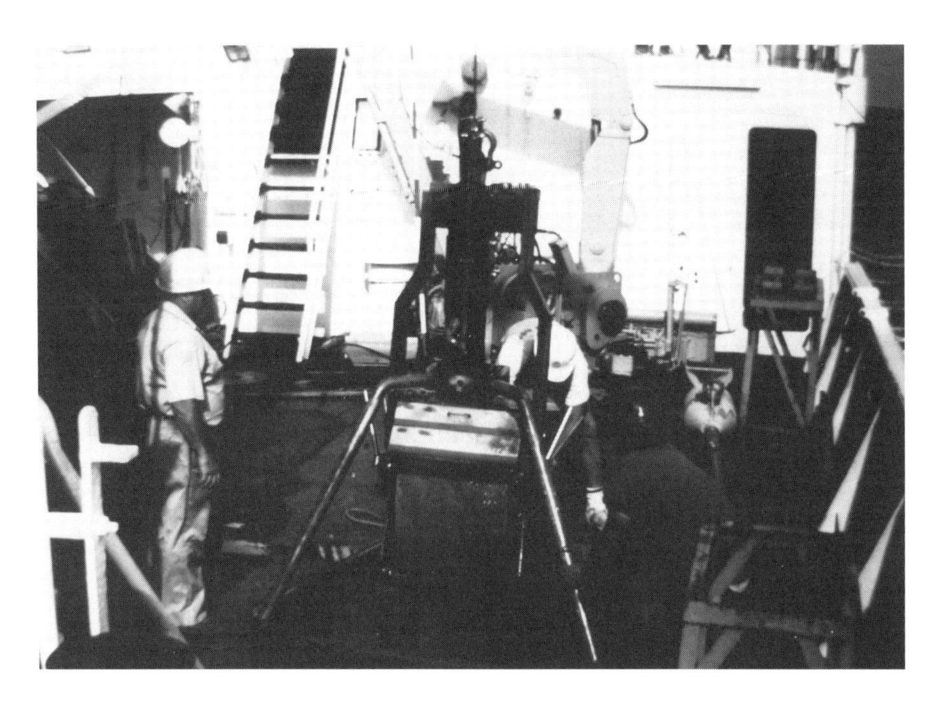

FIGURE 19. Retrieved box corer with the sediment sample.

bottom, the results of any analyses and studies of the sediment carried out at various laboratories can be compared. The lack of disturbance of the sediment recovered in the box corer can be ascertained by the sediment appearance. Hand coring of the sediment from the box corer allows good control of the

FIGURE 20. Inserting core tubes into the sediment collected by the box corer.

compaction of the sediment upon driving the core tubes into the sediment, particularly when clear plastic tubes are used. Recently, a box corer was developed which enables horizontal subsampling of the entire sediment volume recovered by the box corer.[96] An improvement of soft bottom sediments sampling by combined box coring and shipboard sampling using hand-operated piston coring was described by Blomqvist and Boström.[97]

Oriented, undisturbed cores were obtained with an improved German coring device[98] at any water depth. Rectangular sediment samples with an area of about 20 × 30 cm were a maximum of 46 cm long.[92] Construction and use of the box corer and its application in some investigations, such as studies of living organisms and shear strength measurements, were described. A similar box corer[99] was used for retrieval of a large amount of undisturbed sediment for different investigations. The 0.2-m^2 Van Veen grab and the 0.06-m^2 Reineck box corer were tested by comparing numbers and size of the macrofaunal benthos species, and the biomass.[100] The burrowing depth of different species could be observed from the undisturbed box corer samples. It was found that the depth penetration of the grab sampler depended on the type of sediment with small penetration (5 to 6 cm) in relatively fine sand. However, the depth

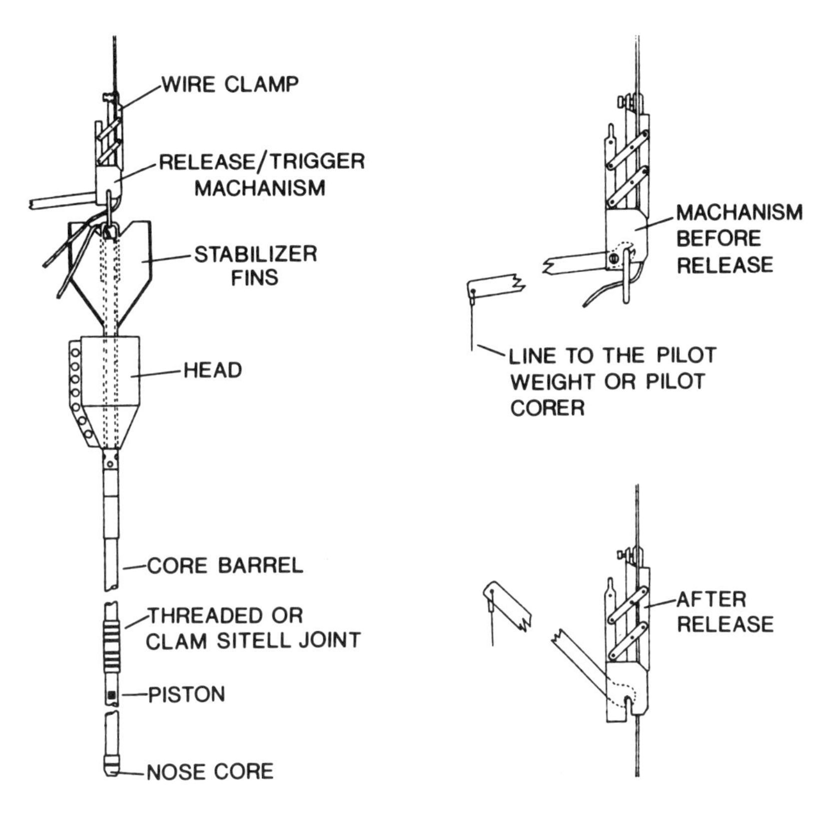

FIGURE 21. Typical parts of a piston corer.

penetration of the box sampler was invariably more than 15 cm, which was sufficient to obtain all infaunal species. Low-cost, plastic box corer liners were developed, and their construction and functioning described by Karl.[101] Dolotov and Zharomskis[102] studied the textural features of recent sediments in the offshore shelf zone with a modified Reinecke box-type sampler. Sundby et al.[94] used a box corer in studies of the distribution of Mn in bottom sediments and the water column in the Gulf of St. Lawrence. The sediment from the box corer was subsampled by scraping off successive layers with a teflon-coated spatula. Carlton and Wetzel[103] described a box corer for sediment-water interface studies of an epipelic microorganism, which permitted sampling of a combined sediment and overlying water volume of about 29 l (cross-sectional area >700 cm^2) with minimal disturbance of sediment structure.

e. Piston Corers

The piston corers are usually used for the studies of bottom sediment stratigraphy in oceans and deep, large lakes and can typically recover relatively undisturbed cores 3 m in length, or up to about 20 m long. A piston corer consists of a weighted stabilized head, a core barrel with a plastic core liner, a piston, a core retainer and cutting head, and a trigger mechanism (Figure 21). Normal deployment (i.e., limited free-fall from the water surface) cannot be undertaken due to the very heavy head weight. The purpose of the triggering mechanism is to permit the main corer to free-fall over a known, but relatively short, distance. The correct adjustment of the length of wire on the trigger-weight and in the main corer enables the core barrel to slide past a piston which remains stationary at the sediment surface. The piston creates a partial

FIGURE 22. Lifting the piston corer before swinging it outboard.

vacuum, facilitating sample entry, reducing core compaction, and promoting sample retention. The core barrel may be forced into the sediment by weight or by vibratory action. The large weight and size of the piston corer require a large vessel with heavy-duty cranes with lifting capacity over 2000 kg, and experienced operators. Figures 22 to 24 show the operation of a piston corer in sampling Lake Ontario sediments.

Sediment sampling by piston corers has been described in many reports and journals. A simple quantitative experiment was conducted to determine possible misinformation on sediment sequence resulting from improperly functioning piston corers.[104] The nonactivated piston appeared to be the major difficulty. Difficulties with the precise adjustment of the position of the piston at the time of impact introduced additional problems in obtaining an undisturbed sediment core. Sources of similar problems causing misinformation from sediment cores obtained by piston corers were discussed by Heezen.[105] Richards and Parker[106] evaluated gravity and piston coring equipment in terms of disturbance affecting the shear strength of the sediment, and concluded that conventional marine piston samplers had undesirable features that disturbed cores more than disturbance encountered in properly designed open-barreled gravity samplers. Ross and Riedel[107] evaluated the effects of the coring process on the mass physical properties of the sediments in three simultaneously collected open-barrel gravity and piston core pairs, and found that sediment cores collected by piston cores were shortened relative to the upper section of simultaneously collected cores by open-barrel gravity corers. Bouma and Boerma[108] found that vertical disturbance of piston cores, most likely originated by the upward motion of the piston during the first stage of pulling up the coring device, led to the sucking-up of sediment. Chmelik et al.[109] described the use of a flexible liner to minimize wall friction in a gravity, free-fall piston corer for recovery of undisturbed sediment cores. An automatic release piston

FIGURE 23. Preparation of the piston corer before lowering to the bottom.

for use in piston coring was developed which eliminates core disturbance due to suction caused by the movement of a conventional piston during pullout of the device.[110] Deep-sea core head camera photography was used to analyze the operation and effects of a piston corer.[111,112] The results of the investigation indicated that piston cores were shortened and disturbed, and often up to 1 m of surface sediment was missing. No consistent relationship was found between the length of the recovered core and the penetration of the device. Recommendations were made for decreasing core disturbance. A giant piston corer was applied in the studies of geotechnical properties of ocean sediments.[113-116] Nine piston and gravity coring devices were tested, compared with *in situ* measurements, and ranked according to the degree of sample disturbance.[117] After correction factors were defined and applied to core samples, the *in situ* strength of fine-grained sediments could be determined within 20%. Prell et al.[118] described a hydraulic piston corer for recovery of virtually continuously undisturbed sections of late Neogen and Quarternary sediments for biostratigraphic, paleoclimatic, and paleoceanographic studies in the Caribbean and equatorial Pacific. A piston corer was developed suitable for sampling peat. The corer was equipped with a serrated cutting edge on the end

FIGURE 24. Cleaning the piston corer pipes before lifting the corer to the ship deck.

of the core tube to cut undecomposed fibers and roots.[119] Kelts et al.[120] described an integrated sediment coring system modular for simple transport, which was successfully used in deep perialpine lakes, deep rift lakes of Africa, and other lakes in the world. Piston core lengths were variable in 2.4- or 5-m sections up to 16 m.

f. Boomerang Corer

The Boomerang corer (Figure 25) is a free-falling sampler designed to take sediment cores from the ocean floor without use of cable connections to the ship. It was designed and developed jointly by Benthos Inc., North Falmouth, MA, and the Woods Hole Oceanographic Institution, U.S. The corer consists of an expendable ballast portion, weighing 75 kg, and a recoverable float portion (11 kg). The ballast portion includes a nose cone, pilot weight, core barrel, ballast, float release mechanism, and a protective shroud for the glass float. The float portion consists of two glass spheres with a nylon net bag tethered to a core assembly composed of a 1.2-m clear plastic tube (liner), I.D. 6.7 cm, with a stainless steel core catcher on one end and a valve/release on the other. One of the two glass spheres contains a battery-powered electronic flash for nighttime recovery (5 km visible range) by a helicopter or a vessel. The glass

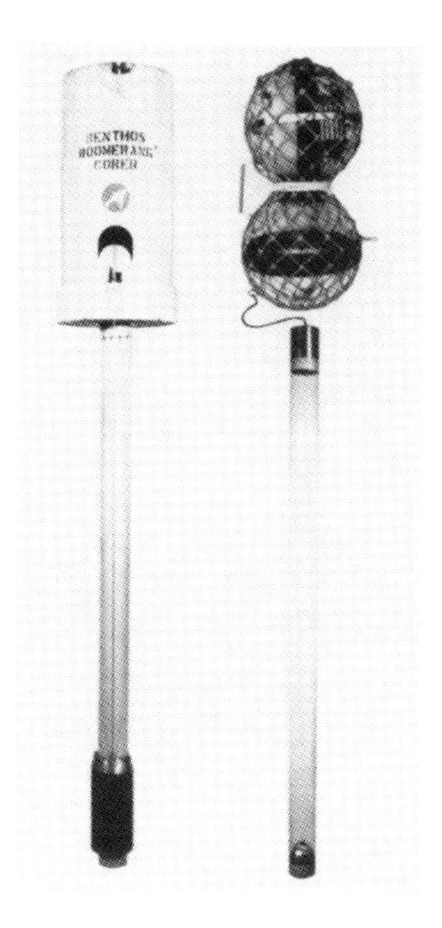

FIGURE 25. Boomerang corer. (Courtesy of Benthos, Inc., North Falmouth, MA.)

spheres allow 1.2-m core liners to float to the surface from depths up to 9000 m. The corer, weighing 86 kg (203 cm overall length with 30 cm O.D. shroud), can be dropped overboard from small or large vessels without stopping. The descent rate is approximately 450 m/min. Round trip time is approximately 15 min/km of depth. When the corer hits the bottom and penetrates the sediment, the release mechanism releases the floats that start to rise, closes a valve at the top of the core liner, and pulls the liner from the barrel. The floats and the liner rise freely to the surface, while the coring device and ballast parts remain on the seafloor. The float valve and core catcher are reusable.

Moore and Heath[11] considered the advantages and disadvantages of a boomerang corer as follows.

Advantages:

- Minimal requirements for shipboard equipment
- To take samples accurately located relative to each other and related precisely to its place on the chart of the bottom topography

Disadvantages:

- Only 1.2 m penetration
- Nighttime recovery when the flash can be seen for several kilometers; daytime recovery can only be made in the calm sea;

FIGURE 26. Vibracorer.

- Initial unit cost which is greater than that of a gravity corer
- Loss rate of 10 to 20%

Bowen and Sachs[121] and James[122] described the use of the boomerang corer in the deep-sea investigations.

g. Vibratory Corers

Typically, corers can only penetrate a few centimeters into sandy unconsolidated sediments. Vibratory corers overcome the resistance factor by a vibration action of the core barrel (either side/side or up/down). Vibratory coring systems are used mainly to assess geotechnical or structural properties of the sediment. They have rarely been used for obtaining sediment samples for the study of environmental pollution.

Sly and Gardener[123] described the construction of a simple vibro-type corer using a standard pneumatic impactor. The corer successfully recovered cores about 5 cm in diameter from gravelly sand, sand, and silty sand sediments. Structure and laminations in recovered sediments appeared to be well preserved, though there was evidence of some repacking of the core. Figure 26 shows a vibracorer similar to that described by Sly and Gardener. McMaster and McClennen[124] developed a vibratory coring system specifically to sample

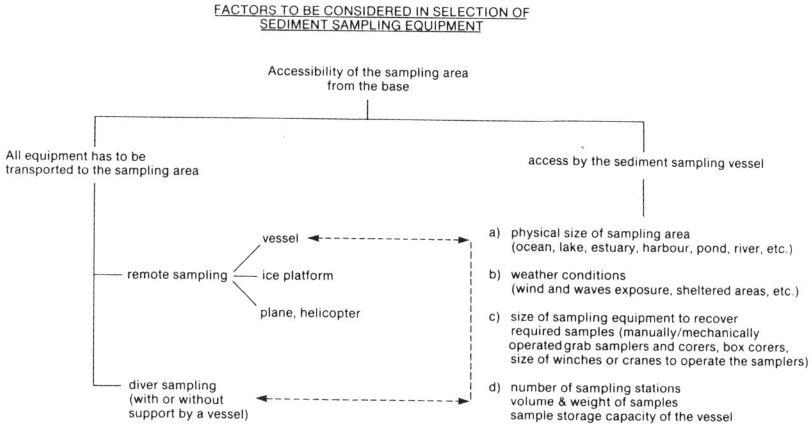

FACTORS TO BE CONSIDERED IN SELECTION OF SEDIMENT SAMPLING EQUIPMENT

Accessibility of the sampling area from the base

All equipment has to be transported to the sampling area

access by the sediment sampling vessel

remote sampling

vessel

ice platform

plane, helicopter

diver sampling (with or without support by a vessel)

a) physical size of sampling area (ocean, lake, estuary, harbour, pond, river, etc.)

b) weather conditions (wind and waves exposure, sheltered areas, etc.)

c) size of sampling equipment to recover required samples (manually/mechanically operated grab samplers and corers, box corers, size of winches or cranes to operate the samplers)

d) number of sampling stations volume & weight of samples sample storage capacity of the vessel

FIGURE 27. Factors to be considered in selection of sediment sampling equipment.

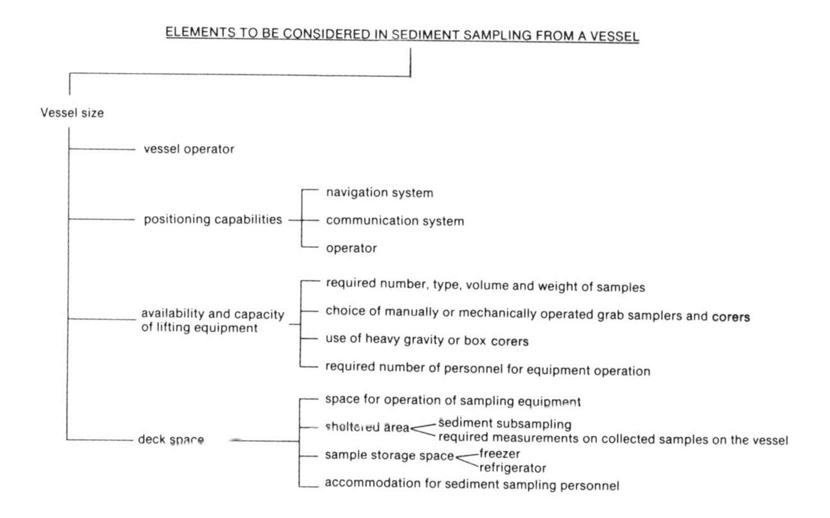

FIGURE 28. Elements to be considered in sediment sampling from a vessel.

sands on continental margins to obtain more information on sedimentary processes. Field tests indicated that the system was operational at all depths on the continental shelf. Dokken et al.[125] described a portable hydraulic vibracorer which recovered 7.5-cm-diameter, 5.4-m-long cores in unconsolidated sediments. The corer was operated by four workers from vessels as small as 10.5 m in length and was easily transported by truck and ship. Assembling and disassembling the corer required scuba divers. An effective and inexpensive method of vibracoring developed for acquisition of continuous cores up to 13 m long in unconsolidated sediments was described.[126] The weight of the system was about 150 kg and it recovered 4- to 6-m-long cores in sediments ranging from clay to sand. Fuller and Meisburger[127] described a simple, lightweight, pneumatic coring device for recovery of 3-m-long sediment cores.

III. SAMPLE COLLECTION

Factors which need to be considered in the selection of sediment sampling equipment are outlined in Figures 27 and 28. Access to the sampling area plays an important role in sampling strategy and logistics and selection of sampling equipment. There are basically two options for the collection of bottom sediment samples: sampling from a platform and sampling by a diver. Sampling platforms could be a vessel, ice, a plane, or a helicopter. Collection by a diver, usually more costly and difficult than sampling from a platform, often yields better quality samples, particularly sediment cores. In areas with a sufficient ice cover over the sampled water body, usually during the winter, sediment samples can be obtained by drilling a hole in the ice and sampling through this hole. The advantage of this sampling is a steady platform and a large space at the sampling station for assembling the equipment and processing the samples.

In areas with no road access, sediments may be collected from a small float plane or from a helicopter. Availability of a plane or a helicopter and cost are factors to be considered.

A. FACTORS TO BE CONSIDERED IN THE SELECTION OF BOTTOM SEDIMENT SAMPLERS

1. Bottom Sediment Sampling from a Vessel

The two primary factors in vessel suitability are the size and the depth of the water body being sampled and the proposed method of sampling: surface sediments or sediment cores. The vessel should offer reasonable on-board working space, particularly for subsampling collected sediments and measurements which need to be carried out immediately after sample retrieval. Large and heavy coring equipment dictates the use of a vessel with sufficient deck space and lifting capacity for sampling operations. Alternatively, the availability of a specific vessel will affect the selection of sampling equipment which can be safely operated from this vessel. Small grab samplers and corers can be hand-operated from a small vessel. Light portable winches, hand reels, or line keepers are commercially sold usually by companies supplying sediment sampling equipment. When more sediment samples need to be collected at an area with water depth greater than 10 m, portable winches can be easily mounted on the vessel for use with a variety of samplers within weight limitations.

Large sampling devices usually weigh between 50 and 400 kg empty. Filled with wet sediment they can weigh 125 to 500 kg (and more), and will require extra winch power to counteract the suction effect of cohesive bottom sediments. For example, a winch or crane with 2000 to 4000 kg capacity is required for a box corer. These samplers need a suitable winch or crane to operate with a lift height of 3 to 5 m above the gunnel of the vessel.

Sampler lines, cables, and depth meters are usually available on large vessels, and are commercially sold by companies supplying sediment samplers. The strength capacity of the lines and cables should be checked regarding combined weights of the sampling equipment to be used and that of the collected sediment.

Sufficient open work area on the vessel's deck and a various number of operators are required for launching and recovering sediment samplers, particularly larger ones. If samples are to be partially or completely processed on-board, laboratory space with running water, distilled water supply, and electric power should be included in the vessel equipment. Special equipment, such as a glove box with inert gas supply necessary for sediment subsampling under anoxic conditions, may also be required. A noncontaminated area for all field work is extremely important in studies of concentrations of trace contaminants in sediments. Lubricants, gasoline, paints, metal parts, and other materials commonly used on vessels can severely contaminate sediment samples and should be avoided in areas designated for handling sediment samples.

Another important factor is the storage facility on the vessel. The size, volume, and weight of sediment samples has to be considered together with the required temperature at which samples need to be stored. Sediment samples collected for the determination of contaminants should be stored frozen or at 4°C. The need for freezers or refrigerators places an extra burden on available space and electricity.

All equipment on the vessel to support a bottom sediment sampling program should be inspected before the departure for the field work.

Great attention should be paid to the maneuvering of the sampling vessel in shallow water. Sly[5] noted the disturbance of grab samples extending to 0.3 or 0.6 cm depth of the loose surface material (soft, silty clays) by maneuvering a 19.5-m-long sampling vessel with a draught of 1.9 m at 5.5 m water depth. In water depths of 4.5 and 3 m, the top 1.9 and 5 cm, respectively, of the sediment was eroded. Generally, the shallower the water depth at the site, the more likely the approach and maneuvering of the sampling vessel will disturb the surface layer of the sediment.

2. Bottom Sediment Sampling from Ice

The advantages of sampling through the ice compared to a floating platform are a steady platform and a large space at the sampling station for assembly of the equipment and sample handling. There are many serious problems with collecting samples from the ice, such as malfunctioning of sampling equipment at low temperatures, the transport of equipment from one sampling station to the other by a snowmobile or sledge, poor weather in the winter, and difficult working conditions which is important particularly for the work in remote areas. Moreover, sampling from the ice can be considered only when there is sufficient thickness of ice and the ice is relatively stable. Information on ice thickness at the sampling area and the area which will be used for equipment transport needs to be obtained before planning the mode of transport for the equipment to the site and to estimate the time for hole drilling. This information can usually be obtained from provincial or federal government agencies. There is no standard procedure or equipment for sampling from the ice. Portability of equipment is the prime factor in equipment selection. Equipment has to be transported to the site, often under difficult and cold operating conditions. Helicopters and small planes can provide transportation of larger equipment and personnel to remote areas. Snowmobiles are suitable for transport over shorter distances.

Other factors which need to be considered include drilling holes in the ice, positioning sampling stations, shelter and safety requirements, equipment (all parts of a sediment sampler including lines, winch and retrieval equipment, power source, sample containers), and all equipment for sample handling at the site.

a. Hole Cutting

Sampling through the ice requires drilling or sawing a hole in the ice at each sampling station. The size of the hole will depend on the equipment chosen for sediment sampling. Small corers without stabilizer fins or small grabs will require a hole of approximately 25 to 30 cm in cross section. Large grabs and corers will require a hole of about 80 to 100 cm cross section. Some devices have been designed specially for sampling from ice. Marine geological investigations in areas of permanently frozen sea involve the lowering of sampling equipment through natural or artificial holes in the ice cover. Practical limits on auger hole diameters prevent the use of piston corers having conventional tripping arms. Marlowe[128] described a device which allowed the collection of piston cores through a hole with a diameter as small as 21 cm. King and Everitt[129] described a sediment and water sampler designed for use through surface ice for Antarctic conditions. The sampler was based on mechanical suction and could be deployed through a hole about 11 cm in diameter which could be drilled easily in thick ice with a SIPRE ice auger.[130]

Commercially available ice augers and power drives are suitable for drilling holes. Gasoline-powered engines, electric motors, or a hydraulic drive can be used for driving ice augers. Usually, hand augers can only drill about 15-cm-diameter holes in thick ice. Drilling up to 45-cm-diameter holes requires a gasoline or electric power auger. Drilling larger holes in ice requires additional power for large diameter augers. Chain saws are necessary for cutting large-size holes at a site where more sediment samples have to be collected. Use of an auger or chain saw takes a considerable amount of time. Sufficient time (and energy) should be allowed in the program for ice cutting. Hot water drills have been developed for the more rapid drilling of holes in ice for studies in the Arctic.[131] The state-of-the-art techniques for making access holes through first- and multiyear Arctic sea ice were reviewed by Mellor.[132] The review includes methods for penetrating the ice, cleaning debris, maintaining the access holes in an open condition, and, particularly, gaining access through the ice to ensure that the surrounding ice remains competent for support of operating equipment and personnel.

b. Sampler Deployment

In shallow water, lightweight samplers, described for use from small vessels, can be lowered by hand to the bottom through a hole in the ice. Typically, however, a hand winch or gasoline-powered winch, mounted on a portable tripod stand, is used for sediment sampling at stations with deeper water or when the sampler is particularly heavy. A larger hole or a few holes in the ice are necessary for collecting more undisturbed samples at one site. Sly and Gardener[123] described a specially designed portable winch and frame built and tested for use in through-ice sampling programs. The complete system, made in modular form, was easily handled by two or three operators at temperatures as low as –40°C, and was transported by a ski plane or snow/ice vehicle. This system was suitable for use with a slightly modified vibratory corer.

Portable gasoline-powered engine drives, power generators, and electric drills are heavy and need to be transported to the sampling site by a snowmobile with a sled, a truck, or an aircraft. A heated shelter should be considered when planning to spend a longer time at the sampling site. A portable insulated shelter, a tent, or a transport vehicle can be used as a shelter. At sites where only small samplers will be used sampling can be carried out in an insulated shelter or a tent built on the ice.

FIGURE 29. Preparation for sediment sampling from ice, Lake Huron, Canada.

FIGURE 30. Hole cutting with a hand saw.

Sampling from the ice demands experience and a strong focus on worker safety and health. First aid and communication equipment, survival gear, food, fuel, etc. need to be provided to a party carrying out sampling in remote areas. All sampling equipment should be thoroughly inspected prior to the departure to the sampling site.

Figures 29 to 34 show different equipment used in sediment sampling from ice in South Bay, Lake Huron, Canada.

3. Collection of Bottom Sediment Samples by Diving

The collection of sediment samples by a diver should be considered when undisturbed samples are required, particularly for the studies of the sediment-water interface. Moreover, the diver can select a suitable area on the bottom at the sampling site and make notes or take photographs/underwater videos of the bottom and control the operation of the sampler. There are limitations due

FIGURE 31. Removing cut ice blocks.

FIGURE 32. Hole cutting with a chain saw.

to water depth, visibility, or currents. The diver's visibility can be obscured if fine-grained sediments are disturbed or the water is turbid.

McIntyre[133] compared sediment samples collected by gravity corers and scuba divers in studies of meiobenthos. His data, which represented the range of variation found throughout the year, showed that the diver-collected cores consistently gave substantially higher counts than the samples collected by

FIGURE 33. Preparation of a multiple corer for sediment sampling from ice.

FIGURE 34. Portable frame for lowering the sampling equipment through the hole in the ice.

gravity corers. This was particularly true for copepods which were largely restricted to the top 1 cm of mud. The results indicated that the downwash, caused by gravity corers during their descent, dispersed a considerable proportion of the superficial sediments and, therefore, these corers failed to collect all the animals associated with the uppermost sediment layer. This deficiency in gravity corers or any other instrument dropped, even gently, onto the sea or lake bed should be considered in selecting sampling techniques in studies of recently deposited organic matter or pollutants concentrated at the sediment-water interface.

Simple corers, core tubes, or boxes which can be sealed at both ends are suitable for surface sediment sampling and coring by a diver. Several corers of this type were described by Hopkins.[2] The penetration of 30 to 120 cm by a diver-operated core can be achieved without assistance from the surface. Several other samplers were described and successfully used for sediment collection by a diver. The Birge-Ekman box corer was modified for use by scuba divers or a deep submergence research vessel, allowing the precision of sampling[134] to be controlled. Testing showed that the abundance of benthic species and individuals was more accurately estimated by a modified Birge-Ekman box corer than by conventional surface ship samplers or small corers used widely for an *in situ* investigation. Gale[135] described a diver-operated corer for sampling benthic macroinvertebrates. Removable drive handles allowed the 9-cm-I.D. barrel to be driven into substrates too firm to be sampled by line-held or conventional diver-operated corers. A diver-operated corer was developed and used for collecting undisturbed cores of unconsolidated coarse sand on submarine canyon slopes.[136] Martin and Miller[137] described a diver-operated corer designed for collecting undisturbed sediment samples, without the loss of the uppermost 5-cm sediment layer. Furthermore, the corer was successfully used for collecting sediments for radiometric dating. Bothner and Valentine[138] described an instrument for collection of fine-grained and flocculated material from the sediment-water interface. The sampler weight was 6.7 kg, and could be used by a diver in shallow water or from a manned research submersible. A sediment sampling device was constructed for operation by scuba divers in the study of meiofauna in muddy sediments covered by a layer of fine detritus.[139] A diver-operated lightweight multiple coring system for collecting undisturbed sediment samples was suitable for studies of small-scale spatial dispersion and sampling patchy habitats.[140]

B. SAMPLER TRANSPORT AND ASSEMBLY AT SAMPLING SITE

The transport of sediment sampling equipment to a remote area is another factor to be considered in the selection of a sampler. The weight and volume of the sampler which needs to be shipped by air usually limit the choice of the equipment. Typically, the sampler has to be dismantled into several parts prior to shipping. Unless the person who assembles the sampler upon its arrival at the sampling location is familiar with the equipment, a detailed description, with a simple drawing of the assembling procedure and a list of all parts, should accompany the sampler. A set of tools essential for the assembly should be included in the shipment. Shipping of critical spare parts with the sampler, particularly those which may become lost or damaged during the sampling, such as springs and pins, often saves time and money.

C. SEDIMENT TYPE

There is no one bottom sediment sampler which can be used for the collection of all physically different sediment types. There are many samplers for collecting surface sediments and sediment columns which will recover an undisturbed sample only in soft, fine-grained sediments. Fewer grabs and corers are available for collecting sediments containing sand, gravel, or firm glaciolacustrine clay or till. It is difficult to choose a proper bottom sediment sampler without knowing the bathymetry and areal distribution of physically different sediment types at the sampling site. Consequently, gathering all reported information on the bathymetry and distribution of physically different sediments should be considered for any area to be sampled. Consultation with personnel experienced in sedimentological studies will help in the final choice of sampling equipment.

Acoustic survey techniques such as echo sounding, seismic reflections, and refraction are able to characterize both the type of surficial sediment layer, such as sand, gravel, or soft silty clay, and the subsurface sediment layers. Acoustic penetration is most effective in unconsolidated sediments consisting of soft, silty clays with high water content. On the other hand, sands or firm, compacted sediments display minimal penetration.

Acoustic techniques have been used in marine sediment classification. For example, Hampton[141] used acoustic techniques to measure the reflection coefficient and sound velocity and to determine porosity and density of sediments. Precision echo sounders have been used by oceanographers and marine geologists for mapping of bottom contours in oceans.[142,143] Echo sounding was successfully used by many researchers in the investigation of subbottom profiling, changes at the sediment-water interface, and distribution of different types of sediments in lakes.[144,145]

Seismic reflection is based on the same principle as echo sounding but uses a lower frequency sound source and allows a deeper penetration, even into a coarse sediment. Seismic refraction is based on the lateral movement of sound between an explosive source and a sound receiver. The time of arrival of specific responses is related to the thickness of sediment layers through which the sound travels. These two techniques require special equipment and a good positioning capability. A correct interpretation of records and observation obtained by acoustic survey techniques often requires the help of a geophysicist.

Information on the physical characteristics of sediments may be obtained by the fall cone technique described by Håkanson and Jansson.[17] A sediment penetrometer, for *in situ* measurement of sediment physical characteristics, can rather simply and rapidly determine sediment types and prevailing bottom dynamics. The penetrometer has a few different cones. The depth of the penetration of these cones reveals the nature of the sediments. A calibration table is supplied with commercially sold sediment penetrometers to convert the penetration depth into several different sediment types ranging from very soft sediments to hard bottom sediments.

From many reviews discussing different methods used for the characterization of the physical properties of sediments only the major ones are provided. Dunsiger et al.[146] examined the correlation between the results obtained by acoustic survey, the bottom sediment samples collected by different sampling devices, and data from a free-fall penetrometer in a study of sediment type at outer Placentia Bay of Newfoundland, Atlantic Ocean. Sly[16] described in detail several techniques developed for making underwater *in situ* measurements of various physical properties of sediments on the seafloor prior to collection of sediment cores. Baldwin et al.[147] investigated the relationship between physical properties and information obtained from processing acoustic reflection signals to yield insight into the subbottom acoustic reflection characteristics.

Briggs et al.[148] studied geoacoustic and related properties of marine sediments in the Caribbean Sea, and Reed et al.[149] investigated the sedimentation pattern by analyzing bathymetry, piston cores, echograms, and seismic reflection data. Bloomberg et al.[150] described the cone penetrometer test used in studies of stratigraphy and soil properties of sinkhole sediments.

In addition to the acoustic techniques and penetrometer tests, several other methods and instruments for geotechnical measurements have been developed, for example, vane shear meters, pressure meters, *in situ* electrical and radiometric probes, etc. Their application has been described in relevant literature (for example, Chaney and Demars[151]).

D. SEDIMENT DEPTH TO BE SAMPLED

There are two primary zones of sediment which are of interest in contaminant studies: the surficial or upper 10 to 15 cm, and the deeper layers. Sampling of the surface layer provides information on the horizontal distribution of parameters or properties of interest for the most recently deposited material, such as particle size distribution or geochemical composition of sediment. A sediment column, which includes the surface sediment layer (10 to 15 cm) and the sediment underneath this layer, is collected to study historical changes in parameters of interest or to define zones of pollution. The "typical" geochemical profile shows an exponential decrease of contaminant concentrations with sediment depth to a "background" concentration, since many chemical compounds of environmental concern are of recent origin. The sediment sampling plan should identify the sampling stations and the sediment depth which needs to be sampled at individual locations. In the evaluation of such sediments, there may be regulatory requirements for the handling or treatment of sediments in excess of a specific concentration of a contaminant in a manner different from the "uncontaminated" underlying sediments, where this handling or treatment entails considerable cost per cubic meter of material. Detailed characterization can closely define the contaminated sediment horizon, thereby limiting such costs. For example, 1 m of sediment needs to be removed by dredging for navigational purposes. Detailed sampling shows that the contamination of the sediment may extend only to a depth of about 30 cm below the sediment surface. Then the upper 30 cm of contaminated sediments could be treated differently than the remaining 70 cm of underlying "uncontaminated" sediments, at a cost saving compared to a special treatment of all of the dredged sediment.

E. POSITIONING

Sediment sampling projects, either reconnaissance surveys or detailed surveys, according to their scope, must start with marking the positions of the planned sampling stations on good quality navigation charts in the horizontal plane in latitude and longitude, grid coordinates, or angles and distances from known control points. Topographic maps of the adjacent shore will be necessary for some positioning techniques. The primary task is to carry out the sampling on these marked positions as near as possible.

Accurate positioning of sediment sampling sites is important in any program, particularly when the sampling is to be subsequently repeated. The required accuracy and precision of the station position will depend on the nature of the study. A 5 to 10% error of the distance between sampling sites can be acceptable for a baseline survey study carried out over a large area (more than 10 km^2). In studies involving monitoring of sediment contaminants over a certain time period, or changes over a small geographic scale, accuracy of positioning within a few meters may be necessary.

The choice of proper positioning methods depends on the project area and the distance between the sampling stations and the shore of the water body. A positioning technique suitable for the sampling program has to be selected before the commencement of the sampling. When the owner or operator of the sampling vessel does not have the necessary expertise and/or positioning equipment, a competent surveyor, a survey company, and/or government agencies should be contacted for advice to assist in positioning.

The distance line or taut wire can be used for direct positioning of sampling stations close to the shore. In this method, a line marked along its length is stretched between the vessel and the shore whereby the distance off may be measured as required.

There are different optical methods for positioning using a sextant, for example, double horizontal sextant angles observed simultaneously to obtain a resention fix. This is a common and versatile method for use at distances of 200 m to 5 km or more from shore marks. Usually, accuracy will be in the region of ±3 to 5 m within the above range.

For small areas, such as harbors, embayments, small lakes, and reservoirs, where sampling is carried out from a small vessel or from ice, the sampling sites can be determined by sextant observations of structures on the shore and shoreline features in combination with navigation/topographic maps. Good visibility is necessary for this positioning technique.

Sampling of large areas, such as large harbors, lakes, and oceans, with a large number of sampling sites requires an electronic positioning system, e.g., Loran C, Mini-Ranger, Trisponder, or radar.

Radio methods include "line-of-sight" microwave electromagnetic position-fixing (EPF) systems. They utilize remote instruments (similar to Tellurometers used in land surveys) ashore at a distance of up to 80 km. Medium range EPF systems have a range capability of 150 to 1200 km. Long-range navigational systems can be used from 150 km to virtually worldwide operations.

The Shoran (for short range system) operates with signal pulses in the 200- to 300-MHz band. A trigger signal from a beacon fixed on the vessel causes two transmitters at fixed shore stations to emit signals, giving two direct-range measurements when the signals are received back on the vessel. The observed time differences represent the transit time from the transmitters to the beacon and back, and are recalculated and expressed as distance. Shoran is limited in range, although it provides high precision (±10 m at maximum range). The Distomat DI 3000 with precision ±0.1 m is suitable for harbors and small areas.

Loran C is one of the microwave electromagnetic position-fixing (EPF) systems with a range up to 2800 km. The name is derived from long-range navigation. The expanded configuration of the Loran C system is operated on a full-time basis by the U.S. Coast Guard as a service to private and commercial navigation around the North Atlantic and North Pacific oceans. However, it is necessary to check the sampling area for availability of reliable continuous Loran C signals. The system functions in chains, each having a master station and several secondary (slave) stations for synchronized transmission. Each pulse transmitted in the Loran C system covers a number of cycles of the 100-kHz signal frequency. A Loran master station transmits nine pulse envelopes spaced 1 ms apart, then waits for a specified time before repeating its nine-pulse pattern. The various slave stations in the chain emit eight-pulse transmissions. The time difference in arrival of pulses from the master station and secondary stations gives hyperbolic lines of position. Travel pulse time must be converted to distances. The accuracy obtained with Loran C is quite variable. Using the best techniques for calibration, accuracy of 20 to 40 m can be achieved. An uncalibrated Loran C system is likely to yield errors of 500 m.

Decca and Omega navigation systems are based on similar principles. In the Decca system, phase comparison of continuous waves gives hyperbolic lines of position. The Omega system is based on phase differences between continuous waves from synchronized transmitters. The range of Decca systems is up to about 460 km, and accuracy from about 1 to 60 m. Omega is a low-frequency system operating at 10 kHz, providing low-order navigation signals that can be received worldwide. With a receiver, small boats and aircrafts can determine their approximate position. However, Omega is not recommended for navigation in geophysical and environmental investigations in oceans, lakes, and rivers because of its poor accuracy of several kilometers.

Satellite navigation systems (SATNAV) consist of a number of navigational artificial satellites in polar orbits and are used extensively for positioning systems worldwide. For example, with the Transit system, a fix is possible during the satellite's pass which lasts about 15 min, at intervals which vary from about 35 to 100 min depending on the observer's latitude. The satellites transmit a navigational message lasting precisely 2 min. From many data and using complex fix computation techniques, the accepted vessel position is calculated. The Transit satellite system was developed as an all-weather tool to provide position fixes for ships at sea, and the geophysical industry was the first among commercial users to apply satellite navigation methods to improve positioning in offshore surveys. The most upgraded system should bring the standard error of the translocated fixes within the range of 5 to 10 m.

However, the continuous calculation of positions is presently not feasible for satellite navigation systems, because the time between fixes is up to 3 h. Consequently, it is desirable to have another navigation system available which will operate between the satellite navigation positions.

Underwater positioning methods utilize different types of underwater acoustic beacons at known positions on the sea or lake floor, and sensors, such as echo sounders, sonars, TV cameras, etc. The beacons can also be used for marking instruments on the lake or ocean floor.

The positioning methods are described in many books in great detail and anyone who needs more information should consult them. Government agencies may provide information on a large-scale positioning system available within the country. Information on a small-scale positioning system may be obtained from different private companies. Many low-frequency, long-range navigation systems are permanently established and operated by national and international agencies, for example, Loran C by the U.S. Coast Guard.

IV. OTHER REQUIREMENTS FOR SAMPLE COLLECTION

A. QUALITY CONTROL

Recently, much attention has been focused on quality assurance/quality control in analyses of sediment samples. It is more complex to measure sampling accuracy of sediments, which are, in most cases, heterogeneous. The following two techniques can be used for quality control in sediment sampling. One technique consists of the collection of more than one sediment sample at selected sampling sites using identical sampling equipment (e.g., multicorers) as well as using identical field subsampling procedures, handling and storage of the samples, and methods for sediment analyses. The results will show variations which are due to sampling and subsampling techniques, but the heterogeneity of the sediment at the sampling site will still affect the test. The sediment sampler must be selected to suit the sediment texture at the test sampling site.

In the other quality control technique the collected sample is divided into a few subsamples and each subsample is treated as an individual sample. The results of geochemical analyses of all subsamples will indicate the variability due to the sampling and analytical techniques and sediment heterogeneity within a single collected sample.

A few control sites should be included in a sampling program for investigation of sediment contamination. They should be selected, after historical data review, at areas where the sediment will most likely not be contaminated. Data obtained at the control sites are important as background values when plotting distribution and concentration gradients of contaminants.

B. CONTAMINATION OF SEDIMENTS FROM SAMPLERS

Sediment samples can be contaminated with pieces of metal paint or surface corrosion products from samplers or equipment used for the operation of the samplers. Most samplers are metallic; some may be electroplated or painted to prevent corrosion, particularly when sampling in salt water. Samplers with metal parts painted with cadmium or lead paints are not suitable for the collection of sediments for the determination of metal concentrations. Similarly, use of oil and grease on the samplers or sampler lifting equipment should be avoided. Sediment samples for the quantitative determination of metals or organic contaminants should always be obtained from the center of the sampler. Plastic liners and core barrels used with gravity corers may be a source of contamination with various organic compounds. However, no data are available about testing contamination of sediment samples collected with plastic liners and core barrels manufactured from different plastic material.

C. FIELD NOTES

Good field notes are the backbone of any sampling program. Poor or incomplete notes can make analytical results impossible to interpret. The following items should be recorded at the time of the sediment sampling:

- Name of sampling site and sample number
- Time and date of sediment collection
- Weather conditions, particularly wind strength and directions, air and water temperature, snow or ice cover, thickness of ice when sampling from the ice
- Positioning information (equipment used for positioning, any problems encountered during the positioning of a station, drawings of sampling site's positions on a chart)
- Type of vessel used (size, power, type of engine)
- Type of sediment sampler used (grab, corer, modifications made on the sampler during the sampling)
- Names of sampling personnel

FIGURE 35. Preparation of Benthos gravity corer for sediment sampling.

- Notes of unusual events which occurred during sampling (for example, a grab sampler not completely closed, the top of the recovered corer smeared with sediment, the loss of part of a sediment sample from a grab sampler or a corer, the loss of a section of a sediment core through the bottom of the core liner before capping the bottom of the tube, problems with the sampling equipment, or observations of possible sample contamination)
- Sediment description including texture and consistency, color, odor, estimate of quantity of recovered samples by a grab sampler, length and appearance of recovered sediment cores
- Notes on further processing of sediment samples in the field, particularly subsampling methods, type of containers and temperature used for sample storage, record of any measurements made in the field, such as pH, Eh

D. SUBSAMPLING OF SEDIMENT CORES

Prior to subsampling, and in case a clear plastic liner was used, the appearance of the sediment core should be recorded along with any obvious features, such as the length of the core, sediment color, texture and structure, occurrence of fauna, etc.

FIGURE 36. Stoppering a sediment core collected by Benthos gravity corer.

Sediment cores have to be stoppered immediately after retrieval to prevent accidental loss of samples (Figures 35 and 36). Recovered sediment cores have to be processed with respect to the type and number of analyses which need to be carried out on each core and as outlined in the sediment sampling plan. The cores should be subsampled as soon as possible after retrieval of the cores.

Sediment cores collected for stratigraphical or geotechnical studies can be stored at 4°C in a humidity-controlled room without any large changes in sediment properties for several months. Long cores, such as those collected by piston coring, can be cut into sections of suitable length for storage, sectioned longitudinally, described, labeled, wrapped to preserve original consistency, and stored in a refrigerated room.

Sediment cores collected for chemical analyses, particularly for the determination of different contaminants, should be extruded from the core liners and subsampled as soon as possible. Cores collected with different gravity corers are usually up to 2 m long, and have to be kept upright to prevent mixing of the uppermost part of the sediment core which consists mostly of very fine, soft, and unconsolidated material. Prior to any transport of these cores, the entire space over the sediment in the core liner needs to be filled with lake or seawater,

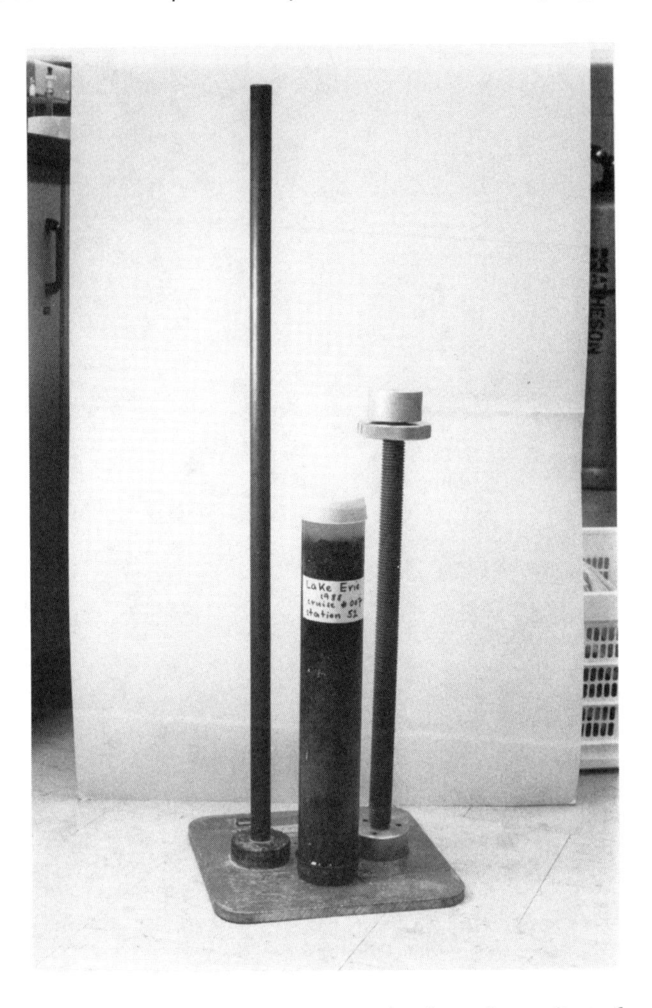

FIGURE 37. Simple piston extruder for subsampling of sediment cores with a metal rod for securing longer cores from falling over, and a labeled sediment core.

and both ends of the core liner have to be completely sealed to prevent mixing of the sediment inside the liner. A refrigerated storage space should be used in case the cores have to be stored even for a short time. The cores should be stored in an upright position and secured from falling over. Sediments with a high content of organic matter often contain large amounts of gasses, which upon recovery can disturb a part of or the entire core.

Sediment cores collected for studies of environmental pollution or sediment dating are usually subsampled into 1-cm sections. By subsampling the sediment into longer sections, for example, 3 to 10 cm thick, the information on the vertical distribution of contaminants can be lost, particularly in the sediments collected from an area with a low sedimentation rate. For example, it would be impossible to assess recent changes in contaminant loadings from 5-cm core sections at an area with a 2-mm annual sedimentation rate.

Often budgetary limitations will determine the number of analyses and thus the decision as to the number of subsampled sediment core sections. To limit a large number of analyses, it is recommended to initially analyze every third

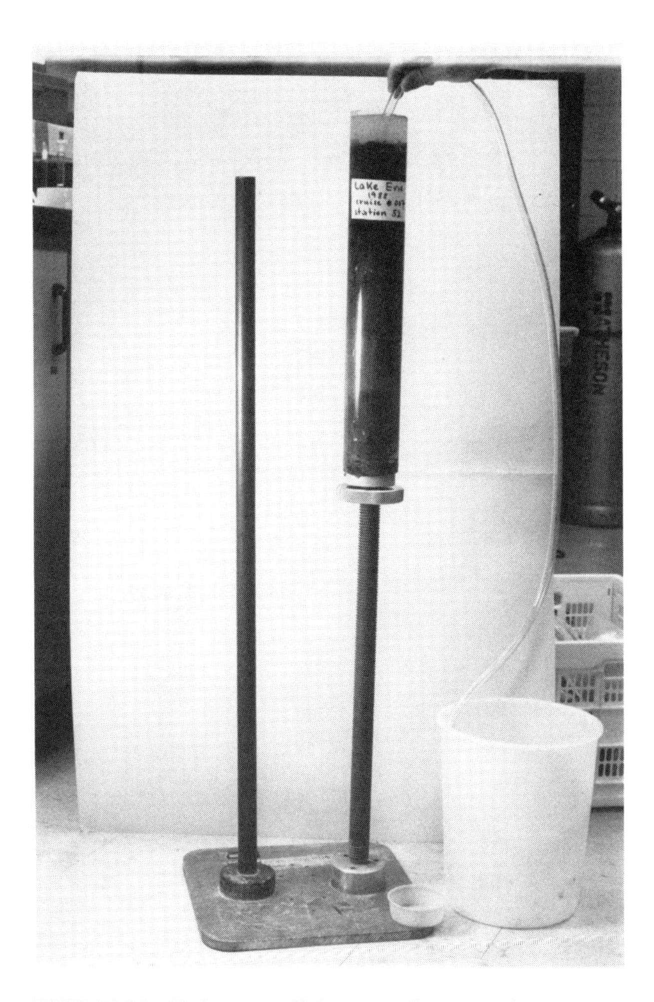

FIGURE 38. Siphoning off the water from a sediment core placed on extruder's piston before starting the subsampling.

section of a core, subsampled into 1-cm sections, while storing the rest for additional analyses as needed.

There are different methods for subsampling sediment cores. A mechanical lightweight device for traverse cutting of soft bottom sediment cores was constructed. The equipment was convenient for operation in the field, and was simple and cheap to manufacture.[152] Håkanson and Jansson[17] described an electro-osmotic knife and/or guillotine, dry-ice freezing method, and a plexiglas slide for subsampling sediment cores. The cores can be extruded by different piston-type extruders.[90] The extruding and sectioning the cores using this simple type of extruder involve the few steps outlined in Figures 37 to 41. The capped core liner containing the sediment and overlying water is uncapped at the lower end and placed vertically on the top of the piston. The top cap is removed and the water is siphoned off to avoid disturbance of the sediment-water interface. The core liner is then pushed slowly down until the surface of the sediment is at the upper end of the liner. Sediment sections are collected by pushing the liner down and cutting the exposed sediment into sections of desired thickness. From each sediment section, a 1- to 2-mm outer layer of sediment which had been in contact with the plastic or metal liner is discarded

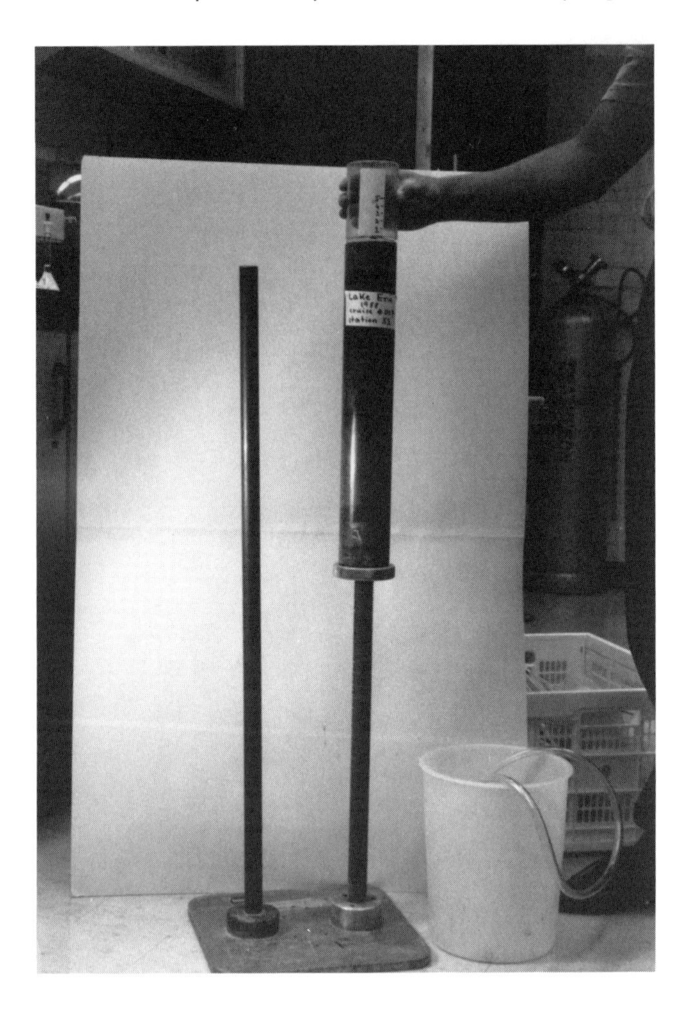

FIGURE 39. Placing a piece of core liner with a scale on the top of the core.

to avoid contamination. A simple device to achieve this side layer removal has been constructed by the National Water Research Institute, Environment Canada, Burlington, Ontario. The device consists of a frying pan with a hole cut in the center to fit the plastic liner. A metal ring with a sharp bottom edge is mounted over the hole in the pan to cut only the center part of the sediment core. The use of the “frying pan” in subsampling of sediment cores is shown in Figures 42 to 45. Individual core sections are collected into precleaned, labeled containers. Because the uppermost sediment layer often consists of very fine sediment with a high water content, it may have to be subsampled using a pipette or large syringe. The extruders shown in Figures 37 and 46 are suitable for the subsampling of about 2-m-long fine-grained sediment cores collected with different gravity corers. Alternatively, cores of more consolidated material can be mounted onto a horizontal U-shaped rail and the liner cut using a saw mounted on a depth-controlling jig. The final cut can then be made with a sharp knife to avoid contamination by liner material, and the core itself sliced with Teflon or nylon string. The core then becomes two “D”-shaped halves which can be easily inspected and subsampled.

The extruder shown in Figures 46 and 47 was designed and constructed at the National Water Research Institute, Environment Canada, Burlington,

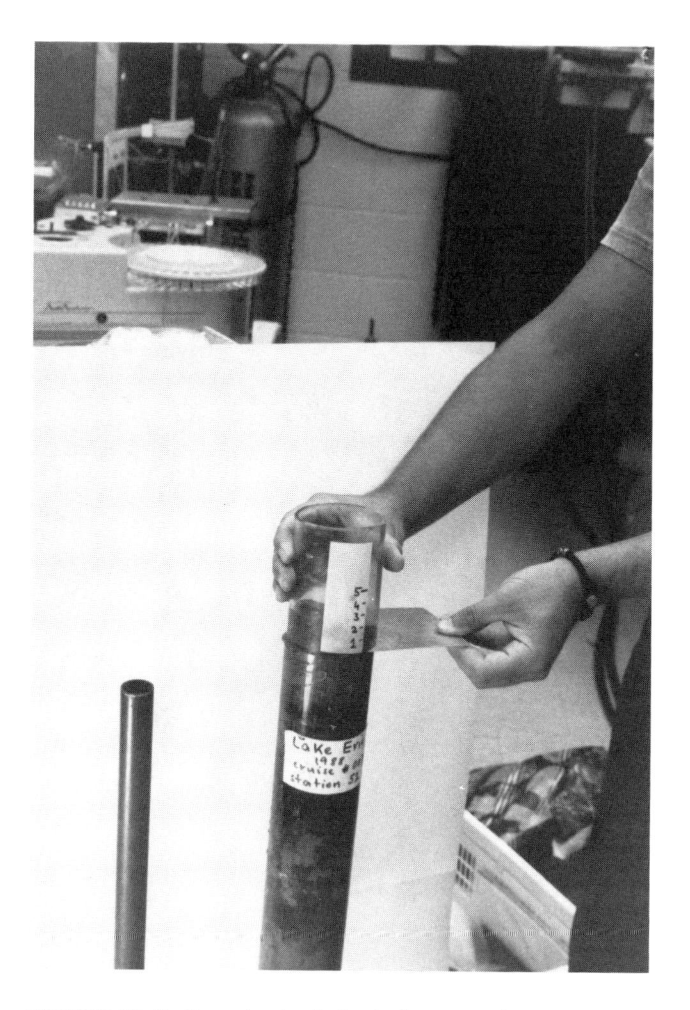

FIGURE 40. Sediment extruded into the piece of core liner with the scale is cut by a metal cutter.

Ontario, Canada, for subsampling sediment cores anywhere in the field. It consists of approximately a 50 × 50 × 0.5-cm-thick aluminum plate with a reinforced aluminum frame and two stainless steel bases mounted on the surface of the plate. The inside diameter of one base corresponds to the outside diameter of the plastic liner commonly used with gravity corers (such as the Benthos corer), and the inside diameter of the second base corresponds to the outside diameter of the plastic liner used with the lightweight corer described by Williams and Pashley.[74] Therefore, this extruder can be used for subsampling sediment cores collected in these two different sizes of core liners. The core liner is clamped in place with two special holders cut to match the outside diameter of the core tube being used. This holds the core upright and secured in the base. A hose is connected to each base to fill the inside space of the base with water. The other end of the hose is connected to a water tap or, at areas with no water supply, to the metal pressurized stainless steel bottle filled up to three-quarters of its volume with water under pressure. A stopper, made from two rubber stoppers joined vertically by a screw, is inserted into the bottom end of the core liner upon retrieval of a sediment core. This rubber stopper fits completely into the plastic liner to prevent the sediment from sliding out. The core is then placed into the base on the extruder's plate, the

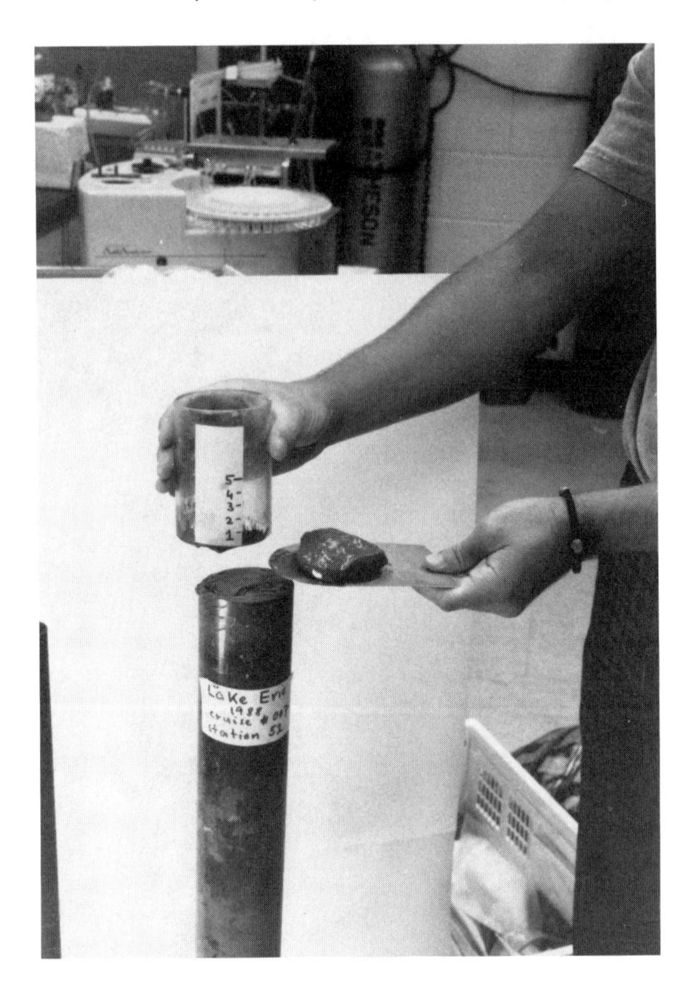

FIGURE 41. Sediment subsample on a metal cutter.

top cap removed, and the sediment pushed upward by the water from the water tap or the pressurized bottle. The water pressure is controlled by a fine needle valve inserted in the supply hose. The sediment extruding from the top of the core liner can be subsampled into sections of desired thickness (minimum about 0.5-cm-thick slices).

Tools, such as spatulas, scoops, cutters, and other utensils used in sediment subsampling, should be made of a suitable material to prevent contamination of the samples. Plastic utensils are suitable for subsampling sediments for the determination of metals, and stainless steel or glass for subsampling sediments for the determination of organic contaminants. Teflon utensils are the best all-round type.

Freezing sediment cores selected for subsampling is not suitable. Sediment freezing changes the sediment volume depending on the water content, and 1 cm of a frozen (or frozen and thawed) sediment section does not equal 1 cm depth of fresh sediment.

1. Procedure for Collection and Preparation of Sediment Cores for Sediment Dating

Sedimentation rates of sediment deposited during the past 100 years are usually determined by different radiometric techniques based on measurements of profiles of either an artificially produced radionuclide, such as ^{137}Cs,

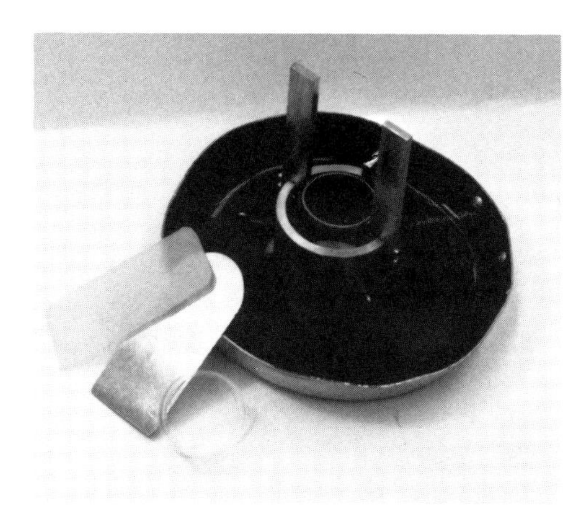

FIGURE 42. Equipment ("frying pan") for subsampling sediment cores with the elimination of the sediment in contact with the core liner.

FIGURE 43. "Frying pan" placed on the top of the sediment core before the subsampling.

FIGURE 44. Cutting a subsample from the sediment core using the “frying pan”.

FIGURE 45. Collecting a sediment subsample into prewashed and labeled glass jar.

or a naturally occurring radionuclide, such as ^{210}Pb.[153-159] Sediment dating by ^{137}Cs and ^{210}Pb has been utilized in assessing the extent of impacts on lakes from human activities, as the input of contaminants into the sediment started in the last century.[160,161] Relative time markers in sediment cores, such as different pollen grains, have also been used in recent sediment dating.

FIGURE 46. Core extruder using pressurized water.

However, these are typically site-specific for a watershed or a larger area.[162] Horizon markers for the determination of age of sediment deposited between the past 300 and about 40,000 years include different fossil invertebrates, paleobotanical remains, and ^{14}C.[163]

Collection of sediment cores for the determination of sediment age and sedimentation rates needs to meet several conditions:

- The selection of a proper location in freshwater and marine environment for collecting cores is critical. Only locations with undisturbed, fine-grained sediment accumulation are suitable.
- The core tube should be of a sufficient diameter to yield an adequate amount of sediment for analyses.
- The corer should gently penetrate the sediment-water interface without any disturbance of the interface by a shock wave preceding the corer, and any loss of fine-grained particles. Sediment cores with disturbed top sediment are not suitable for dating.
- The sample interval, typically 1 cm thick, is selected for core subsampling. The top sections of the sample, generally up to 3 cm, consist usually of soft, fine-grained, unconsolidated material with a high water content (up to 95%), and must be subsampled carefully, for example, using a large syringe or a pipette with the wide bottom opening. The rest of the sediment core can be subsampled by the method described above ("subsampling of sediment cores") into 1-cm sections. Alternatively, each section around the circumference can be trimmed to minimize the risk of contamination by sediment particles which may have been carried from the surface down the inside of the core tube during the penetration into the sediment. However, the quantity of the material adhering to the walls of the core tube is so small that this contamination was found negligible.[164]

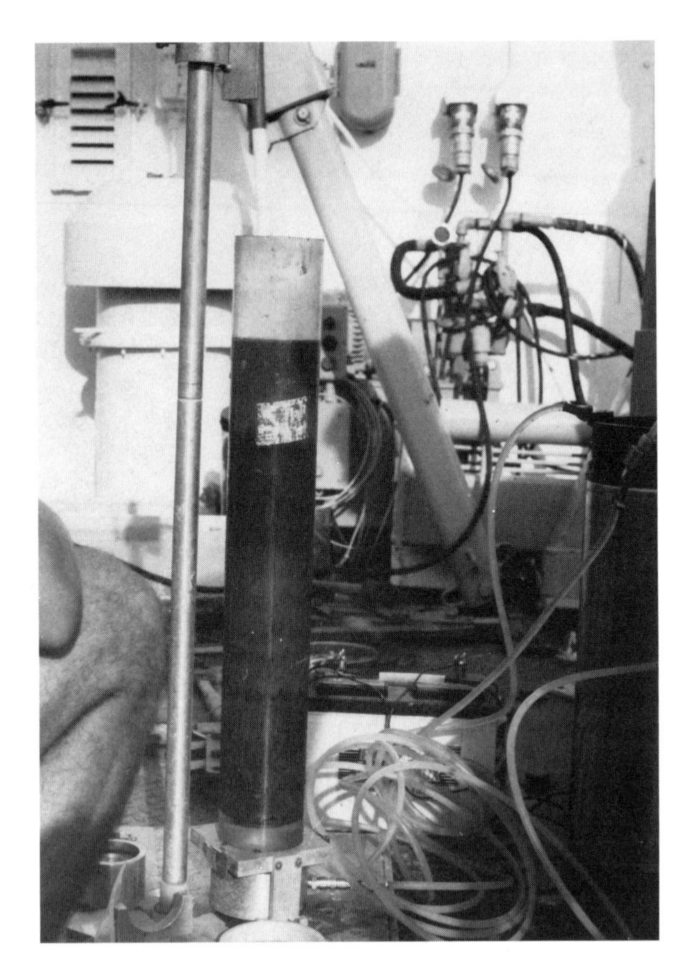

FIGURE 47. Sediment core secured in the extruder.

- Dry and wet weigths of each sediment section have to be recorded for the determination of water content.
- The ratio water content/dry weight of the deepest section of the sediment core is typically the reference value to which all other ratios are normalized.
- The effect of sediment compaction must be allowed for.[155] Normalization procedure converting the measured length of an uncompacted sediment section to a theoretical compacted length was described.[156]

Procedures for the determination of sediment age can be carried out only by specialized laboratories and institutions, usually at universities or government research laboratories. Consequently, these laboratories will issue final detailed instruction to the sampling personnel about the handling and preparation of the sediment samples.

Procedures before and after sediment collection and preparation for dating by a radiometric method (^{137}Cs or ^{210}Pb) usually involve the following steps:

- Preparation of an appropriate number of labeled, numbered plastic vials or bottles with lids (in measuring radionuclide concentrations it is recommended to contact the personnel of the laboratory equipped with detectors which will be used, and follow their instruction on the size of the vials or bottles which may be placed directly in the detector chamber, saving the time for transferring the sediment samples from the sampling vials into other containers)

- Weighing the vials without lids and recording their weights
- Selection of the sampling interval of the recovered sediment core
- Extruding and subsampling the sediment core into selected intervals
- Placing each sediment interval into a preweighed, numbered vial or bottle, and applying tightly the lid to prevent moisture loss
- In the laboratory, removing the lids and weighing the vials with wet sediment intervals, and recording the weights
- Drying the sediment after removing the lids in an oven or a freeze dryer (see Chapter 6) to a constant weight
- Weighing the vials with dry sediment intervals without lids, and recording the dry weights
- Preparation of a sediment data file in the form of a spreadsheet
- Application of the selected method for the determination of sediment age and sedimentation rates

E. STRATIGRAPHIC ANALYSES OF UNCONSOLIDATED SEDIMENT CORES

Sediment cores are sometimes used for stratigraphical studies. Geological studies of sediment cores recovered from ocean basins include early diagenetic processes, the general nature of climatic fluctuations, the correlation of the marine record to the classical glacial sequence on land, etc.[11] Bouma[3] described in detail many different techniques used in studies of sedimentary structures. Stratigraphical analyses of sediment cores are a useful tool in the studies of historical changes in sedimentary processes and sediment geochemistry, and in the interpretation of results obtained by studies of the input of contaminants into the aquatic environment. Procedures used in such stratigraphic analyses of unconsolidated sediment cores usually include X-radiography of the cores, extruding or splitting the core, and photographing and logging the core for color, texture, and structure. The X-radiography of the cores is a nondestructive technique that reveals internal structures and particles that may not be visible to the naked eye, and provides a permanent record of these internal structures of the cores. Details of the principles and use of radiography in sediment studies were described by Krinitzski.[165] The equipment for X-radiography is commercially sold, for example, by different companies dealing with X-ray instruments.

For visual inspection and different tests, the sediment in a plastic core liner can be split lengthwise by a core liner cutting device similar to that described by Mallik.[166] Also, the core can be extruded on a plastic sheet using different core extruders selected with respect to consistency of the material and length of the core. The surface of the extruded or split core should be cleaned by gently scraping the oxidized surface material (usually a different color than that of the inside of the core) with a wet spatula. Scraping should be done across the core, not lengthwise, to prevent smearing the sediment over the entire core. For taking a photograph of the core, identification labels should be placed showing the top and the bottom of the core, core number, sampling station, time of collection, and a scale, for example, a meter stick placed along the core. For

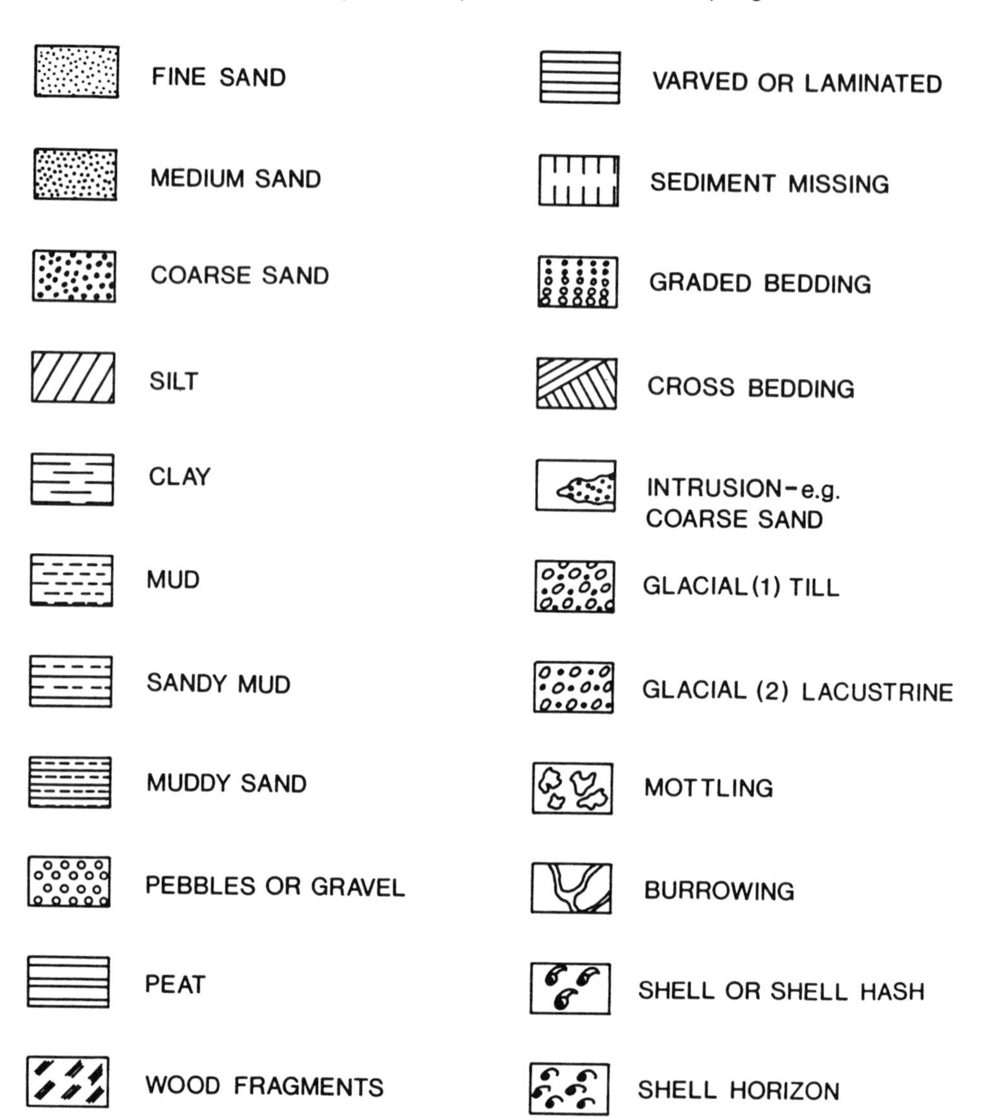

FIGURE 48. Lithological symbols used in core logging.

example, a 40- to 45-cm section of the core can be photographed at one time at a distance of 80 cm. A label with the number of the section of the core should be placed near the core when the core is divided into more sections for photographing.

A visual description of the core should contain the following information: length of retained core; equipment used for core collection; name of the operator who collected, handled, and split (or extruded) the core; description of the splitting (or extruding) of the core; thickness of the sediment units in the core, which may be based on changes in color using, for example, a Munsell color chart; consistency, for example, described as soupy, soft, medium firm, firm, stiff, loose, packed, etc.; texture (estimated particle size, for example, gravel, sand, silt, clay — or the principal component and then the modifier, such as silty clay, sandy clay, etc.); structure (graded bedding, cross bedding, laminates, lenses, varves, etc.), recorded in centimeters; presence of organic

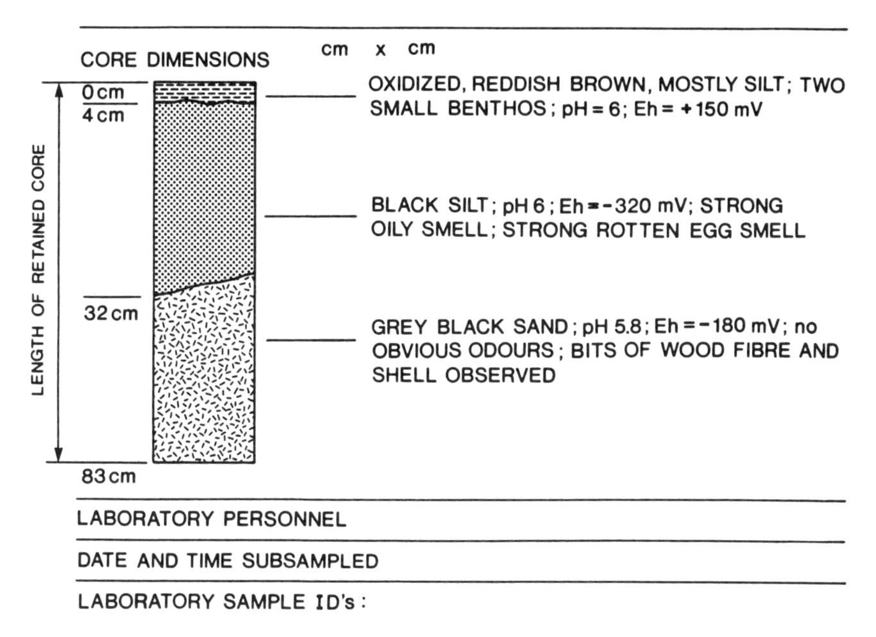

FIGURE 49. Core logging sheet.

matter, shells and coarse fragments with a description of their type and size; sediment odor, for example, odorless (clean material), chemical odor (chlorine, petroleum, sulfurous), or decaying organic odor (manure, sewage); appearance of oil, coal dust, ash, etc.; presence of carbonates (tested by a drop of 10% hydrochloric acid which generates effervescence in the presence of carbonates).

Figure 48 shows an example of lithologic symbols for core logging[167] and Figure 49 gives examples of core logging sheets. The advantage of logging is the standardization of the description of sediment stratigraphy by different workers and easier interpretation of results.

ACKNOWLEDGMENTS

We would like to thank many colleagues from the National Water Research Institute, Environment Canada, Burlington, Ontario, Canada, for valuable comments, particularly Mr. M. R. Mawhinney for comments and advise on the description of the sediment sampling equipment, Messr. J. Van Nynatten and M. L. Donnelly for preparation of the illustrations for this chapter, and Ms. Dianne Crabtree for suggestions and comments on the style and format. Last but not least, we thank Mr. O. Mudroch for his help with preparing this chapter.

REFERENCES

1. **Holme, N.A.**, Methods of sampling the benthos, in *Advances in Marine Biology*, Vol. 2, Russel, F.S., Ed., Academic Press, New York, 1964, 171.
2. **Hopkins, T.L.**, A survey of marine bottom samplers, in *Progress in Oceanography*, Vol. 2, Sears, M., Ed., Pergamon-Macmillan, 1964, 213.
3. **Bouma, A.H.**, *Methods for the Study of Sedimentary Structures*, John Wiley & Sons, New York, 1969, 458.
4. **Kajak, Z.**, Benthos of standing water: survey of samplers, in *A Manual on Methods for the Assessment of Secondary Productivity in Fresh Water*, Edmonson, W.T., Ed., IBP Handbook No. 17, Blackwell Scientific, Oxford, 1969, 25.
5. **Sly, P.G.**, Bottom sediment sampling, in *Proc. 12th Conf. Int. Assoc. Great Lakes Res.*, International Association of Great Lakes Research, Ann Arbor, MI, 1969, 883.
6. **Martinais, J.**, L'instrumentation scientifique au Centre Oceanologique de Bretagne, Colloque Inter. sur l'Exploit des Oceans, Theme V, Vol. 1, 1971, 11.
7. **McIntyre, A.D.**, Deficiency of gravity corers for sampling meiobenthos and sediments, *Nature*, 231, 260, 1971.
8. **Collins, V.G.**, Methods in sediment microbiology, in *Advances in Aquatic Microbiology*, Vol. 1, Droop, M.R. and Jannasch, H.W., Eds., Academic Press, New York, 1977, 219.
9. **Elliot, J.M. and Tullett, P.A.**, A bibliography of samplers for benthic invertebrates, Occasional Publ. No. 4, Freshwater Biological Association,1978.
10. **Elliot, J.M. and Tullett, P.A.**, A supplement to a bibliography of samplers for benthic invertebrates, Freshwater Biological Association, Occasional Publication No. 20, 1983.
11. **Moore, T.C. and Heath, G.R.**, Sea-floor sampling techniques, in *Chemical Oceanography*, Vol. 7, 2nd ed., Riley, J.P. and Chester, R., Eds., Academic Press, New York, 1978, 75.
12. **Bascom, W.N.**, Instruments for measuring pollution in the sea, *Progr. Water Technol.*, 4, 99, 1979.
13. **Fowler, G.A. and Kingston, P.F.**, The use of bottom operating sampling equipment for investigation of the continental shelf, in *Proc. 1st Can. Conf. Mar. Geotech. Engineering*, 1979, 238.
14. **Le Tirant, P.**, Seabed exploration coring devices and techniques, in *Seabed Reconnaissance and Offshore Soil Mechanics for the Installation of Petroleum Structures* (transl. from the 1976 French ed. by J. Chilton-Ward), Techniq, Paris, 1979, 133.
15. **LeTirant, P.**, Seabed exploration by in situ measurements, in *Seabed Reconnaissance and Offshore Soil Mechanics for the Installation of Petroleum Structures* (transl. from the 1976 French ed. by J. Chilton-Ward), Techniq, Paris, 1979, 218.
16. **Sly, P.G.**, Equipment and techniques for offshore survey and site investigations, *Can. Geotech. J.*, 18, 230, 1981.
17. **Håkanson, L. and Jansson, M.**, *Principles of Lake Sedimentology*, Springer-Verlag, Berlin, 1983, 32.
18. **Wigley, R.L.**, Comparative efficiencies of Van Veen and Smith-McIntyre grab samplers as revealed by motion pictures, *Ecology*, 48, 168, 1967.
19. **Håkanson, L.**, Modifications of the Ekman sampler, *Int. Rev. Ges. Hydrobiol.*, 71, 719, 1986.
20. **Birkett, L.**, A basis for comparing grabs, *J. Cons. Cons. Perm. Int. Explor. Mer.*, 23, 202, 1958.
21. **Gallardo, V.A.**, Observations on the biting profiles of three 0.1 m^2 bottom samplers, *Ophelia*, 2, 319, 1965.
22. **Lassig, J.**, An improvement to the Van Veen bottom grab, *J. Cons. Cons. Perm. Int. Explor. Mer.*, 29, 352, 1965.
23. **Lie, U. and Pamatmat, M.M.**, Digging characteristics and sampling efficiency of the 0.1 m^2 Van Veen grab, *Limnol. Oceanogr.*, 10, 379, 1965.
24. **Stirn, J.**, Manual of methods in aquatic environment research. VIII. Ecological assessment of pollution effects, Fish. Tech. Paper 209, Fisheries and Oceans, Canada, 1981, 70.
25. **Kennedy, V.S.**, A summer benthic survey in Conception Bay, Newfoundland, emphasizing zoogeography of annelids and amphipods, *Can. J. Zool.*, 63, 1863, 1985.
26. **LeRoy, L.W.**, Pressure-disc sediment sampler, *J. Sediment. Petrol.*, 33, 944, 1963.
27. **Kutty, M.K. and Desai, B.N.**, A comparison of the efficiency of the bottom samplers used in benthic studies off cochin, *Mar. Biol.*, 1, 168, 1968.
28. **Holme, N.A.**, Macrofaunal sampling: grab sampling and suction sampling, in *International Biology Handbook No. 16*, Holme, N.A. and McIntyre, A.D., Eds., Blackwell Scientific, Oxford, 1971.

29. **Smith, K.L. and Howard, J.D.**, Comparison of a grab sampler and a large volume corer, *Limnol. Oceanogr.*, 17, 142, 1972.
30. **Christie, N.D.**, Relationship between sediment texture, species richness, and volume of sediment sampled by a grab, *Mar. Biol.*, 30, 89, 1975.
31. **Hunter, B. and Simpson, A.E.**, A benthic grab designed for easy operation and durability, *J. Mar. Biol. Assoc. U.K.*, 56, 951, 1976.
32. **Tyler, P. and Shakley, S.E.**, Comparative efficiency of the Day and Smith-McIntyre grabs, *Estuarine Coastal Mar. Sci.*, 6, 439, 1978.
33. **Bascom, W.N.**, Instruments for measuring pollution in the sea, *Progr. Water Technol.*, 4, 99, 1979.
34. **Dall, P.C.**, A new grab for the sampling of zoobenthos in the upper stony littoral zone, *Arch. Hydrobiol.*, 92, 396, 1981.
35. **Elliot, J.M. and Drake, C.M.**, A comparative study of seven grabs used for sampling benthic macroinvertebrates in rivers, *Freshwater Biol.*, 11, 99, 1981.
36. **Jensen, K.**, Comparison of two bottom samplers for benthos monitoring, *Environ. Technol. Lett.*, 2, 81, 1981.
37. **Hulle, M. and Jestin, J.M.**, L'echantillonnage des peuplements benthiques: mise au point d'une methodologie fondee sur l'Analyse Factorielle des Correspondances; son application aux Oligochetes du lac de Cretil (France), *Hydrobiologia*, 99, 95, 1983.
38. **Vidakovic, J.**, Meiofauna of silty sediments in the coastal area of the North Adriatic, with special reference to sampling methods, *Hydrobiologia*, 118, 67, 1984.
39. **Grizzle, R.E. and Stegner, W.E.**, A new quantitative grab for sampling benthos, *Hydrobiologia*, 126, 91, 1985.
40. **Dorgelo, J. and Hengst, P.A.**, Two modifications of the quantitative hydraulic lift sampler for benthic invertebrates, *Water Res.*, 20, 825, 1986.
41. **Blomqvist, S.**, Sampling performance of Ekman grabs - *in situ* observations and design improvements, *Hydrobiologia*, 206, 245, 1990.
42. **Rosfelder, A.M. and Marshall, N.F.**, Obtaining large, undisturbed, and oriented samples in deep water, in *Marine Geotechnique*, Richards, A.F., Ed., University of Illinois Press, Urbana, 1967, 243.
43. **Baxter, M.S., Farmer, J.F., McKinley, I.G., Swan, D.S., and Jack, W.**, Evidence of the unsuitability of gravity coring for collecting sediment in pollution and sedimentation rate studies, *Environ. Sci. Technol.*, 15, 843, 1981.
44. **Pettersson, H. and Kullenberg, B.**, A vacuum core-sampler for deep sea sediments, *Nature*, 145, 306, 1940.
45. **Emery, K.O. and Dietz, R.S.**, Gravity coring instruments and mechanics of sediment coring, *Bull. Geol. Soc. Am.*, 52, 1685, 1941.
46. **Piggot, C.S.**, Factors involved in submarine core sampling, *Bull. Geol. Soc. Am.*, 52, 1513, 1941.
47. **Hvorslev, M.J. and Stetson, H.C.**, Free-fall coring tube: a new type of gravity bottom sampler, *Bull. Geol. Soc. Am.*, 57, 935, 1946.
48. **Phleger, F.B.**, Ecology of foraminifera, northwest Gulf of Mexico, I. Foraminifera distribution, *Geol. Soc. Am. Memoir*, 46, 1, 1951.
49. **Heezen, B.C.**, Methods of exploring the ocean floor: a discussion, in *Oceanographic Instrumentation*, Nat. Res. Coun. Publ. No. 309, National Academy of Sciences, Washington, D.C., 1954, 200.
50. **Mackereth, F.J.H.**, A portable core sampler for lake deposits, *Limnol. Oceanogr.*, 3, 181, 1958.
51. **Hersey, J.B.**, Acoustically monitored bottom coring, *Deep Sea Res.*, 6, 170, 1960.
52. **Richards, A.F. and Keller, G.H.**, A plastic-barrel sediment corer, *Deep Sea Res.*, 8, 306, 1961.
53. **Emery, K.O. and Hulsemann, J.**, Shortening of sediment cores collected in open barrel gravity corers, *Sedimentology*, 3, 144, 1964.
54. **Van den Bussche, H.K.J. and Houbolt, J.H.C.**, A corer for sampling shallow-marine sands, *Sedimentology*, 3, 155, 1964.
55. **Willermoes, M.**, A ball-stoppered quantitative sampler for the microbenthos, *Ophelia*, 1, 235, 1964.
56. **McManus, D.A.**, A large-diameter coring device, *Deep Sea Res.*, 12, 227, 1965.
57. **Burns, R.E.**, Free-fall behaviour of small, lightweight gravity corers, *Mar. Geol.*, 4, 1, 1966.
58. **Kermabon, A., Blavier, P., Cortis, V., and Delauze, H.**, The "sphincter" corer: a wide-diameter corer with watertight core-catcher, *Mar. Geol.*, 4, 149, 1966.

59. **Kermabon, A. and Cortis, U.**, A new "sphincter" corer with a recoilless piston, *Mar. Geol.*, 7, 147, 1969.
60. **Burke, J.C.**, A sediment coring device of 21-cm diameter with sphincter core retainer, *Limnol. Oceanogr.*, 13, 714, 1968.
61. **Inderbitzen, A.L.**, A study of the effects of various core samples on mass physical properties in marine sediments, *J. Sediment. Petrol.*, 38, 473, 1968.
62. **Isaacs, J.D. and Brown, D.M.**, "Bootstrap" corer, *J. Sediment. Petrol.*, 38, 159, 1968.
63. **Menzies, R.J. and Rowe, G.T.**, The LUBS, a large undisturbed bottom sampler, *Limnol. Oceanogr.*, 13, 708, 1968.
64. **Barnett, P.R.O.**, A stabilizing framework for the Knudsen bottom sampler, *Limnol. Oceanogr.*, 14, 648, 1969.
65. **Mackereth, F.J.H.**, A short core sampler for subaqueous deposits, *Limnol. Oceanogr.*, 14, 145, 1969.
66. **Holme, N.A.**, Macrofaunal sampling: grab sampling and suction sampling, in *International Biology Handbook No. 16*, Holme, N.A. and McIntyre, A.D., Eds., Blackwell Scientific, Oxford, 1971.
67. **Kannenworf, E. and Nicolaisen, W.**, The "haps": a frame-supported bottom corer, *Ophelia*, 10, 119, 1973.
68. **Keegan, B.F. and Konnecker, G.**, *In situ* quantitative sampling of benthic organisms, *Helgol. Wiss. Meeresunters.*, 24, 256, 1973.
69. **Meischner, D. and Rumohr, J.**, A lightweight, high-momentum gravity corer for subaqueous sediments, *Senckenbergiana Marit.*, 6, 105, 1974.
70. **Thayer, G.W., Williams, R.B., Price, T.J., and Colby, D.R.**, A large corer for quantitatively sampling benthos in shallow water, *Limnol. Oceanogr.*, 20, 474, 1975.
71. **Axelsson, V. and Håkanson, L.**, A gravity corer with a simple valve system, *J. Sediment. Petrol.*, 48, 630, 1978.
72. **Barton, C.E. and Burden, F.R.**, Modifications to the Mackereth corer, *Limnol. Oceanogr.*, 24, 977, 1979.
73. **Hongve, D. and Erlandsen, A.H.**, Shortening of surface sediment cores during sampling, *Hydrobiologia*, 65, 283, 1979.
74. **Williams, J.D.H. and Pashley, A.E.**, Lightweight corer designed for sampling very soft sediments, *J. Fish. Res. Board. Can.*, 36, 241, 1979.
75. **Avilov, V.I. and Trotsyuk, V.Y.A.**, Improved means for obtaining hermetically sealed samples of bottom sediment, *Oceanology*, 20, 365, 1980.
76. **Lebel, J., Silverberg, N., and Sundby, B.**, Gravity core shortening and pore water chemical gradients, *Deep Sea Res.*, 29, 1365, 1982.
77. **Frithsen, J.B., Rudnick, D.T., and Elmgren, R.**, A new, flow-through corer for the quantitative sampling of surface sediments, *Hydrobiologia*, 99, 75, 1983.
78. **De Ressequier, A.**, A portable coring device for use in the intertidal environment, *Mar. Geol.*, 52, M19, 1983.
79. **Satake, K.**, A small handy corer for sampling of lake surface sediment, *Jpn. J. Limnol. Rikusuizatsu*, 44, 142, 1983.
80. **McCoy, F.W. and Selwyn, S.**, The hydrostatic corer, *Mar. Geol.*, 54, M33, 1983/1984.
81. **Ali, A.**, A simple and efficient sediment corer for shallow lakes, *J. Environ. Qual.*, 13, 63, 1984.
82. **Evans, H.E. and Lasenby, D.C.**, A comparison of lead and zinc sediment profiles from cores taken by diver and a gravity corer, *Hydrobiologia*, 108, 165, 1984.
83. **Twinch, A.J. and Ashton, P.J.**, A simple gravity corer and continuous-flow adaptor for use in sediment/water exchange studies, *Water Res.*, 18, 1529, 1984.
84. **Blomqvist, S.**, Reliability of core sampling of soft bottom sediment - an *in situ* study, *Sedimentology*, 32, 605, 1985.
85. **Fleischack, P.C., de Freitas, A.J., and Jackson, R.B.**, Two apparatuses for sampling benthic fauna in surf zones, *Estuarine Coastal Shelf Sci.*, 21, 287, 1985.
86. **Kuehl, S.A., Nittrouer, Ch.A., DeMaster, D.J., and Curtin, T.B.**, A long, square-barrel gravity corer for sedimentological and geochemical investigation of fine-grained sediments, *Mar. Geol.*, 62, 365, 1985.
87. **Pedersen, T.F., Malcolm, S.J., and Sholkowitz, E.R.**, A lightweight gravity corer for undisturbed sampling of soft sediments, *Can. J. Earth Sci.*, 22, 133, 1985.
88. **Brinkhurst, R.O., Chua, K.E., and Batoosingh, E.**, Modification in sampling procedures as applied to studies on the bacteria and tubificid oligochaetes inhabiting aquatic sediments, *J. Fish. Res. Board. Can.*, 26, 2581, 1969.

89. **Hamilton, A.L., Burton, W., and Flannagan, J.F.**, A multiple corer for sampling profoundal benthos, *J. Fish. Res. Board. Can.*, 27, 1867, 1970.
90. **Kemp, A.L.W., Saville, H.A., Gray, C.B., and Mudrochova, A.**, A simple corer and method for sampling the mud-water interface, *Limnol. Oceanogr.*, 16, 689, 1971.
91. **Jones, A.R. and Watson-Russell, C.**, A multiple coring system for use with scuba, *Hydrobiologia*, 109, 211, 1984.
92. **Bouma, A.H. and Marshall, N.F.**, A method for obtaining and analyzing undisturbed oceanic sediment samples, *Mar. Geol.*, 2, 81, 1964.
93. **Bruland, K.H.**, Pb-210 Geochronology in the Coastal Marine Environment, Ph.D. thesis, University of California, San Diego, 1974.
94. **Sundby, B., Silverberg, N., and Chesselet, R.**, Pathways of manganese in an open estuarine system, *Geochem. Cosmochem. Acta*, 45, 293, 1981.
95. **Mawhinney, M.R. and Bisutti, C.**, Common Corers and Grab Samplers Operating Manual, Report, Technical Operations, National Water Research Institute, Environment Canada, Burlington, Ontario, 1987.
96. **Brunskil, G.J.**, personal communication, 1989.
97. **Blomqvist, S., and Boström, K.**, Improved sampling of soft bottom sediments by combined box and piston coring, *Sedimentology*, 34, 715, 1987.
98. **Reineck, H.E.**, Kastengreifer und Lotrohre "Schnepfe", *Senckenbergiana Lethaea*, 39, 45, 1958.
99. **Bouma, A.H. and Sheppard, F.P.**, Large rectangular cores from submarine canyons and fan valleys, *Bull. Am. Assoc. Pet. Geol.*, 48, 225, 1964.
100. **Beukema, J.J.**, The efficiency of the Van Veen grab compared with the Reineck box sampler, *J. Cons. Cons. Perm. Int. Explor. Mer.*, 35, 319, 1974.
101. **Karl, H.A.**, Box core liner system developed at the Sedimentary Research Laboratory, University of Southern California, *Mar. Geol.*, 20, M1, 1976.
102. **Dolotov, Y.S. and Zharomskis, R.B.**, Method for studying coastal deposits from monoliths of undisturbed texture, *Oceanology*, 19, 338, 1979.
103. **Carlton, R.G. and Wetzel, R.G.**, A box corer for studying metabolism of epipelic microorganisms in sediment under *in situ* conditions, *Limnol. Oceanogr.*, 30, 422, 1985.
104. **Burns, R.E.**, A note on some possible misinformation from cores obtained by piston-type coring devices, *J. Sediment. Petrol.*, 33, 950, 1964.
105. **Heezen, B.C.**, Discussion of "A note on some possible misinformation from cores obtained by piston-coring devices", *J. Sediment. Petrol.*, 34, 699, 1964.
106. **Richards, A.F. and Parker, H.W.**, Surface coring for shear strength measurements, in *Proc. Civ. Eng. in the Oceans (ASCE)*, p. 445, 1967.
107. **Ross, D.A. and Riedel, W.R.**, Comparison of upper parts of some piston cores with simultaneously collected open-barrel cores, *Deep Sea Res.*, 14, 285, 1967.
108. **Bouma, A.H. and Boerma, J.A.K.**, Vertical disturbances in piston cores, *Mar. Geol.*, 6, 231, 1968.
109. **Chmelik, F.B., Bouma, A.H., and Bryant, W.R.**, Influence of sampling on geological interpretation, *Trans. Gulf Coast Assoc. Geol. Soc.*, 18, 256, 1968.
110. **Winterhalter, B.**, An automatic-release piston for use in piston coring, *Deep Sea Res.*, 18, 361, 1971.
111. **McCoy, F.W., Jr. and Von Herzen, R.P.**, Deep-sea corehead camera photography and piston coring, *Deep-Sea Res.*, 18, 361, 1971.
112. **McCoy, F.W., Jr.**, An analysis of piston coring through corehead camera photography, in *Underwater Soil Sampling, Testing, and Construction Control*, Am. Soc. Testing and Materials Spec. Tech. Publ. 501, 1972, 90.
113. **Hollister, C.D., Silva, A.J., and Driscoll, A.H.**, A giant piston-corer, *Ocean Eng.*, 2, 159, 1973.
114. **Silva, A.J. and Hollister, C.D.**, Geotechnical properties of ocean sediments recovered with giant piston corer. I. Gulf of Maine, *J. Geophys. Res.*, 78, 3597, 1973.
115. **Driscoll, A.H. and Hollister, C.D.**, The W.H.O.I. giant piston core: state of the art, in *Proc. Mar. Tech. Soc. 10th Conf.*, 1974, 663.
116. **Silva, A.J., Hollister, C.D., Laine, E.P., and Beverly, B.E.**, Geotechnical properties of deep sea sediments: Bermuda Rise, *Mar. Geotech.*, 1, 195, 1976.
117. **Lee, H.J.**, Offshore soil sampling and geotechnical parameter determination, *J. Pet. Technol.*, 32, 891, 1980.
118. **Prell, W.L., Gardner, J.V., Adelseck, Ch., Blechschmidt, G., Fleet, A.J., Keigwin, L.D., Jr., Kent, D., Ledbetter, M.T., Mann, U., Mayer, L., Reidel, W.R., Sancetta, C., Spariosu, D., and Zimmerman, H.B.**, Hydraulic piston coring of late Neogen and Quartenary sections in the Caribbean and equatorial Pacific: preliminary results of deep-sea drilling project Leg 68, *Geol. Soc. Am. Bull.*, 91 (Part 1), 433, 1980.

119. **Wright, H.E., Jr., Mann, D.H., and Glasser, P.H.**, Piston corers for peat and lake sediments, *Ecology*, 65, 657, 1984.
120. **Kelts, K., Briegel, U., Ghilardi, K., and Hsu, K.**, The limnogeology-ETH coring system, *Schweiz. Z. Hydrol.*, 48, 104, 1986.
121. **Bowen, V.T. and Sachs, P.L.**, New instrument — the free corer, *Oceanus*, 9, 1964.
122. **James, P.N.**, Cable-free data retrieval from the deep sea, *Sea Technol.*, 1976.
123. **Sly, P.G. and Gardener, K.**, A vibro-corer and portable tripod-winch assembly for through ice sampling, in *Proc. 13th Conf. Great Lakes Res.*, International Association of Great Lakes Research, Ann Arbor, MI, 1970, 297.
124. **McMaster, R.L. and McClennen, C.E.**, A vibratory coring system for continental margin sediments, *J. Sediment. Petrol.*, 43, 550, 1973.
125. **Dokken, Q.R., Circe, R.C., and Holmes, C.W.**, A portable, self supporting, hydraulic vibracorer for coring submerged, unconsolidated sediments, *J. Sediment. Petrol.*, 49, 658, 1979.
126. **Lanesky, D.E., Logan, B.W., Brown, R.G., and Hine, A.C.**, A new approach to portable vibracoring underwater and on land, *J. Sediment. Petrol.*, 49, 654, 1979.
127. **Fuller, J.A. and Meisburger, E.P.**, A simple, ship-based vibratory corer, *J. Sediment. Petrol.*, 52, 642, 1982.
128. **Marlowe, J.I.**, A piston corer for use through small ice holes, *Deep Sea Res. Oceanogr. Abstr.*, 14, 129, 1967.
129. **King, E.W. and Everitt, D.A.**, A remote sampling device for under-ice water, bottom biota, and sediments, *Limnol. Oceanogr.*, 25, 935, 1980.
130. Instruments and methods: ice drills and corers, *J. Glaciol.*, 3, 30, 1958.
131. **Verrall, R.J. and Baade, D.**, A hot-water drill for penetrating the Elsemere Island ice shelves, DREP Technical Memorandum 82-9, Defence Research Establishment Pacific, 1982.
132. **Mellor, M.**, Equipment for making access holes through arctic sea ice, Special Report 86-32, U.S. Army Corps of Engineers, Cold Regions Research and Engineering Laboratory, Hanover, NH, 1986.
133. **McIntyre, A.D.**, Efficiency of benthos sampling gear, in *International Biology Handbook No. 16*, Holme, N.A. and McIntyre, A.D., Eds., Blackwell Scientific, Oxford, 1971, 140.
134. **Rowe, G.T. and Clifford, C.H.**, Modification of the Birge-Ekman box corer for use with scuba or deep submergence research vessel, *Limnol. Oceanogr.*, 18, 172, 1973.
135. **Gale, W.F.**, A floatable benthic corer for use with scuba, *Hydrobiologia*, 77, 273, 1981.
136. **Anima, R.J.**, A diver operated reverse corer to collect samples of unconsolidated coarse sand, *J. Sediment. Petrol.*, 51, 653, 1981.
137. **Martin, E.A. and Miller, R.J.**, A simple, diver operated coring device for collecting undisturbed shallow cores, *J. Sediment. Petrol.*, 52, 641, 1982.
138. **Bothner, M.H. and Valentine, P.C.**, A new instrument for sampling flocculent material at the water/sediment interface, *J. Sediment. Petrol.*, 52, 639, 1982.
139. **Jensen, P.**, Meiofaunal abundance and vertical zonation in a sublittoral soft bottom, with a test of the Haps corer, *Mar. Biol.*, 74, 319, 1983.
140. **Jones, A.R. and Watson-Russell, C.**, A multiple coring system for use with scuba, *Hydrobiologia*, 109, 211, 1984.
141. **Hampton, L.**, *Physics of Sound in Marine Sediments*, Plenum Press, New York, 1974.
142. **Kirby, R. and Parker, W.R.**, Seabed density measurements related to echo sounder records, *Dock Harbour Auth.*, 54, 423, 1974.
143. **MacIsaak, R.R. and Dunsiger, A.D.**, Ocean sediment properties using acoustic sensing, in *Proc. 4th Int. Conf. Port and Ocean Engineering under Arctic Conditions*, 1977, 1074.
144. **Thomas, R.L., Jaquet, J.-M., and Kemp, A.L.W.**, Surficial sediments of Lake Erie, *J. Fish. Res. Board Can.*, 33, 385, 1976.
145. **Finckh, P., Kelts, K., and Lambert, A.**, Seismic stratigraphy and bedrock forms in perialpine lakes, *Geol. Soc. Am. Bull.*, 95, 1118, 1984.
146. **Dunsiger, A.D., Chari, T.R., Fader, G.B., Peters, G.R., Simpkin, P.G., and Zielinski, A.**, Ocean sediments — a study relating geophysical, geotechnical, and acoustic properties, *Can. Geotech. J.*, 18, 492, 1981.
147. **Baldwin, K.C., LeBlanc, L.R., and Silva, A.J.**, An analysis of 3.5 kHz acoustic reflections and sediment physical properties, *Ocean Eng.*, 12, 475, 1985.
148. **Briggs, K.B., Richardson, M.D., and Young, D.K.**, Variability in geoacoustic and related properties of surface sediments from the Venezuela Basin, Caribbean Sea, *Mar. Geol.*, 68, 73, 1985.

149. **Reed, D.L., Meyer, A.W., Silver, E.A., and Prasetyo, H.**, Contourite sedimentation in an intraoceanic forearc system: Eastern Sunda Arc, Indonesia, *Mar. Geol.*, 76, 223, 1987.
150. **Bloomberg, D., Upchurch, S.B., Hayden, M.L., and Williams, R.C.**, Cone penetrometer exploration of sinkholes: stratigraphy and soil properties, *Environ. Geol. Water Sci.*, 12, 99, 1988.
151. **Chaney, R.C. and Demars, K.R.**, Strength Testing of Marine Sediments: Laboratory and *In-Situ* Measurements, American Society for Testing and Materials Special Technical Publication 883, Publication Code Number 04-883000-38, ASTM, Philadelphia, PA, 1985.
152. **Blomqvist, S. and Abrahamsson, B.**, A device for rapid sectioning of soft bottom sediment cores, *Schweiz. Z. Hydrol.*, 49, 393, 1987.
153. **Krishnaswamy, S., Lal, D., Martin, J.M., and Meybeck, M.**, Geochronology of lake sediments, *Earth Planet. Sci. Lett.*, 11, 407, 1971.
154. **Koide, M., Soutar, A., and Goldberg, E.D.**, Marine geochronology with Pb-210, *Earth Planet. Sci. Lett.*, 14, 442, 1972.
155. **Robbins, J.A. and Edgington, D.N.**, Determination of recent sedimentation rates in Lake Michigan using Pb-210 and Cs-137, *Geochim. Cosmochim. Acta*, 39, 286, 1975.
156. **Farmer, J.G.**, The determination of sedimentation rates in Lake Ontario using the Pb-210 method, *Can. J. Earth Sci.*, 15, 431, 1978.
157. **Durham, R.W. and Joshi, S.R.**, Recent sedimentation rates, Pb-210 fluxes, and particle settling velocities in Lake Huron, *Chem. Geol.*, 31, 53, 1980.
158. **Joshi, S.R.**, Recent sedimentation rates and Pb-210 fluxes in Georgian Bay and Lake Huron, *Sci. Total Environ.*, 41, 219, 1985.
159. **Turner, L.J. and Delorme, L.D.**, Pb-210 dating of lacustrine sediments from Hamilton Harbour (Cores 137, 138, 139, 141, 142, 143), Technical Note LRB-88-9, Environment Canada, National Water Research Institute, Burlington, Ontario, 1988.
160. **Durham, R.W. and Oliver, B.G.**, History of Lake Ontario contamination from the Niagara River by sediment radiodating and chlorinated hydrocarbon analysis, *J. Great Lakes Res.*, 9, 160, 1983.
161. **Mudroch, A., Joshi, S.R., Sutherland, D., Mudroch, P., and Dickson, K.M.**, Geochemistry of sediments in the Back Bay and Yellowknife Bay of the Great Slave Lake, *Environ. Geol. Water Sci.*, 14, 35, 1989.
162. **McAndrews, J.H.**, Fossil history of man's impact on the Canadian flora: an example from southern Ontario, *Can. Bot. Assoc. Bull.*, Suppl. 9, 1, 1976.
163. **Haworth, E.Y. and Lund, J.W.G.**, *Lake Sediments and Environmental History*, University of Minnesota Press, Minneapolis, 1984, 411.
164. **Robbins, J.A.**, personal communication.
165. **Krinitzski, E.L.**, *Radiography in the Earth Sciences and Soil Mechanics*, Plenum Press, New York, 1970, 47.
166. **Mallik, T.K.**, An inexpensive hand-operated device for cutting core liners, *Mar. Geol.*, 70, 307, 1986.
167. **Duncan, G.A.**, Manual on procedures for stratigraphic analysis of unconsolidated sediment cores, Report No. 82-24, Hydraulics Division, National Water Research Institute, Burlington, Ontario, Canada, 1982.

Chapter 5
SAMPLING THE SETTLING AND SUSPENDED PARTICULATE MATTER (SPM)

Fernando Rosa, Jürg Bloesch, and David E. Rathke

I. INTRODUCTION

A myriad of living organisms and particles of nonliving matter free floating in water is the very basis of the life support system in aquatic environments. The suspended particles are an integral component for both their nutritive qualities and their ability to adsorb organic and inorganic pollutants.[1] Simultaneously, these minute particles are the main transport media for the dispersion and translocation of these contaminants. In the past decade, research into the role of suspended particles has been increasing, in recognition of their importance in the cycling of nutrients and pollutants. Proper sampling techniques and proper sampling devices are of equal importance to obtain a representative sample which would adequately describe the ambient concentration and flux of settling and suspended particulate matter (SPM). This chapter emphasizes the definition, origin, and fate of particulate material (Sections I.A and B), the sampling strategy dependent thereof (Section I.C), the state of the art of different sampling devices (Sections II.A to D) and sample processing prior to analysis (Section III).

A. DEFINITION OF SUSPENDED PARTICULATE MATTER (SPM)

Seston (suspended sediments) is defined by Hutchinson[2] as all the SPM in the free water of a lake, including bioseston (plankton and nekton) and abioseston (tripton). The size spectrum of SPM can range from fibrils or colloids, <0.05 μm,[3] on the lower end, up to particles >2 mm, at the upper end of the scale. In natural waters dissolved as well as particulate chemical compounds are present. However, the boundary between dissolved and particulate matter is not so distinct as illustrated by the presence of colloids. Bacteria and viruses, representing the smallest "biological particles", are in the size range of large organic molecules.

There are many definitions used to delimit the soluble/particulate phase and, for the most part, are dependent on the nature of the experiment. In practice, 0.45-μm filters are commonly but arbitrarily accepted to delineate the soluble/particulate fractions. However, recent research has shown that the 0.45-μm membrane filter does *not* serve as a proper cutoff filter for distinguishing dissolved from particulate organic carbon.[4] Burnison and Leppard[4] used ultrafiltration (see Figure 1) to separate the colloidal fraction that passed through the 0.45-μm filter, and found that the colloidal fraction contained 33% of the total "dissolved" organic carbon. Furthermore, size selective filtration is not possible,[5-7] mainly due to filter clogging and heterogeneity in the different shapes and sizes of particles.

1-56670-027-2/94/$0.00+$.50

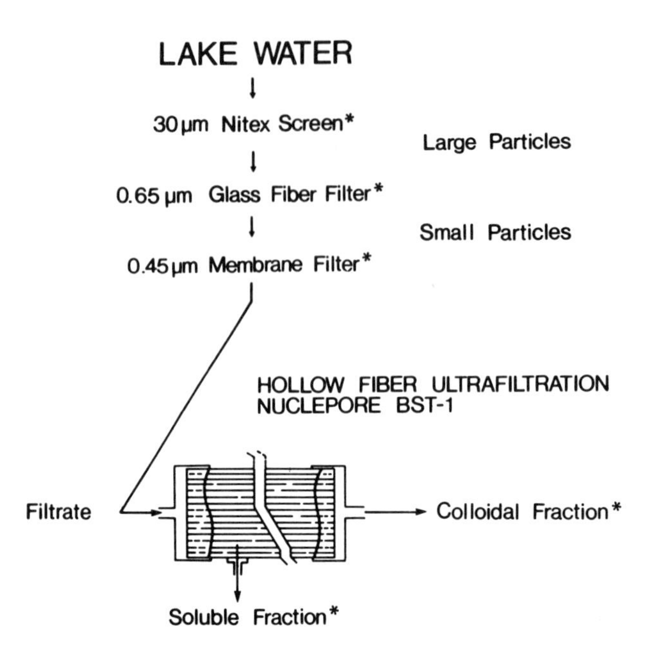

FIGURE 1. Separation of colloidal fibrils from lake water. *Observed by transmission electron microscopy. (Modified after Burnison, B. K. and Leppard, G. G., *Can. J. Fish. Aquat. Sci.*, 40, 373, 1983. With permission.)

Most of the research done on the classification of SPM has used water subsamples obtained from instantaneous/point and integrating water samplers described in detail in Section II.B, and by Fay.[8] Many researchers have used filters and screens for quantifying and qualifying different types of SPM through chemical/physical analyses on the total fraction,[5,9-12] or on the different size fractions.[13] A filter is normally used to remove particles from the liquid phase, while the screen is generally used to quantitatively separate two particle size fractions.[5] Although the mesh size of a screen may be a good measure of particle size separation, the stated pore size of a filter may not be as precise.[14] (This may primarily relate to glass fiber and not membrane filters.) For example, Sheldon[5] reported that different filters had different particle retention capacity (Figure 2), and that the manufacturer's stated pore size was not necessarily the effective pore size of the filter.

In addition there are two significant problems encountered in using either filters or screens. Apart from the problem of selecting the proper pore or mesh size for each individual application, the sample size that can be processed by filters or screens is often limited by clogging. Also, the amount of SPM required for performing multiple analyses on the same sample, especially for organic contaminants, is not always achieved.

B. SOURCE, ROLE, TRANSPORTATION, AND TRANSFORMATION OF SUSPENDED PARTICULATE MATTER

Particulate matter in lakes and oceans can originate from allochthonous or autochthonous sources.[15,16] Riverine input is the major allochthonous source for lakes, whereas primary production contributes significantly to the autochthonous bioseston. Aeolian input and hydrothermal activity provide other important particulate matter sources in oceans.[17]

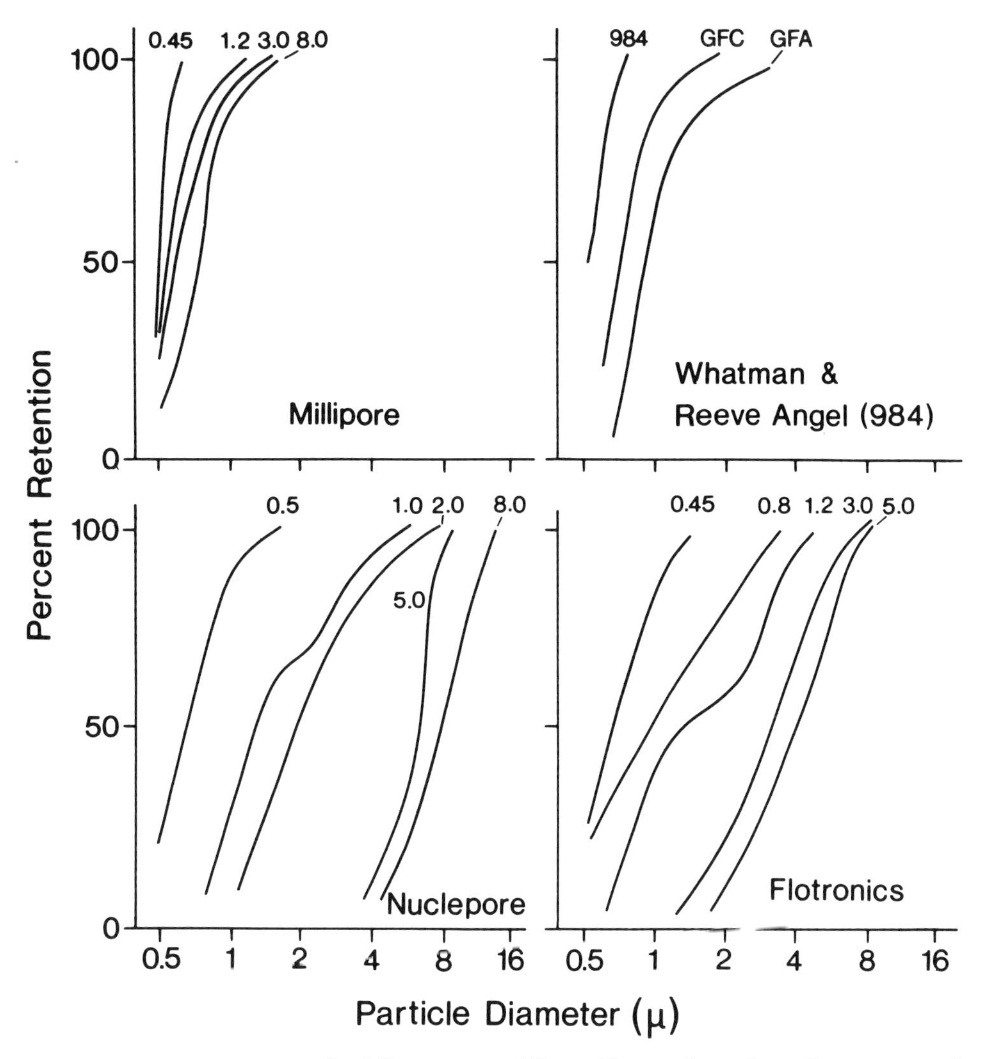

FIGURE 2. Retention curves for different types of filters. The numbers above the curves are the pore sizes given by the manufacturers. (After Sheldon, R. W., *Limnol. Oceanogr.*, 7, 494, 1972. With permission.)

Sediment transport, distribution, deposition, erosion, and resuspension are processes mainly driven by meteorological forces and currents.[18,19] Particles are transformed by physical, chemical, and biological processes such as decomposition, mineralization, dissolution, adsorption, coagulation, precipitation, fecal pellet production, and bacterial degradation.[20-24] The residence time of particles in the water column is of crucial importance in transformation processes, and is usually limited by the downward flux rather than loss through river outlets. Settling flux of SPM is recognized as one of the fundamental processes for the understanding of aquatic ecosystems, especially with regard to primary production and loss of organic matter from the trophogenic layers,[25] scavenging and transport of pollutants,[1] and particulate organic carbon input to the benthic food chain.[26]

The sinking mechanism of particles, not yet fully understood, was comprehensively reviewed, for example, by Hutchinson[2] and Lerman.[17] The sinking mechanism is controlled largely by the various size, shape, and density of

TABLE 1
Vertical Sinking Velocities of Particles vs. Horizontal Movements in Air and Water

Falling or settling particle	Diameter (μm)	Vertical sinking velocity (cm/s)	Wind speed or horizontal current velocity (cm/s)	Proportional difference in velocity
Raindrops	50—5000	0.2—1000	<10^3	<10
Snowflakes	—	50		
Small particles	1	10^{-4}	<20—200 (oceans)	10—10^6
Large particles	40	10^{-1}	<5—30 (lakes)	
Fecal pellets	—	0.04—1		

According to Gardner;[79] modified from Bloesch, J. and Burns, N. M., *Schweiz. Z. Hydrol.*, 42, 15, 1980.

suspended particles, e.g., planktonic cells, detritus, mineral crystals and clay minerals, aggregates, as well as the physico-chemical or biological mechanisms of particle dissolution and formation. In addition, phytoplankton are known for their ability to regulate sinking rates.[27,28]

The fundamental question — To what extent does physical turbulence influence particle settling velocity? — is still being debated. Disagreement among researchers on increasing effects,[29] as well as decreasing effects,[30] in settling velocity has been reported. There is also some controversy whether Stokes' law can be applied to settling particles.[31,32] In turbulent water, small particles are carried passively in eddies with a small constant component of downward movement, rather than sinking vertically or at a certain angle as do raindrops or snowflakes in the air. This apparent downward movement is due to the striking differences between vertical sinking velocities of particles and horizontal movements in air and water (Table 1). Thus, for most sestonic particles we have to dismiss the concept of a steady "rain" sedimentation. In oceans, however, large fecal pellets and large amorphous aggregates ("marine snow") may be the exception;[33] also, nearshore turbidites occurring in lakes with large tributaries/deltas and steep slopes, such as the Rhone River delta in Lake Geneva, Switzerland, may have occasional and time-limited "fallout" of particles.[34]

C. SAMPLING STRATEGIES

The purpose of sampling a lake is to make a judgment about the whole lake by examining only a small portion of the whole. This judgment will only be valid if the sampling strategy/design used will be representative of the whole water body. If field data are to be useful, the survey design must be planned according to established statistical procedures (see Chapter 3).

Because of the chemical/physical heterogeneity in the distribution of SPM that exists within water bodies — whether they may be lakes, oceans, or reservoirs — sampling procedures are a very important consideration in order to acquire representative samples. The concentration of SPM and settling sediments changes in space and time, and is dependent on the unique characteristics of the particular system. There are two categories of SPM and settling sediment samplers: (1) instantaneous samplers (water samplers) which collect a volume of water-SPM mixture passing a stationary sampling point at a specific time (see Section II.B) and (2) integrating samplers which collect material over an extended time or space/distance. Space integrating samplers collect a volume of water and associated SPM continuously at each point, while descending at a constant speed, over a vertical distance (depth) in the water column (see Section II.C). Time integrating samplers, e.g., sediment traps, collect suspended and settling sediments at a specific point in space, over a time period, so that temporal fluctuations are averaged (see Section II.D). These samplers act as SPM and settling sediment interceptors, collectors, or accumulators.

A complete citation of work dealing with SPM sampling prior to 1969 is given by the Australian Water Resources Council.[35] The report reviews reasons for collecting data on sediment in reservoirs and streams. Examples are provided on the effects of sedimentation both in Australia and worldwide, and the need for collecting data to assist in the solution of specific engineering and scientific problems caused by sediments. In addition, the report reviews methods of data collection and analytical techniques used by water resource organizations in Australia, and covers suspended sediment transport, bed load transport, and reservoir sedimentation.

II. DESCRIPTION AND USE OF DIFFERENT SAMPLING DEVICES: THE STATE OF THE ART

A. OPTICAL METHODS AND TURBIDITY MEASUREMENTS

Low and high concentration zones of SPM can, for example, be identified by optical measurements using a multiband transmissometer (Figure 3, bottom). Lake water color can be directly related to water quality indicators such as chlorophyll and suspended minerals once the absorption and scattering properties of the water mass and the particulate constituents have been determined.[36,37] Optical measurements therefore should be a prerequisite to field sampling surveys for SPM analysis. Such measurements could substantially reduce sampling and analytical time/costs, by only sampling the depths of interest based on the optical (depth-transmission) trace.

Reasonable approximation of the concentrations of suspended matter at various sites and water depths may also be derived from the measurement obtained with a turbidity meter. Turbidity meters require careful calibration but may be used together with a fluorometer (to approximate chlorophyll content) to provide an estimate of the living and nonliving components. This information, however, in no way approaches the derivation which is possible from the analysis of actual water samples.[38]

B. SAMPLING SUSPENDED PARTICULATE MATTER (SPM) AT DISCRETE DEPTHS

SPM is usually collected with different types of water samplers (see Table 2).[8,39,40] Most sampling devices collect water vertically in the water column, integrating the water column equal to the length of the sampler, usually ranging from 0.5 to 1.0 m. Sampling at a close interval is important to assess the variability of SPM in the water column, especially close to the sediment-water interface where large gradients in SPM concentrations can exist.[41-44] These samplers can also be used anywhere in the water column to determine gradients in particle concentration, e.g., within the thermocline or chemocline where sharp gradients also occur.[45]

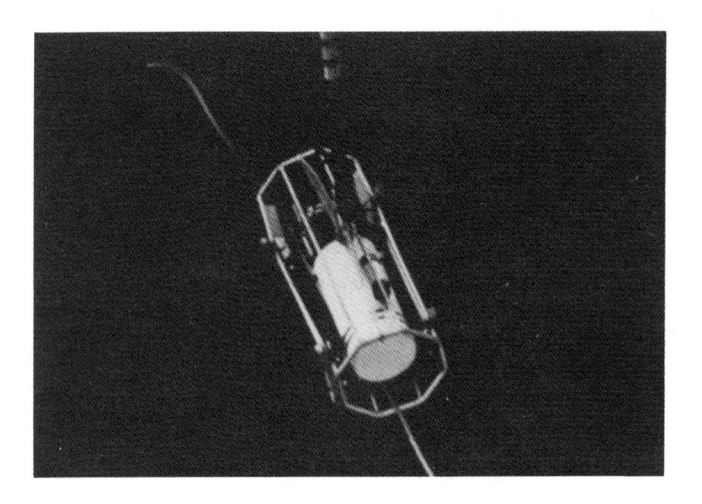

FIGURE 3. Integrating samplers and an optical sensor.

1. Instantaneous Single Point Water Samplers

All instantaneous water samplers (Table 2) basically work on the same principle, that is, they collect and retain water by closing and sealing the lids at each end of the sampler by use of a remote messenger.

The **Knudsen bottle** (Figure 4) was designed for the collection of water samples and water temperature data using up to three reversing thermometers.

TABLE 2
Comparison of Different Types of Water Samplers[a]

Water Sampler; Volume (l); Sampler Suitability; and Construction

Knudsen;[b,c] 1.2; capable of holding three reversing thermometers; nickel-plated brass construction
Nansen;[c] 1.5; similar to Knudsen bottle, with the same brass construction; tin-plated or Teflon lined
Van Dorn-Vertical;[b,c] 2,3,6; good for general water sampling; PVC construction
Van Dorn-Horizontal;[c] 2,3,6; good for sampling in thin mesolimnion and hypolimnion situations, and could be used to obtain sediment-water interface samples; PVC construction.
Kemmerer;[c] 0.5—8; good for general water sampling; available in copper, brass, nickel-plated brass, and PVC
Niskin;[d] 1.2—30; can be used individually, in series or in a Rosette sampler, a multidimensional sampler; available in nonmetallic PVC and also Teflon lined for organics sampling
Go-Flo;[d] 1.7—100; recommended for deep sampling of trace metals due to the close-open-close design of the bottles which eliminates surface contamination; PVC construction
Rosette;[b,d] 1.2—30; multibottle sampler capable of holding and sampling up to 12 bottles per cast; the optional temperature and depth sensors are recommended; Rosette frame can utilize either Go-Flo or Niskin bottles
Bacti-Bulb;[b,e] 0.8—1; spherical shape to withstand pressure for deep sampling down to 366 m; made of glass
Ruttner;[f] 1—3; Perpex (plexiglass); the open lids are in the horizontal position
Friedinger;[148,f] 1—3; light metal or PVC; the open lids are in the vertical position
Depth Integrator;[b,g] variable; designed to collect an equally weighted depth sample; useful for portraying water quality without the expense of multiple depth samples, available in 0—10 m (8 l), 9—20 m (4 l), and 0—50 m (1 l) sizes; Al and PVC construction
Submersible Pump (MARSH);[h] unlimited, pump is capable of supplying an unlimited volume of sample; available in brass or stainless steel

[a] Revised after Fay, L. A., Great Lakes Methods Manual, Field Procedures, Report D-4857W, International Joint Commission, Windsor, Ontario, 1988.
[b] These samplers are shown on Figures 3 and 4.

SAMPLER MANUFACTURERS (Suppliers)

[c] Wildco (Egetec Enterp. Inc., 265 Canarctic Dr., Downsview, Ontario, Canada M3J 2N7).
[d] General Oceanic.
[e] John Scientific Inc. (175 Hanson St., Toronto, Ontario, Canada M4C 1A7).
[f] Züllig, Apparatebau für die Wasserwirtschaft (CH-9424 Rheineck, Switzerland).
[g] Canada Centre for Inland Waters, National Water Research Institute, Engineering Services (867 Lakeshore Rd., Box 5050, Burlington, Ontario, Canada L7R 4A6).
[h] Miller Plastics (19 Advance Rd., Toronto, Ontario, Canada M8Z 2S6).

The sampler is constructed of nickel-plated metal with an average length and capacity of 0.5 m and 1.2 l, respectively.

The **Nansen bottle** is similar to the Knudsen bottle, but is designed to sample depths of 6100 m or more. The end valves are made of bronze, and the cylinder is made of brass.[46] It is available either with a tin-plated or Teflon-lined cylinder. It has a capacity of 1.3 l and an overall length of 0.7 m.

The **Van Dorn bottle** (Figure 4), also known as the Alpha or Beta bottle, is constructed of polyvinyl chloride (PVC) and sealed at either end by force cups pulled together by a length of strong, flexible, rubber tubing. The sampler is available in 2-, 3-, and 6-l capacities, and two configurations: vertical or horizontal. The horizontal position is useful for collecting water at the sediment-water interface or sampling a thin layer of the water column. The

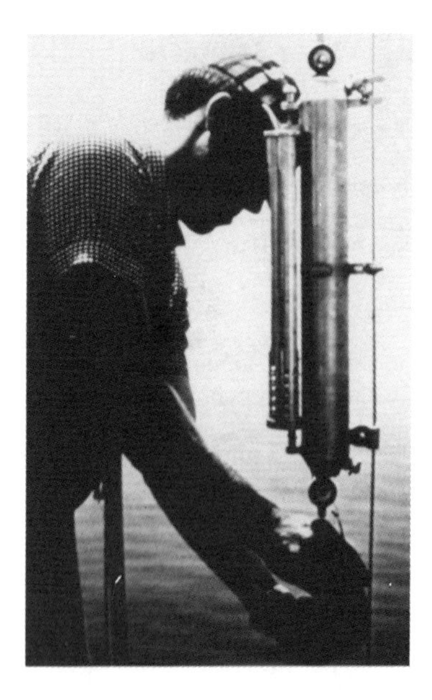

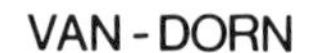

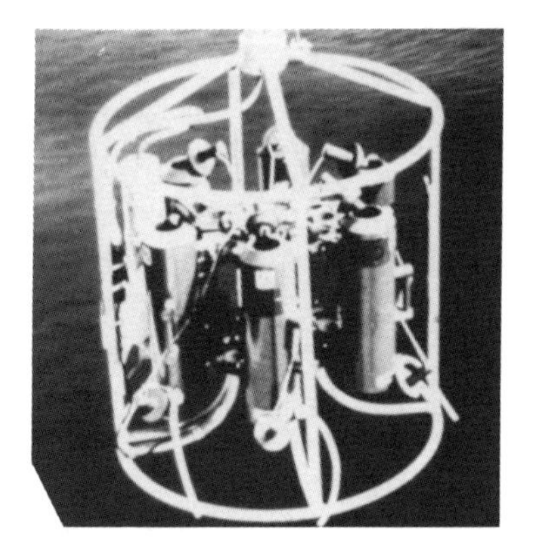

FIGURE 4. Instantaneous water samplers.

sampling procedure varies for the horizontal and vertical bottles although the bottles are functionally the same.[8]

The **Bacti-Bulb** (Figure 4) was specifically designed to collect water samples for bacteriological analyses,[47] but it can be used for collecting SPM. For bacteriological sampling it is of utmost importance that the water samples are collected as aseptically as possible to accurately reflect microbiological conditions at the time of sample collection. To collect water samples, the air in the bulb must first be removed by using a vacuum pump, and the bulb is then placed on a special stainless steel holder and attached to the sampling line.

The **Rosette Water Sampler** (Figure 4) consists of a cylindrical shaped frame capable of holding 6 or 12 GO-FLO Model 1800 water sampling bottles or Model 1010 Niskin water bottles (Table 2). Bottle size may vary from 1.2- to 30-l capacity. The deck command unit is capable of remote actuation of the water bottles in sequence. The Rosette sampler is also capable of actuating three or four reversing thermometer assemblies. The submersible bottle array has optional temperature and depth sensors. The operator of the sampler can verify that the individual water bottles collect water at selected depth and temperature.

The investigator has to give careful attention to the selection of the water sampler, depending on the required analyses of the sample. Bacon et al.[48] found that in analyzing for hydrogen sulfide (H_2S), there was a systematic difference between samples collected with the Nansen bottles and those taken with the Niskin bottles. The Nansen samples consistently gave lower results by about 15%. The authors attributed this loss to a consumption of H_2S by reaction with the brass structure of the Nansen sampler, due to imperfections in the Teflon coating. Nansen bottles have also been shown to produce systematically low results for dissolved oxygen analyses,[49] but the effects were much smaller than for H_2S. Consequently, if organochlorine, hydrocarbons, and trace metals are to be analyzed, glass or Teflon-coated materials must be used to prevent contamination.

2. Instantaneous Multiple Point Water Samplers

The close interval water sampler (CIWS, Figure 5) is an instantaneous multiple water sampler that was specifically developed[50] to aid in the interpretation and understanding of sediment resuspension in the Laurentian Great Lakes.[41] (See also Section II.D.3.) This sampler, with overall dimensions of ≈2.7 m high by 2.7 m wide, is designed to allow discrete, 3-l, undisturbed water samples to be taken at multiple heights between 20 cm and 1.8 m above the sediment-water interface or anywhere in the water column where gradients in particle concentration are suspected to exist. Samples are taken in horizontally operating, stainless steel, and PVC piston-type bottles, driven by pneumatic cylinders. The supporting structure is an aluminum central column, mounted on an aluminum Y-shaped structure, with aluminum pads at the center and outer ends (Figure 5). The structure is braced with stainless steel cables, tensioned with turnbuckles. The system is operated by an electronic timer, controlling a solenoid valve. Power for the cylinder motion is supplied by a SCUBA cylinder through a pressure reducing regulator and safety valve.

3. Single Point Particle Concentrators

Before starting any project with SPM, the separation of the required quantity of SPM from the water (dewatering) should be carefully considered. The concentration of SPM in the water column varies significantly between water bodies and within a water body (see also Table 9). For example, open water concentrations of SPM in the Laurentian Great Lakes can range from less than 0.5 up to 10 mg/l, with some tributary concentrations exceeding 100 mg/l.[51] Consequently, the quantity of water needed to collect 1 g of SPM from these two water masses is significantly different. In this respect, the water samplers are not satisfactory for obtaining the SPM quantities necessary for specific chemical analyses such as for trace elements or organic compounds, which generally require 0.5 to 10 g of dry sample. In order to optimize the effort it is advisable to predetermine the quantity of SPM prior to extensive sampling.

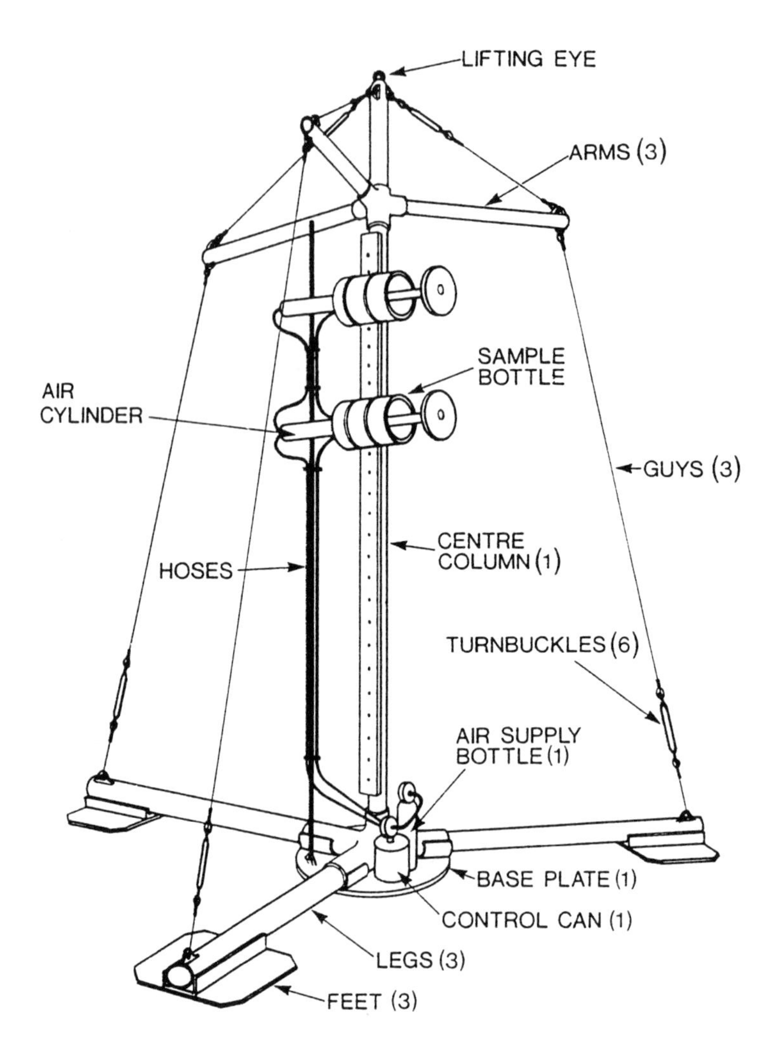

FIGURE 5. Close-interval water sampler. (Design of H. Savile and F. Rosa.)

Several methods have been employed to collect, dewater, and concentrate SPM with various degrees of success (Table 3). Two procedures have proven to be most effective: the flow-through centrifuge system (FTCS) (e.g., Westfalia four-bowl continuous flow centrifuge, Model KDD 605[52-54] (WFTC), Figure 6) and the tangential flow filtration system (TFFS) (Pellicon — manufactured by Millipore,[55] Figure 7). Both systems can be used in the field, thus eliminating the need for transportation, storage, and preservation of large volumes of water.

Recently, two FTCS systems and the TFFS were extensively studied, in river and stream waters with SPM concentrations ranging from 30 to 350 mg/l. The two systems were compared with respect to SPM recovery, effluent quality, efficiency, and sample contamination problems associated with trace metal analyses.[56] In nearly all cases, the Westfalia FTCS was found to be most efficient, with most recoveries exceeding 95%. Similar efficiencies have been reported by Williams et al.[57] The TFFS recoveries were lower (87%) and were attributed to problems associated with removing (backwashing) the SPM from the membrane pack within the TFFS. In contrast, the final effluent from the

TABLE 3
Procedures Used to Dewater and Concentrate SPM

Method	Advantage	Disadvantage
Conventional filtration	Quick, field adaptable	Limited volume, and material recovery difficult
Settling	Controlled conditions	Storage and transport of large water volumes, degradation, long settling time for fine particles
Centrifugation (laboratory)	Controlled conditions	Storage and transport of large water volumes, degradation
Centrifugation flow-through (field)	Field, quick, efficient, very large volume	Recovery <100%, expense
Tangential-flow filtration	Field, quick, efficient large volumes	Recovery <100%

TFFS was nearly 100% free from SPM (0.45 μm), while effluent from the FTCS averaged 95%. The removal efficiencies using the TFFS were the greatest of all systems tested and hence the effluent water from the TFFS was considered appropriate for use in the analysis of "dissolved" substances.

In addition to this study, a series of tests were conducted to determine if any differences existed in the qualitative or quantitative characteristics of SPM collected via the FTCS and TFFS, in the central basin of Lake Erie[57-59] (Table 4). Using a submersible pump, water from 5 m was transferred from the lake to a large container (600 l) on board ship, subsequently subsampled from the holding tank, and processed by the two methods. Although, the particulate phosphorus (PP) values were similar, total phosphorus (TP) concentrations from the centrifuge effluent were nearly double those obtained from the TFFS effluent (Table 4). It is inferred that the higher values encountered from the centrifuging process resulted from an increase of the total filtered phosphorus (TFP), i.e., dissolved phase, in the centrifuge effluent as a result of cell breakage and disruption during the centrifuge process. Other discrepancies were apparent in the effluent concentrations of chlorophyll and suspended organic matter. Concentrations of these parameters determined in the centrifuge effluent were very similar to those in the water prior to processing, while the effluent from the TFFS contained no measurable concentration of chlorophyll and a 95% reduction in total suspended matter over the ambient concentrations.

In conclusion, the FTCS operates less efficiently in lakes than in rivers, since SPM concentrations in open lake waters are considerably lower than in rivers (between 1 and 20 mg/l), and SPM in lakes is composed of smaller particles (less than 10 μm). Moreover, lake water has larger quantities of phytoplankton and zooplankton (subject to breakage during centrifugation or filtration) and a different ratio of organic to inorganic material than found in rivers.

It seems clear from the above comparison that the FTCS would be preferable for collecting SPM for bulk analyses, but if detailed information is required for

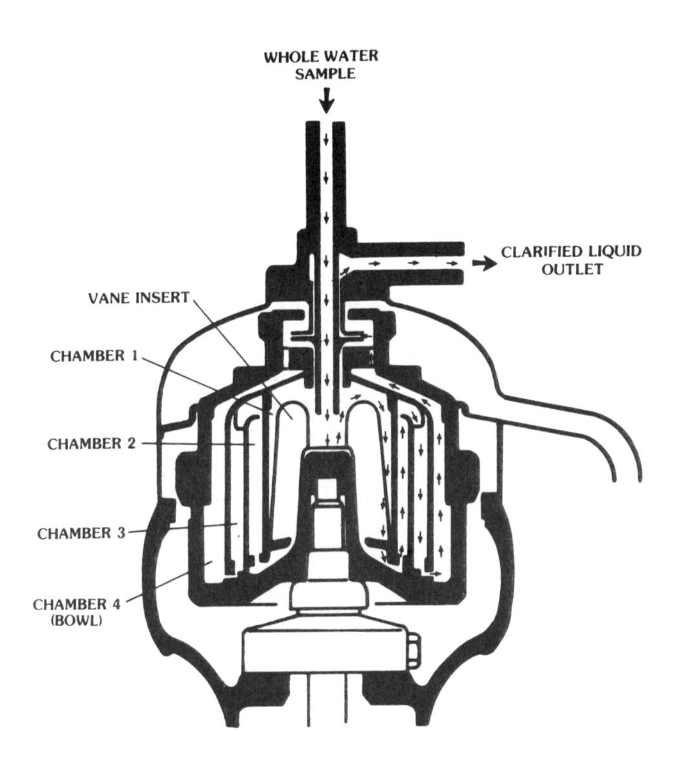

FIGURE 6. (Upper) Westfalia, flow-through centrifuge (WFTC) system. (Lower) Exploded view of the WFTC bowl (Savile).[52] (From Horowitz, A. J., Elrick, K. A., and Hooper, R. C., *Hydrol. Process.*, 2, 163, 1989. With permission.)

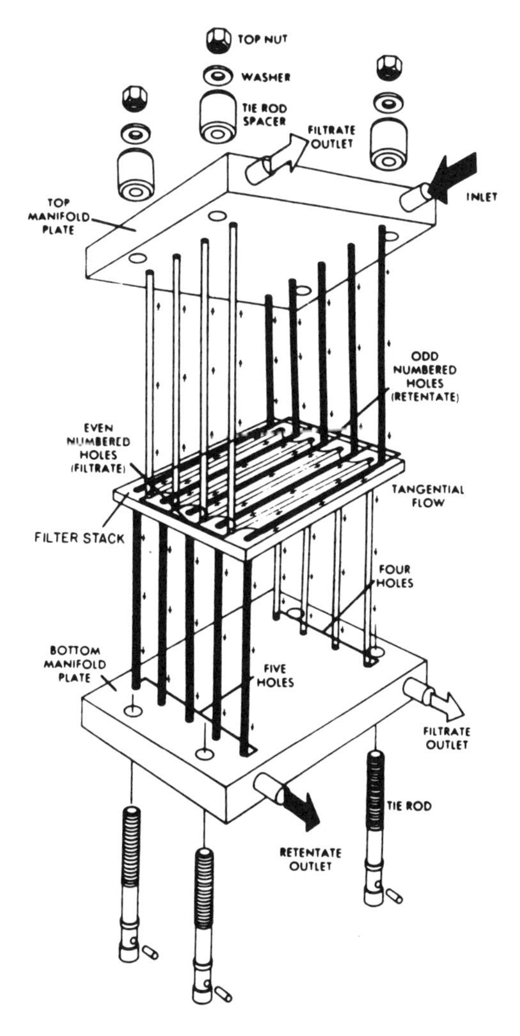

FIGURE 7. (Upper) Pellicon, tangential-flow filtration system (TFFS). (Lower) Schematic diagram of the PTFFS filtration stack (membrane pack) assembly (Millipore).[55] (From Horowitz, A. J., Elrick, K. A., and Hooper, R. C., *Hydrol. Process.* 2, 163, 1989. With permission.)

TABLE 4
Concentrations of Parameters in Lake Erie Obtained by Analyzing Centrifugation (FTCS) and Filtration (TFFS) Effluents

	WFTC[a] effluent	PTFFS[b] effluent	Total ambient water
Total phosphorus (mg/l)	5.8	2.7	13.0
Total filtered phosphorus (mg/l)	5.8	2.7	3.7
Particulate phosphorus (mg/l)	0.9	0.7	9.3
Chlorophyll a (mg/l)	2.90	0.0	2.72
Total suspended matter (mg/l)	1.66	0.09	1.69
Inorganic suspended matter (mg/l)	0.43	0.04	0.62
Organic suspended matter (mg/l)	1.23	0.05	1.07

[a] Westfalia flow-through centrifuge.
[b] Pellicon tangential-flow filtration system.

fractionation and for dissolved chemical analysis, the TFFS would be more suitable.

During the initial testing of the TFFS, it was discovered that the SPM determined from the material collected from the 0.45-μm membrane pack (Figure 7, bottom) was generally greater than that measured by conventional filtration. In order to determine the cause for this discrepancy several samples were examined in detail by using the following procedure: Once the water sample had been processed through the TFFS, the collected material was flushed (backwashed) from the membrane pack using 1 l of the effluent water from the process. This backwash water and 1 l of the TFFS effluent water were frozen and later freeze-dried to determine the *actual* and *effluent* concentrations, respectively (Table 5). It was determined that the dissolved solids recovered from freeze-drying the effluent water, which was used to flush the membrane pack, significantly added to the SPM sample weight. Consequently, the total freeze-dried weight needs to be corrected to obtain the true weight of the SPM recovered by the TFFS (*corrected* concentration, Table 5).

As previously discussed, the most significant problem with the TFFS lies in recovering the material from the membrane pack. As is evident from these results, preliminary studies on harvesting procedures and recovery efficiencies need to be conducted prior to the actual use of the TFFS.

C. SAMPLING SUSPENDED PARTICULATE MATTER (SPM) INTEGRATING OVER DEPTH

Certain studies require that discrete samples must be obtained to describe concentration changes of different constituents throughout the water column. The timely and expensive discrete multiple sampling can be avoided by the use of a depth integrator (Figure 3). The integrator is designed to collect approximately equal volumes of water-SPM mixture from each layer/point in the water column it traverses. While the sampler is being lowered, water enters the cylinder and interior cone through a one-way valve. To maintain equal volumes of water passing through the valve as the sampler is lowered at a near-uniform speed of 1 m/s, the trapped air inside the sampler is compressed in relation to the increasing depth and prevailing ambient pressure according to the equation:

TABLE 5
Total Suspended Matter (SPM) Recovery by Using Conventional Filtration, and Freeze-Drying Methods

Location	Concentration (mg/l) Actual (A)	Effluent (E)	Corrected (A—E)	Total suspended matter
Lake Erie west	3.46	2.44	1.02	1.30
Lake Erie central	3.81	2.96	0.85	0.72
Lake Erie central	4.07	3.17	0.90	0.58
Lake Erie central	5.16	3.69	1.47	1.69
Lake Erie east	4.74	3.51	1.23	2.06
Lake Huron	2.38	0.87	1.51	0.53
Lake Huron	1.96	1.20	0.76	0.59

$$h = hz/d + 1 \quad (1)$$

where h is the compressibility factor, hz is the overall cylinder height, and d is the hydrostatic pressure.[60] After the sampler is retrieved to the surface, the water inside the cone remains trapped by the one-way valve, while excess water that has entered the cylinder, but not the cone, is pushed out by the compressed air inside the cone and cylinder or flows out by itself.[61] Besides the Schröder sampler,[61] there are three integrators (built by the Engineering Section, Research Support Division, NWRI) that have been specifically designed to integrate three different depth ranges of 10, 20, and 50 m with a resulting sample volume of 8, 4, and 1 l, respectively. Due to existing concentration gradients in the water column, the sample needs to be completely drained into a container and well mixed before subsampling for SPM analysis.

SPM may also be collected *in situ* by means of plankton nets (Figure 3). Plankton nets can be used for vertical, horizontal, or oblique sampling, the vertical haul being the simplest. The most common type of plankton net is the conical net sampler, consisting of a simple support loop, towing bridle, filtration net, and straining (filtering) bucket. Also, nets with tilt-closing mechanisms for quantitative plankton sampling have been used.[62] The most popular mouth-diameter-to-net-length ratio is 1:3, however, nets can be ordered in a wide variety of ratios and sizes depending on user requirements. An ideal plankton net must collect SPM or plankton with known effectiveness. For some purposes the qualitative aspects of sampler selectivity are relatively more important than the quantitative ones, or vice versa, but neither can ever be completely ignored in the design and selection of the sampling gear. Sampler selectivity has usually been evaluated empirically, by comparing results obtained with one sampler under different conditions or by comparing various sampling methods.[63]

Plankton nets may be lowered to the lake bottom, or any specified depth, and hauled vertically at a steady speed to the surface. The approximate volume of water filtered depends on the distance from where the net-haul starts to the surface, the hauling speed, the mesh size, the size of the net opening, and the ratio of opening area to filtration area. Usually the volume filtered is estimated by multiplying the distance integrated by the area of the mouth of the opening.

If the net is hauled too quickly or if the net is clogged, the resistance of the net will allow a smaller volume of water to be filtered than calculated above.[64] The use of flow meters mounted at the net opening is recommended for more accurate volume measurements.[47]

Under ideal conditions, this technique can provide a uniform sample of all depth strata over an area equivalent to that of the net opening. The sampling accuracy of these net-hauls operated from a vessel is best during calm conditions. During rough conditions, the net tends to ascend in a series of rapid rises followed by sharp decelerations and possibly short reversals in directions.[65] These fluctuations tend to introduce errors in the results due to uncertainty in the volume filtered. Plankton nets with mesh sizes from 5 µm to >1 mm can collect large quantities of SPM but mesh size can affect the apparent patterns of particle composition/dispersion.[66]

D. SAMPLING SETTLING SEDIMENTS INTEGRATING OVER TIME: SEDIMENT TRAPS AND SETTLING CHAMBERS

1. Sediment Trap Design and Trapping Efficiency

In order to understand processes associated with SPM which aid in establishing correct functional lake models, it is of crucial importance to provide methods which allow for the correct measurement of the downward settling flux of particulate matter. Sediment traps are a unique tool that can be used to investigate particle settling flux throughout the water column, whereas other methods such as sediment dating and mass balance (input-output models, for lakes) can only measure accumulation rates of bottom sediments.[67-69] A critical review of the ^{210}Pb method (in comparison with the trap method) given by Bloesch and Evans[70] was further discussed by Binford and Brenner,[71] Benoit and Hemond,[72] and Binford and Brenner.[73] The statement of Binford and Brenner[71] that "traps tend to misestimate sedimentation" cannot be substantiated as illustrated by the extensive methodological studies on sediment trap efficiency reviewed below. The methodological errors of the trap technique (being in the acceptable range of ±10 to 20%) must be compared with those of the ^{210}Pb method, which may be biased by some underlaid assumptions,[74] dilution effects,[70] or bioturbation.[18] Bottom sediment resuspension can interfere with both methods (see Section II.D.3). In view of this debate, it must be stressed that neither of these two methods is "right or wrong", because they apply to different measurements of different processes.

Sediment traps, which were first used at the turn of the century, were "reintroduced" in the 1950s and have become increasingly popular in the last decade. The literature on sediment trap methodology, up to about 1980, has been comprehensively reviewed by Bloesch and Burns,[75] Reynolds et al.,[76] and Blomqvist and Håkanson.[77] The misconception concerning the "snow-fence" effect, which states that any collector tends to overtrap material by decreasing the turbulence in the vicinity of the trap, and also the fear of undertrapping which loses collected material, led to the use of many different trap designs. These include flat containers, bottles, jars, plastic bags, funnels, and cylinders, which often contain lids or collars (Figure 8). Recent work on trapping efficiency[75,78-85] has shown that cylindrical traps with appropriate dimensions, mainly aspect ratio (height/diameter), are the best instruments to correctly measure the settling downward flux of particulate matter.

The following discussion on trap efficiency is based on the assumption that turbulence affects particle distribution rather than particle sinking velocity,

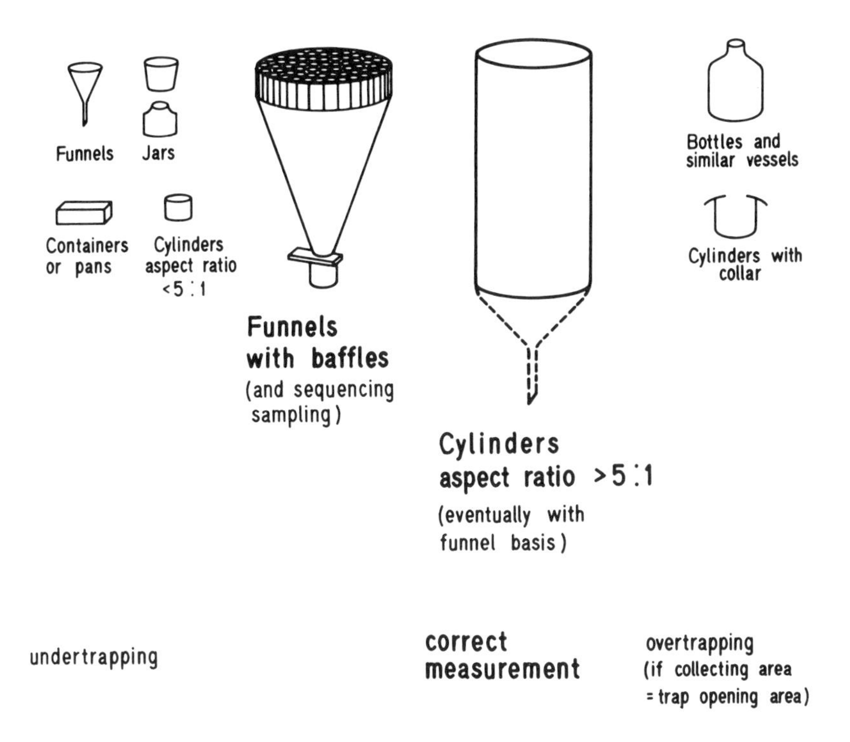

FIGURE 8. Different geometrical forms of sediment traps. (After Bloesch, J. and Burns, N. M., *Schweiz. Z. Hydrol.*, 42, 15, 1980. With permission.)

provided that particle size is <250 μm and particle Reynolds number (Re) is <0.5,[75] according to the equation:

$$Re = vd/\nu \tag{2}$$

where v = particle sinking velocity, d = particle diameter, and ν = kinematic viscosity of fluid. Particle trapping was transcendently described by Butman et al.[84] The collecting efficiency can be experimentally determined by exposing traps in tanks, where the uninfluenced settling flux at the tank bottom provides the reference value.[78,80,83] *In situ* sediment trap efficiency has also been tested recently by using a free drifting trap as the control.[85]

If a trap is to measure the correct settling flux in both calm and turbulent conditions the following conditions need to prevail: (1) the particle concentration must be equal outside and inside the upper, turbulent part of the trap at all times; (2) the sedimented material on the trap bottom must not be resuspended by turbulence. Theoretical considerations (Figure 8, from Bloesch and Burns[75] and Butman et al.[84]) and experimental evaluations with different trap designs[78-81,83] have shown that only a cylindrical trap of appropriate length and diameter (aspect ratio of at least 5:1) can fulfill both requirements. Bottle-like traps tend to overtrap because requirement (1) is not fulfilled, funnel-shaped traps and flat containers undertrap because both requirements (1) and (2) are not met (Figure 9 and Table 6). Baffles, funnels, or bottles placed at the bottom of traps have no influence on trapping efficiency; whereas the baffles are of no use,[75] the funnels and bottles may be useful for harvesting purposes, and where sequencing traps[86] are used. The choice of the aspect ratio is dependent on *in situ* turbulences[82] (Table 7). Using the experimentally found relationship between aspect ratio and resuspension of oil droplets (Figure 10,

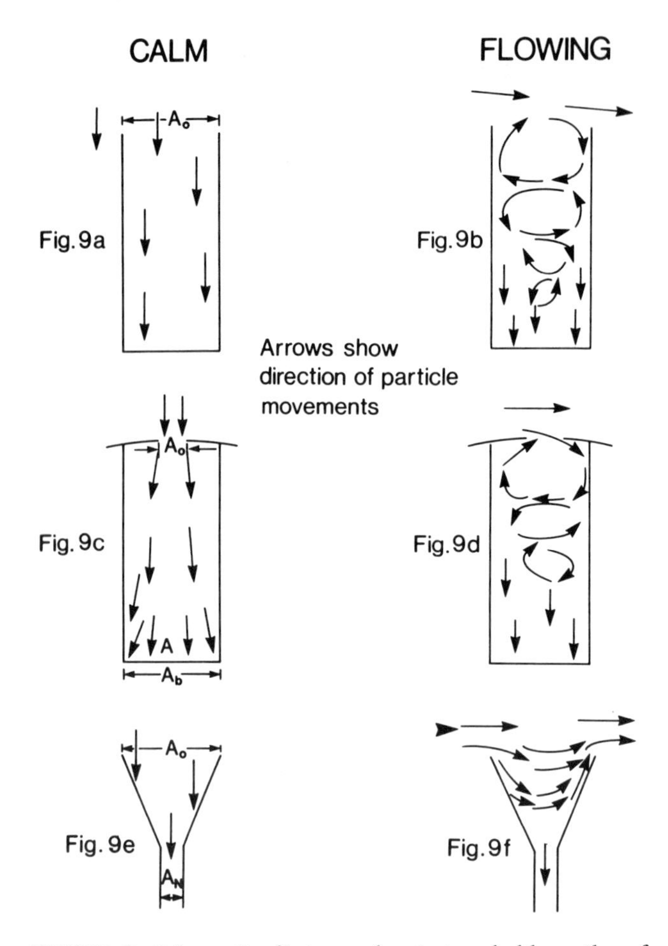

FIGURE 9. Schematic diagram showing probable paths of particles around different traps, under calm and turbulent conditions. (From Bloesch, J. and Burns, N. M., *Schweiz. Z. Hydrol.*, 42, 15, 1980. With permission.)
Figure 9a, c, e: $F_i = A_oF_o = A_oV_oC_o = A_{b,n}V_iC_i$; $F_{i,o}$ = particle settling flux inside, outside the trap.
Figure 9a, c, e, b: $V_o = V_i$; $A_o = A_b$; $C_o = C_i$; $V_{i,o}$ = mean particle settling velocity inside, outside the trap.
Figure 9d: $V_o = V_i$; $A_o < A_b$; $C_o > C_i$; $C_{i,o}$ = mean particle concentration inside, outside the trap.
Figure 9f: $V_o = V_i$; $A_o > A_n$; $C_o < C_i$; $A_{o,b,n}$ = area of trap opening, bottom, neck.
$F_{i,o}$ = particle settling flux inside, outside the trap.
$V_{i,o}$ = mean particle settling vilocity inside, outside the trap.
$C_{i,o}$ = mean particle concentration inside, outhe trap.
$A_{o,b,n}$ = area of trap opening, bottom, neck.

from Lau[87]), the maximum horizontal flow velocity above which resuspension at the trap bottom begins can be estimated (Table 8, from Bloesch and Burns[75] and Butman[83]), and thus the appropriate ratio can be determined.

An ideal trap should have an aspect ratio between 5:1 and 10:1; hence, in practice, the trap diameter may range between 5 to 20 cm, and the trap height between 0.25 and 2 m. Trap catches have been reported to be directly proportional to the collecting area if the trap diameter is >5 cm (see references in Bloesch and Burns[75]). Blomqvist and Kofoed[82] found traps with a narrower diameter to undertrap small, dense mineral particles and to overtrap large,

TABLE 6
Collecting Efficiency of Sediment Traps of Various Designs

Trap design	Collecting efficiency (%)
Cylinders	95—100
Funnels	25—60
Funnels with baffles	60—90
Bottles[a]	230—1000
Flat containers	2—12

[a] The collecting efficiency is similar to that of cylinders, if the bottom of the bottle is taken as the collecting area.[78]

Compiled after Gardner, W. D., Fluxes, Dynamics and Chemistry of Particulates in the Ocean, Ph.D. thesis, MIT/WHOI, Woods Hole, MA, 1977.

TABLE 7
Lake Erie Field Data Showing the Influence of Aspect Ratio upon Entrapped SPM Retention

	Trap diameter (cm)	6.6	6.6	6.6	6.6	6.6
Exposure time	**Trap length (cm)**	15.2	33	66	91.4	132
(2 weeks)	**Aspect ratio**	2.3:1	5:1	10:1	14:1	20:1
Period of high turbulence	Retention (%)[a]	7	57	100	96	104
Period of low turbulence	Retention (%)[a]	21	103	97	107	97

[a] Mean values of total SPM, POC, PP, and PN retention.

Modified after Bloesch, J. and Burns, N. M., *Schweiz. Z. Hydrol.*, 42, 15, 1980.

light organic particles. These authors attribute the density-selective catch to the catapulting centrifugal force, thus carrying heavier (mineral) particles to the center of a vortex. However, this explanation is probably based on a physical misconception, as circular (vortex) motion cannot be the sorting mechanism for particles having the density range found in natural waters. However, the observed phenomenon may be explained by the changed ratio between trap wall surface area and total trap volume; if this ratio exceeds 1, "wall effects" occur, i.e., the natural eddies cannot enter the small trap, and the turbulence inside the trap is drastically reduced, hence increasing the settling velocity of large (organic) particles, whereas small (inorganic) particles are not affected.

Modern traps are made of either transparent or nontransparent PVC or plexiglass which are readily available and easily used. However, Teflon-coated materials must be used to prevent contamination, if organochlorine and hydrocarbons are to be analyzed. Similarly, if trace metals are to be analyzed, metallic material on trap frames and moorings must be avoided.

2. Sediment Traps: Exposure and Moorings

Although there is a striking difference in turbulence and currents between lakes and oceans, the trap dynamics are essentially the same in both systems.

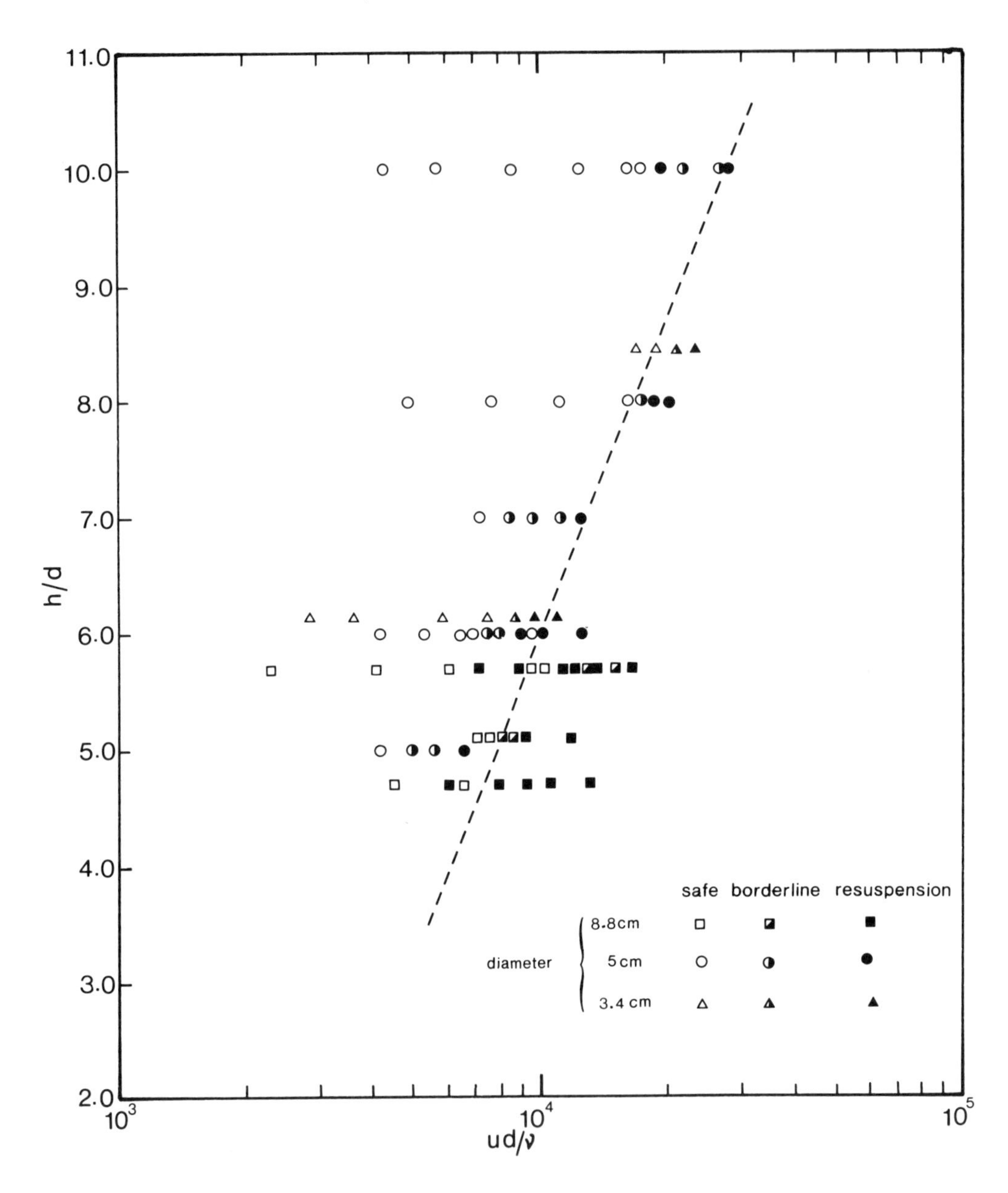

FIGURE 10. Dependence of resuspension of oil droplets from the bottom of a sediment trap on aspect ratio and flow Reynolds number (according to Lau).[87] h = length of cylinder, d = diameter of cylinder, u = velocity of water flowing past the cylinder, n = kinematic viscosity of water. (After Bloesch, J. and Burns, N. M., *Schweiz. Z. Hydrol.*, 42, 15, 1980. With permission.)

Recent work has shown the importance of high and variable current speeds with regard to trapping efficiency, as typically found in oceans and large lakes.[84,85] Thus, trap catches in these systems may be biased. The *in situ* experiments by Baker et al.,[85] performed in a deep estuarine tidal passage, showed that at speeds exceeding 12 cm/s the collection efficiency of moored traps decreased due to increasing trap Reynolds number and decreasing particle settling velocity. Since particle concentration and settling flux are extremely small in open oceans (Table 9), settling flux measurements require a large collection area, so that cylinders with respective aspect ratios may not

TABLE 8
Calculation of the Maximum Horizontal Flow Velocity (cm/s) above which Resuspension Begins

Trap diameter (cm)	6.6	6.6	6.6	6.6	6.6
Aspect ratio (height/diameter)	2.5:1	5:1	10:1	14:1	20:1
Re-number at which resuspension starts[a]	4,500	7,000	20,000	35,000	70,000
Mean catch (%)[b]	5—20	65—100	100	100	100
Critical horizontal flow velocity (cm/s)					
At 4°C	10.7	16.6	47.5	83	166
At 10°C	8.9	13.9	39.6	69	139
At 20°C	6.8	10.6	30.3	53	106

[a] From Figure 3, in Lau, Y. L., *J. Fish. Res. Board Can.*, 36, 1288, 1979.
[b] From basic data of Table 7.

Modified after Bloesch, J. and Burns, N. M., *Schweiz. Z. Hydrol.*, 42, 15, 1980.

be used. This situation is usually not found in nearshore areas and fjords. Thus, for practical reasons funnel-shaped traps are often utilized.[86,88] However, the use of baffles on top is recommended because they significantly decrease the undertrapping tendency of the funnel (Table 6). Special attention should be given to particulate material adhering to the funnel walls.

Equipment used for trap deployment such as cables, lines, trap frame, attachment of traps, and mooring, shown in Figure 11, can vary. In principle, any disturbance resulting from turbulence around the trap opening which results from the mooring configuration should be minimized. The trap should always be oriented perpendicular to the current direction, since tilting can cause significant overtrapping by increasing collection area.[89] Usually, a vertical position of the sediment trap is adequate. In large lakes and oceans, traps often have a fin attached to the mooring in such a manner as to allow rotation for orientation with the current. Where strong currents do not cause tilting of the mooring, either a single trap or a frame carrying one or more replicate traps can be used. Recently, free-drifting traps were successfully used in oceans to minimize the effect of tilting by strong currents.[85,90]

3. Vertical Flux Studies Using Sediment Traps and Settling Chambers

In situ experiments with replicate traps, either positioned in parallel moorings (in open waters) or fixed at the same site, usually result in coefficients of variation <15%, including the errors of subsampling and chemical analyses.[75,91] The sediment trap methodology yields settling flux with an acceptable precision.

Overtrapping is the result of measuring gross settling flux instead of net settling flux. Hence, some restrictions are necessary for trap exposure in order to measure net downward settling flux. Traps positioned in the upper layers of lakes and oceans will collect excess particles due to Langmuir cells. However, they may effectively undertrap (yield lower sedimentation rates) in warm epilimnetic layers, where decomposition of abundant organic material exceeds the overtrapping effect caused by these currents (see Section II.D.4).

Also, traps positioned near bottom sediments (<3 m)[41] will collect resuspended particles in addition to the downward settling particles,[41,75] and hence

TABLE 9
Primary Production, SPM Concentration, Settling Flux and Accumulation Rates in Open Oceans, Neritic Oceans (Fjords, Bays), and Lakes

Parameter	Open oceans	Fjords, bays	Lakes
Primary production (g C assimilated/m²/year)	25—75	100—400	100—500
SPM concentration			
Dry weight (g/m³)	0.05—0.2	0.7—2	0.1—10
Particulate organic carbon (mg/m³)	0.5—5	10—1000	100—1500
Particulate nitrogen (mg/m³)	0.1—0.5	10—200	10—1500
Particulate phosphorus (mg/m³)	0.1—1	1—10	1—100
Settling flux			
Dry weight (g/m²/d)	0.01—0.3	0.1—3	0.1—30
Particulate organic carbon (mg/m²/day)	0.1—30	10—1000	50—2500
Particulate nitrogen (mg/m²/day)	0.01—2	4—100	5—200
Particulate phosphorus (mg/m²/day)	0.1—2	1—10	1—60
Accumulation rates (mm/year)	$(3—5) \times 10^{-3}$	0.01—1	3—5

Note: The given average figures represent order of magnitudes, and specific sites may differ significantly.

invalidate some of the rates obtained from sediment trap experiments.[92,93] Resuspension of bottom sediments is caused by peripheral wave action, leading to horizontal sediment transport and sediment focusing.[94] This latter process is of fundamental importance for the food supply to benthic organisms, the preservation of historical records in the sediments, and the geochemical cycling and early diagenesis of the various elements. Resuspension may explain some of the puzzling features pertaining to the distribution of pollutants in lake sediments.[41,95] Bloesch and Uehlinger[96] have reviewed the sparse literature on differences in horizontal sedimentation rates in lakes and have shown that sediment resuspension is of minor importance in the profundal of small deep lakes, whereas it may be important in shallower more turbulent lakes such as Lake Erie[97] or Frains Lake.[98]

It is debated whether significant horizontal sedimentation differences in lakes exist.[96] Obviously, this is dependent on each lake's morphology and geographical position. However, Dillon et al.[69] have shown that within the same lake, horizontal sedimentation differences can change with time. The nearshore settling flux is often increased due to bottom sediment resuspension.[19,94] With respect to the "correct" measurement of net settling flux, the midlake sampling site conventionally chosen may generally be considered representative for the whole lake. Since the shores are steep and small when compared to the whole lake area, and wind stress is not strong, studies in these littoral regions provide information as to conditions only for that specific area.

The upward flux of tripton has so far received only modest attention.[42,45,97,99] The first quantitative measurement of upward vs. downward flux of particulates at the bottom of Lake Ontario has been reported by Rosa et al.[41] The authors reported that resuspension rates of TSM, at a site greater than 70 m

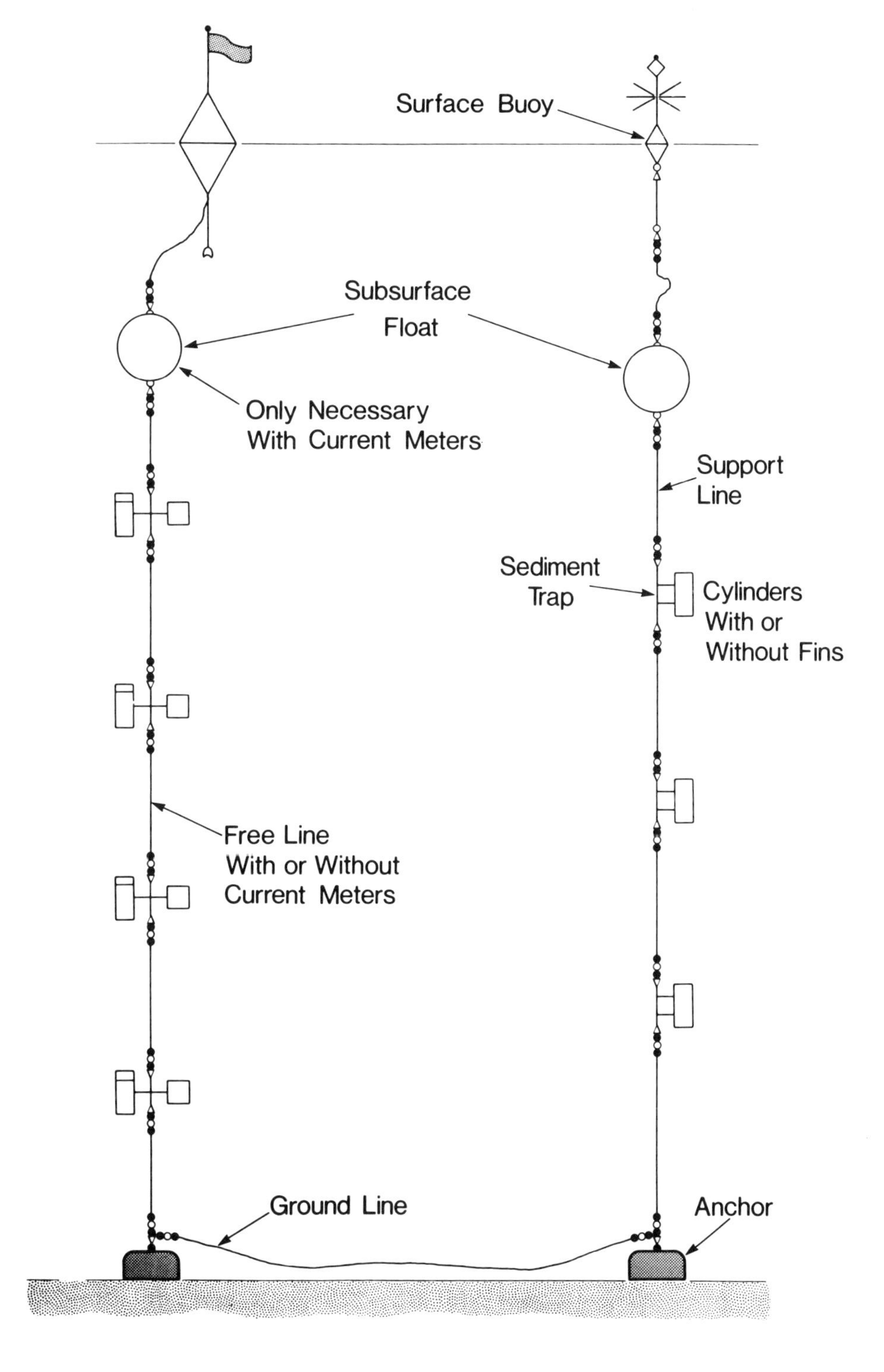

FIGURE 11. Typical trap mooring system: the free line with current meters is optional. (After Bloesch, J. and Burns, N. M., *Schweiz. Z. Hydrol.*, 42, 15, 1980. With permission.)

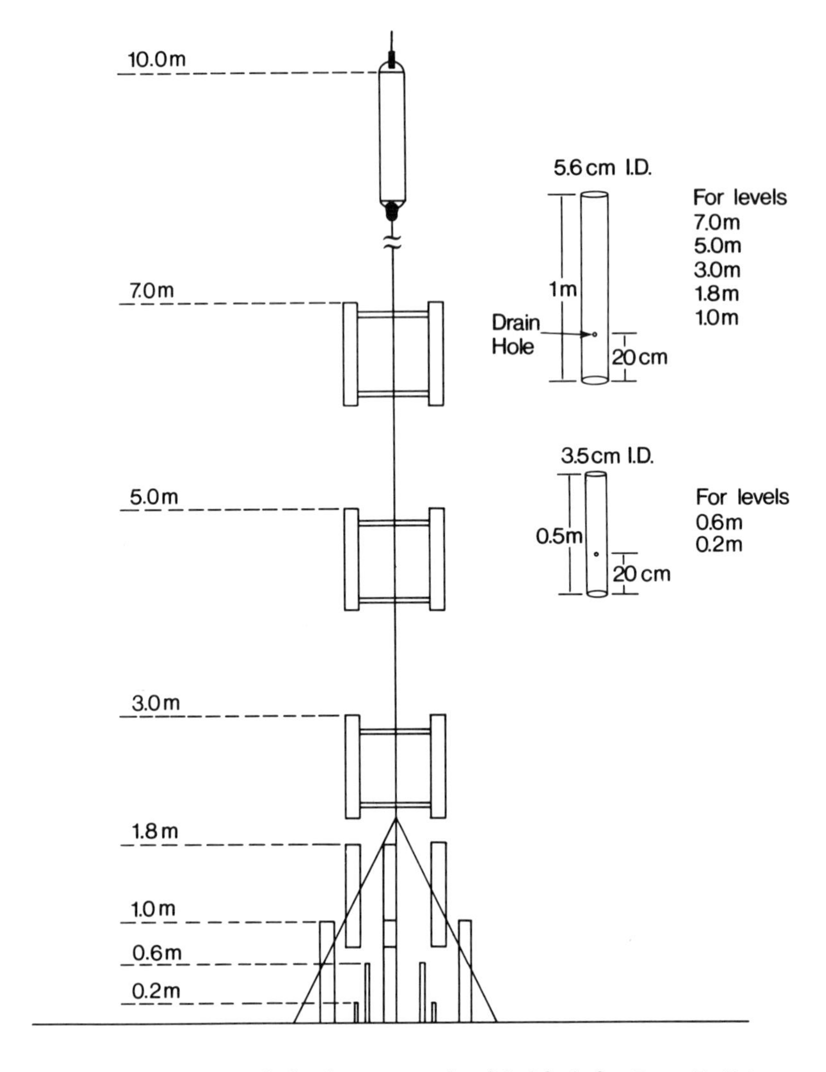

FIGURE 12. Resuspended sediment sampler. (Modified after Rosa, F., Nriagu, J. O., Wong, H. K. T., and Burns, N. M., *Chemosphere*, 12, 1345, 1983.)

deep, were 85 and 45% at 0.2 and 7 m above the sediments, respectively. A special resuspension sediment sampler (RSS) was built with an array of cylindrical traps at seven depths varying from 0.2 to 7 m above the sediment water interface (Figure 12). To avoid the collection of any bottom sediment material stirred up during installation, trap tubes positioned lower than 1.8 m above the lake bottom were capped. The caps were subsequently removed with the aid of a self-contained electronic timer, after a predetermined time period.

The settling fluxes measured by traps can be used to calculate mean sinking velocities of any bulk particulate material and phytoplankton, if compared to the mean particle/plankton concentration in the water above or near the traps.[21,100-102] Such calculations, however, provide no precise value, but rather results within an order of magnitude. Moreover, the precision of these calculated velocities is proportional to the frequency of sampling the SPM/plankton concentration in the water. Daily SPM sampling is recommended throughout the trap exposure period, in order to calculate a representative mean particle concentration.

An alternative method for measuring settling flux and sinking velocities of particles is the use of settling chambers.[33,103,104] An *in situ* method using settling chambers[102] for measuring the settling velocities of different size particles has been found adequate for measuring the diurnal flux of organic carbon out of the epilimnion.[104] When settling chambers and traps are exposed *in situ*, they will not necessarily yield the same results.[106]

Settling chambers are also a valuable alternative to traps in estimating the sinking velocity of phytoplankton,[104] although they cannot fully overcome the problems with extrapolating sinking velocities experimentally obtained in tanks[107] or with settling chambers exposed *in situ*.[104] Alternative methods and the problems of determining intrinsic sinking rates of algae have been comprehensively reviewed by Bienfang et al.,[108] Walsby and Reynolds,[109] and Reynolds.[28]

4. Problems Related to the Sediment Trap Technique: Active Swimmers, Poisons, and Preservatives

Although the best trap configuration has been determined during the last decade (see Section II.D.1), some physical aspects on trapping efficiency are still debated.[83-85,110] Moreover, two major problems in the trap methodology remain: (1) zooplankton (and other swimmers) actively entering traps[111-113] and (2) bacterial decomposition of particulate organic material within the traps.[75,114,115]

The problems related to the fate of entrapped material are presently recognized as a major concern in regard to the correct measurement of settling fluxes.[25,110] For instance, the impact of grazing and excretion may be significant during periods of high zooplankton abundance such as the clear water phase observed in lakes in late spring or early summer.[23] In any case, entrapped zooplankton should be picked or removed by screening prior to analysis.[110,116] Further, the accelerated bacterial decomposition of organic matter accumulated within the traps is dependent on the composition of entrapped materials, the adjacent temperature, and the exposure time of the traps. Thus, in oligotrophic lakes at temperate zones with little primary production and high input of inorganic allochthonous material, loss of particulate organic carbon through decomposition is negligible,[117] and in hypolimnetic traps exposed to cold temperatures (below 6°C), metabolic and chemical processes of particle decomposition and dissolution are minimized. Various experiments comparing short-term with long-term exposure times have shown that in many situations, loss through accelerated decomposition is kept in an acceptable range of <10%, if the traps are not exposed longer than 1 to 2 weeks.[75,118,119] Thus, a 2-week exposure has been recommended by Bloesch and Burns[75] and adapted by most researchers. During colder winter periods, a 3-week exposure is acceptable. In tropical lakes and oceans, the bias of artificial decomposition inside the traps is of more concern and may not be easily overcome.

Despite any precautions taken, the problem of accelerated changes of entrapped material needs to be recognized, investigated, and overcome. The use of poisons and preservatives was believed to keep entrapped material unchanged. Poisons only kill bacteria and planktonic organisms, whereas preservatives also conserve, at least partly, the dead cells. However, formalin does not prevent leaching of dissolved organic matter from entrapped planktonic cells.[120] The poisons and/or preservatives so far used, together with the known and possible interfering effects of these chemicals, are presented in Table 10. This list makes it clear that the introduction of any substance may introduce additional bias to that which would occur in an unpreserved trap sample, through oxidation, dissolution, adsorption, disintegration of swimmers, or analytical problems. For instance, chloroform, a poison and an organic solvent, did not stop microbial degradation, but killed entering zooplankton, increased their disintegration, and considerably biased POC, chlorophyll, and lipid flux measurements.[113,121] However, the mode of chloroform applications may play a crucial role on its bactericide effectiveness.[120] Therefore, any poison and/or preservative should be carefully checked in laboratory experiments prior to being used *in situ*. For the time being, and without knowing the ideal poison/preservative, we recommend to deploy parallel treated and untreated traps; this may help to enable proper data interpretation.

In addition, before using poisons and/or preservatives to prevent changes of entrapped material either by the grazing zooplankton or bacterial degradation, one should consider the parameters to be analyzed according to the aims of the study and the extent of possible interferences. At present, there is no perfect poison and/or preservative available. Sometimes, several different additives must be used in parallel trap tubes to overcome the specific chemical and analytical problems. The concentration of the poison and/or preservative should be a compromise between the effectiveness (lower end of concentration range) and analytical interference (upper end of concentration range) established by laboratory tests. Whether or not the concentration of poisons and/or preservatives applied is maintained throughout exposure *in situ* should be checked, for example, by measuring ATP uptake, H_3-thymidine incorporation by bacteria, growth of bacteria (MPN), and measuring concentrations of dissolved nutrients or poisons and/or preservatives in overlying trapwater. In addition to the problematic selection of additives, the application mode of poisons and/or preservatives is not yet fully standardized.

The application of additives in the solid phase (with low solubility) is more difficult than that of soluble additives. Injection of dissolved additives into the trap through a siphon is recommended, but injection through the trap wall, or introduction of diffusion chambers to the trap bottom may cause additional complications. The establishment of density gradients by adding NaCl solution or ice cubes to keep the additive at the trap bottom are unnecessary if the trap aspect ratio is correct (5:1), since only negligible loss of poison and/or preservative can occur through diffusion and not through turbulence.

III. SUBSAMPLING, SAMPLE HANDLING, AND PROCESSING

Between the time of sample collection and sample analysis, the samples obtained will be subject to subsampling, sample handling, and processing. This is the most crucial time for the possibility of sample contamination to occur, due to improper procedures (see also Chapter 6).

Sample handling both in SPM sampling and subsampling requires extreme caution to obtain successful results, but in many cases this caution is not exercised. The problem of contamination is always present, whether due to handling or improper use of containers. Sample handling then should always be considered as a potential source of error, which will add to the scatter of the data. Subsampling of SPM can be performed either on *wet* samples (before filtering) or on *dry* samples (when processing dried particulate matter for analysis). When subsampling, the major problem encountered is to ensure

TABLE 10
Poisons and Preservatives Used in Sediment Traps to Minimize Degradation of Organic Matter

Poisons and preservatives	Interfering effects[a]	Ref.
Formaldehyde	Changes of organic bindings of organic matter[a]	122
Para-formaldehyde (crystals)	Chemical reactions with POM (dissolution)[a]	64
Glutaraldehyde	Artificial increase of POC flux[a]	123
	Decreasing particle size[a]	124
	If not buffered by $CaCO_3$, diminishing DW flux[a]	125
	Increasing number of dead zooplankton (which may be removed by picking or prescreening the samples)	115, 126, 127
Lugol	General bias[a]	128
I_2 (crystals)	—	129
Chloroform	Dissolves POM, pigments, lipids, amino acids, etc.	122
	Does not kill all bacteria	64
	Disintegrates swimmers (zooplankton)	113, 130—134
Na-N_3 (sodium azide)	Oxidizes PN to NO_3[a]	135
	Bias of N-analysis	136
	Does not kill all bacteria[a]	88, 115, 137—139
$HgCl_2$ (mercurium)	Bias of P-analysis	140
	Bias of Hg (and other trace metals)[a] analysis	114, 115, 134
$KMnO_4$	MnO_2 is formed and precipitated	141
Polyacrylamide	May influence settling velocity of small particles[a]	86
Phenol	Does not kill all bacteria[a]	142
Pentachlorophenol (PCP)[a]	Increasing P and N flux by increased sulfur bacteria	—
Antibiotics (thymol; merphen = phenyl-Hg-borate)	May not kill all bacteria, as wide ecological spectrum of species is beyond the range of chemical	143—146
Solid copper[a]	—	—
Tributyl tin (TBT) (antifouling)[a]	—	—
NaCl	Does not kill all bacteria	147
	Density gradient prevents smallest particles from settling to trap bottom (undertrapping effects)	141

[a] Not definitely proven.

Modified after Wassman, P. and Heiskanen, A.S., Eds., *Sediment Trap Studies in the Nordic Countries, 1*, Workshop, Tvärminne Zoological Station, Finland, 1988, 190; Fisch, M., Experimentelle Untersuchungen über die Anwendung von Bakteriziden zur Verhinderung der kunstlichen Mineralisation in Sedimentfallen, M.Sc. thesis, ETH, Zürich, 1985.

homogeneity of the sample which can be achieved by thorough shaking/mixing of the *wet* samples, and by extensive grinding of the *dried* material. Ultrasound may also be used to dislodge conglomerates of particles to achieve homogeneity, but care should be taken not to produce smaller particles which may pass through the filter, and thus become lost from the particulate phase.

Dry particulate matter can be obtained either by filtration or centrifugation of SPM, and subsequent freeze-drying of the collected matter, or by oven-drying at 50 to 105°C for about 2 h (crystal water may not be removed with temperatures below 100°C). Subsequent combustion at 450 to 550°C yields organic matter; the appropriate temperature for organic matter combustion must be chosen according to the calcium/magnesium content of carbonates: in calcium-rich lake sediments, 550°C, in magnesium-rich ocean sediments, 450°C is suitable to avoid combustion of carbonates; too-low temperatures would result in insufficient combustion of organic matter. In an additional step, the inorganic residue can be combusted at ~880°C to get ash-free residue (clay minerals and silica).

Any analysis can be performed on both *wet* and *dry* samples. Filtration and subsequent analysis of the filter is less time consuming, but is associated with filtration problems (see Section I.A). Centrifugation and subsequent freeze- or oven-drying yield a sample which can be efficiently stored (see Chapter 6) for a long period of time, and hence conveniently saved for future analysis.

The factors associated with subsampling, sample handling, and processing are important. An effective quality control program must be an integral part of any scientific research program, so as to test at random that results are not severely affected by improper sampling, subsampling, and sampling procedures.

ACKNOWLEDGMENTS

The authors would like to thank Ms. Ellie Kokotich for her excellent work in word processing in the preparation of this chapter. We acknowledge the reviews of P. Wassmann, A. Mudroch, R. Gächter, and R. D. Evans. Also, many thanks to the publishers who have granted us copyright permission for previously published work. Last, but not least, we thank the editors for giving us the opportunity to write this chapter.

REFERENCES

1. **Allan, R. J.,** *The Role of Particulate Matter in the Fate of Contaminants in Aquatic Ecosystems*, NWRI Report, Scientific Series No. 142, Canada Centre for Inland Waters, Burlington, Ontario, 1986.
2. **Hutchinson, G. E.,** *A Treatise on Limnology*, Vol. 2, John Wiley & Sons, New York, 1967, 235.
3. **Leppard, G. G., Buffle, J., De Vitre, R. R., and Perret, D.,** The ultra and physical characteristics of a distinctive colloidal iron particulate isolated from a small eutrophic lake, *Arch. Hydrobiol.*, 114, 405, 1988.
4. **Burnison, B. K. and Leppard, G. G.,** Isolation of colloidal fibrils from lake water by physical separation techniques, *Can. J. Fish. Aquat. Sci.*, 40, 373, 1983.

5. **Sheldon, R. W.,** Size separation of marine seston by membrane and 6 lens-fiber filters, *Limnol. Oceanogr.*, 7, 494, 1972.
6. **Runge, J. A. and Ohman, M. D.,** Size fractionation of phytoplankton as an estimate of food available to herbivores, *Limnol. Oceanogr.*, 27, 570, 1982.
7. **Murphy, L. S. and Haugen, E. M.,** The distribution and abundance of phototrophic ultraplankton in the North Atlantic, *Limnol. Oceanogr.*, 30, 47, 1985.
8. **Fay, L. A.,** *Great Lakes Methods Manual, Field Procedures*, Report D-4857W, International Joint Commission, Windsor, Ontario, 1988.
9. **Montegut, C. C. and Montegut, G. C.,** Chemical analysis of suspended particulate matter collected in the Northeast Atlantic, *Deep Sea Res.*, 19, 445, 1972.
10. **Cranston, R. E. and Buckley, D. E.,** *The Application and Performance of Microfilters in Analyses of Suspended Particulate Matter*, Report BIO-R-72-7, Bedford Institute of Oceanography, Dartmouth, Nova Scotia, 1972.
11. **Lal, D.,** The oceanic microcosm of particles, *Science*, 198, 997, 1977.
12. **Zabawa, C. F.,** Microstructure of agglomerated suspended sediments in northern Chesapeake Bay estuary, *Science*, 202, 49, 1978.
13. **Kondratieff, P. F. and Simmons, G. M., Jr.,** Nutritive quality and size fractions of natural seston in an impounded river, *Arch. Hydrobiol.*, 101, 401, 1984.
14. **Sheldon, R. W. and Sutcliffe, W. H., Jr.,** Retention of marine pesticides by screens and filters, *Limnol. Oceanogr.*, 14, 441, 1969.
15. **Kelts, K. and Hsü, J. J.,** Freshwater carbonate sedimentation, in *Lakes — Chemistry, Geology, Physics*, Lerman, A., Ed., Springer-Verlag, New York, 1978, 295.
16. **Bloesch, J., Armengol, J., Giovanoli, F., and Stabel, H. H.,** Phosphorous in suspended and settling particulate matter of lakes, *Arch. Hydrobiol. Beih. Ergebn. Limnol.*, 30, 84, 1988.
17. **Lerman, A.,** *Geochemical Processes: Water and Sediment Environments*, John Wiley & Sons, New York, 1979.
18. **Håkanson, L. and Jansson, M.,** *Principles of Lake Sedimentology*, Springer-Verlag, Berlin, 1983.
19. **Hilton, J., Lishman, J. P., and Allen, P. V.,** The dominant processes of sediment distribution and focusing in a small, eutrophic, monomictic lake, *Limnol. Oceanogr.*, 31, 125, 1986.
20. **Paerl, H. W.,** Detritus in Lake Tahoe: structural modification by attached microflora, *Science*, 180, 496, 1973.
21. **Sommer, U.,** Sedimentation of principal phytoplankton species in Lake Constance, *J. Plankton Res.*, 6, 1, 1984.
22. **Stumm, W., Ed.,** *Chemical Processes in Lakes*, John Wiley & Sons, New York, 1985.
23. **Bloesch, J. and Bürgi, H. R.,** Changes in phytoplankton and zooplankton biomass and composition reflected by sedimentation, *Limnol. Oceanogr.*, 34, 1048, 1989.
24. **Wassmann, P.,** Relationship between primary and export production in the Boreal Coastal Zone of the North Atlantic, *Limnol. Oceanogr.*, 35, 464, 1990.
25. **Wassmann, P.,** Primary production and sedimentation, in *Sediment Trap Studies in the Nordic Countries, 1*, Wassmann, P. and Heiskanen, A. S., Eds., Workshop, Tvärminne Zoological Station, Finland, 1988, 100.
26. **Smetacek, V.,** The supply of food to the benthos, in *Flows of Energy and Materials in Marine Ecosystems: Theory and Practice*, Fasham, M. J., Ed., Plenum Press, New York, 1984, 517.
27. **Smayda, T. J.,** The suspension and sinking of phytoplankton in the sea, *Oceanogr. Mar. Biol. Annu. Rev.*, 8, 353, 1970.
28. **Reynolds, C. S.,** *The Ecology of Freshwater Phytoplankton*, Cambridge Studies in Ecology, Cambridge University Press, Cambridge, 1984.
29. **Jobson, H. E. and Sayre, W. W.,** Vertical transfer in open channel flow, *J. Hydraul. Eng.*, American Society of Civil Engineers, Hydraulics Division, 703, 1970.
30. **Murray, S. P.,** Settling velocities and vertical diffusion of particles in turbulent water, *J. Geophys. Res.*, 75, 1647, 1970.
31. **Smith, I. R.,** Turbulence in lakes and rivers, *Freshwater Biol. Assoc. Sci. Publ.*, 29, 79 pp., 1975.
32. **Chase, R. R. P.,** Settling behaviour of natural aquatic particulates, *Limnol. Oceanogr.*, 24, 417, 1979.
33. **Shanks, A. L. and Trent, J. D.,** Marine snow: sinking rates and potential role in vertical flux, *Deep Sea Res.*, 27A, 137, 1980.
34. **Vernet, J. P.,** personal communication, 1989.

35. *Sediment Sampling in Australia, Hydrological Series No. 3*, Department of National Development, Australian Water Resources Council, Australia, 1969.
36. **Bukata, R. P., Jerome, J. H., Bruton, J. E., Jain, S. C., and Zwick, H. H.,** Optical water quality model of Lake Ontario. 1. Determination of the optical cross sections of organic and inorganic particulates in Lake Ontario, *Appl. Opt.*, 20, 1696, 1981.
37. **Bukata, R. P., Bruton, J. E., and Jerome, J. H.,** *Application of Direct Measurements of Optical Parameters to the Estimation of Lake Water Quality Indicators*, IWD Report, Scientific Series No. 140, Canada Centre for Inland Waters, Burlington, Ontario, 1985.
38. **Golterman, H. L., Sly, P. G., and Thomas, R. L.,** Study of the relationship between water quality and sediment transport, in *Technical Papers in Hydrology, Volume 26*, UNESCO, Paris, 1983, 1.
39. **Singh, S. A.,** Uppal instantaneous suspended sediment sampler, *Indian J. Lower River Valley Dev.*, 12, 7, 1962.
40. **Golterman, H. L., Clymo, R. S., and Ohnstad, M. A. M.,** *Methods for Physical and Chemical Analysis of Fresh Waters*, IBP Manual No. 8, Blackwell Scientific, Oxford, 1978, 1.
41. **Rosa, F., Nriagu, J. O., Wong, H. K. T., and Burns, N. M.,** Particulate flux at the bottom of Lake Ontario, *Chemosphere*, 12, 1345, 1983.
42. **Chambers, R. L. and Eadie, B. J.,** Nepheloid and suspended particulate matter in southeastern Lake Michigan, *Sedimentology*, 28, 439, 1981.
43. **Wahlgren, M. A. and Nelson, D. M.,** *Factors Affecting the Collection Efficiency of Sediment Traps in Lake Michigan*, Report ANL-76-88, Argonne National Laboratory, Argonne, IL, 1976.
44. **Eadie, B. J., Chambers, R. L., Gardner, W. S., and Bell, G. L.,** Sediment trap studies in Lake Michigan: resuspension and chemical fluxes in the southern basin. *J. Great Lakes Res.*, 10, 307, 1984.
45. **Rosa, F.,** Sedimentation and sediment resuspension in Lake Ontario, *J. Great Lakes Res.*, 11, 13, 1985.
46. Wildco Catalog, Wildlife Supply Company, Saginaw, MI, 1988.
47. *Methods for Microbiological Analysis of Waters*, IWD/NWRI, Scientific Operations Divisions, Canada Centre for Inland Waters, Burlington, Ontario, 1978.
48. **Bacon, M. P., Brewer, P. G., Spencer, D. W., Murray, J. W., and Goddard, J.,** Lead-210, polonium-210, manganese and iron in the Cariaco Trench, *Deep Sea Res.*, 27A, 119, 1980.
49. **Spencer, D. W.,** GEOSECS II, the 1970 North Atlantic station: hydrographic features, oxygen and nutrients, *Earth Planet. Sci. Lett.*, 16, 91, 1972.
50. **Savile, H.,** *Operating Manual for the Close Interval Water Sampler (CIWS)*, Report ES-89001, NWRI, Research Support Division, Engineering Service Section, Canada Centre for Inland Waters, Burlington, Ontario, 1989.
51. **Rathke, D.,** unpublished data, 1989.
52. **Savile, H.,** *CCIW/NWRI Operating and Maintenance Manual for Westphalia Clarifiers*, Report ES-1087, NWRI, Canada Centre for Inland Waters, Burlington, Ontario, 1980.
53. **Thomas, R. L.,** unpublished method, 1980.
54. **Nriagu, J. O., Wong, H. K. T., and Coker, R. D.,** Particulate and dissolved trace metals in Lake Ontario, *Water Res.*, 15, 91, 1981.
55. Millipore Corporation, Operator's Manual OM-141, Millipore Corporation, MA, 1986, chap. 3 and 4.
56. **Horowitz, A. J., Elrick, K. A., and Hooper, R. C.,** A comparison of instrumental dewatering methods for the separation and concentration of suspended sediment for subsequent trace element analysis, *Hydrol. Process.*, 2, 163, 1989.
57. **Williams, J. D. H., Shear, H., and Thomas, R. L.,** Availability to *Scenedesmus quadricauda* of different forms of phosphorous in sedimentary materials from the Great Lakes, *Limnol. Oceanogr.*, 25, 1, 1980.
58. **Mudroch, A.,** Chemistry, mineralogy, and morphology of Lake Erie suspended matter, *J. Great Lakes Res.*, 10, 286, 1984.
59. Beak Consultants, *Study of Toxic Substances in the Great Lakes: Trace Organic Analyses*, Report No. 45, Department of Fisheries and Environment, PRD, Burlington, Ontario, 1979.
60. **Carew, T. J. and Williams, D. J.,** *Surveillance Methodology*, IWD Technical Bulletin No. 92, Canada Centre for Inland Waters, Burlington, Ontario, 1975, 1.
61. **Schröder, R.,** Ein summierender Wasserschöpfer, *Arch. Hydrobiol.*, 66, 241, 1969.
62. **Bürgi, H. R.,** Eine neue Netzgarnitur mit Kipp-Schliessmechanismus für quantitative Zooplanktonfänge in Seen, *Schweiz. Z. Hydrol.*, 45, 505, 1983.
63. **Barkley, R. A.,** Selectivity of towed-net samplers, *Fish. Bull.*, 70, 799, 1972.

64. **Edmondson, W. T. and Winberg, G. G.,** *A Manual on Methods for the Assessment of Secondary Productivity in Freshwater*, IBP Handbook No. 17, Blackwell Scientific, Oxford, 1971, 1.
65. **Smith, P. E. and Richardson, S. L.,** Standard Techniques for Pelagic Fish Egg and Larva Studies, Fisheries Technical Paper 175, Food and Agriculture Organization of the United Nations, Rome, 1977.
66. **Leslie, J. K.,** Nearshore contagion and sampling of freshwater larval fish, *J. Plankton Res.*, 8, 1137, 1986.
67. **Krishnaswami, S. and Lal, D.,** Radionuclide limnochronology, in *Lakes — Chemistry Geology, Physics*, Lerman, A., Ed., Springer-Verlag, New York, 1978, 153.
68. **Imboden, D. M. and Lerman, A.,** Chemical models of lakes, in *Lakes — Chemistry, Geology, Physics*, Lerman, A., Ed., Springer-Verlag, New York, 1978, 341.
69. **Dillon, P. J., Evans, R. D., and Molot, L. A.,** Retention and resuspension of phosphorus, nitrogen and iron in a central Ontario lake, *Can. J. Fish. Aquat. Sci.*, 47, 1269, 1990.
70. **Bloesch, J. and Evans, R. D.,** Lead-210 dating of sediments compared with accumulation rates estimated by natural markers and measured with sediment traps, *Hydrobiologia*, 92, 579, 1982.
71. **Binford, M. W. and Brenner, M.,** Dilution of ^{210}Pb by organic sedimentation in lakes of different trophic states, and application to studies of sediment-water interactions, *Limnol. Oceanogr.*, 31, 584, 1986.
72. **Benoit, G. and Hemond, H. F.,** Comment on "Dilution of ^{210}Pb by organic sedimentation in lakes of different trophic states, and application to studies of sediment-water interactions" (Binford and Brenner[71]), *Limnol. Oceanogr.*, 33, 299, 1988.
73. **Binford, M. W. and Brenner, M.,** Reply to comment on "Dilution of ^{210}Pb by organic sedimentation in lakes of different trophic states, and application to studies of sediment-water interactions" (Benoit and Hemond[72]), *Limnol. Oceanogr.*, 33, 304, 1988.
74. **Evans, R. D. and Rigler, F. H.,** The measurement of whole lake sediment accumulation and phosphorus retention using lead-210 dating, *Can. J. Fish. Aquat. Sci.*, 37, 817, 1980.
75. **Bloesch, J. and Burns, N. M.,** A critical review of sedimentation trap technique, *Schweiz. Z. Hydrol.*, 42, 15, 1980.
76. **Reynolds, C. S., Wiseman, S. W., and Gardner, W. D.,** An annotated bibliography of aquatic sediment traps and trapping methods, *Freshwater Biol. Assoc. Occas. Publ.*, 11, 1, 1980.
77. **Blomqvist, S. and Håkanson, L.,** A review on sediment traps in aquatic environments, *Arch. Hydrobiol.*, 91, 101, 1981.
78. **Hargrave, B. T. and Burns, N. M.,** Assessment of sediment trap collection efficiency, *Limnol. Oceanogr.*, 24, 1124, 1979.
79. **Gardner, W. D.,** Fluxes, Dynamics and Chemistry of Particulates in the Ocean, Ph.D. thesis, MIT/WHOI, Woods Hole, MA, 1977.
80. **Gardner, W. D.,** Sediment trap dynamics and calibration. A laboratory evaluation, *J. Mar. Res.*, 38, 17, 1980.
81. **Gardner, W. D.,** Field assessment of sediment traps, *J. Mar. Res.*, 38, 41, 1980.
82. **Blomqvist, S. and Kofoed, C.,** Sediment trapping — A subaquatic *in situ* experiment, *Limnol. Oceanogr.*, 26, 585, 1981.
83. **Butman, Ch. A.,** Sediment trap biases in turbulent flows: results from a laboratory flume study, *J. Mar. Res.*, 44, 645, 1986.
84. **Butman, Ch. A., Grant, W. D., and Stolzenbach, K. D.,** Predictions of sediment trap biases in turbulent flows: a theoretical analysis based on observations from the literature, *J. Mar. Res.*, 44, 601, 1986.
85. **Baker, E. T., Milburn, H. B., and Tennant, D. A.,** Field assessment of sediment trap efficiency under varying flow conditions, *J. Mar. Res.*, 46, 573, 1988.
86. **Jannasch, H. W., Zafiriou, O. C., and Farrington, J. W.,** A sequencing sediment trap for time-series studies of fragile particles, *Limnol. Oceanogr.*, 25, 939, 1980.
87. **Lau, Y. L.,** Laboratory study of cylindrical sedimentation trap, *J. Fish. Res. Board Can.*, 36, 1288, 1979.
88. **Honjo, S., Connell, J. F., and Sachs, P. L.,** Deep-ocean sediment trap; design and function of PARFLUX Mark II, *Deep Sea Res.*, 27, 745, 1980.
89. **Gardner, W. D.,** The effect of tilt on sediment trap efficiency, *Deep Sea Res.*, 32, 349, 1985.
90. **Staresinic, N., Bröckel, K. von, Smodlanka, N., and Clifford, C. H.,** A comparison of moored and free-drifting sediment traps of two different designs, *J. Mar. Res.*, 40, 273, 1982.

91. **Botnen, H. B.,** Horizontal variation in sedimentation in a land-locked fjord: preliminary results, in *Sediment Trap Studies in the Nordic Countries, 1*, Wassmann, P. and Heiskanen, A. S., Eds., Workshop, Tvärminne Zoological Station, Finland, 1988, 128.
92. **Davis, M. B. and Brubaker, L. B.,** Differential sedimentation of pollen grains in lakes, *Limnol. Oceanogr.*, 18, 635, 1973.
93. **Gasith, A.,** Tripton sedimentation in eutrophic lakes — simple correction for the resuspended matter, *Verh. Int. Verein. Limnol.*, 19, 116, 1975.
94. **Hilton, J.,** A conceptual framework for predicting the occurrence of sediment focusing and sediment redistribution in small lakes, *Limnol. Oceanogr.*, 30, 1131, 1985.
95. **Nriagu, J. O., Kemp, A. L. W., Wong, H. K. T., and Harper, N.,** Sedimentary record of heavy metal pollution in Lake Erie, *Geochim. Cosmochim. Acta*, 43, 247, 1979.
96. **Bloesch, J. and Uehlinger, U.,** Horizontal sedimentation differences in a eutrophic Swiss lake, *Limnol. Oceanogr.*, 31, 1094, 1986.
97. **Bloesch, J.,** Inshore-offshore sedimentation differences resulting from resuspension in the eastern basin of Lake Erie, *Can. J. Fish. Aquat. Sci.*, 39, 748, 1982.
98. **Davis, M. B.,** Redeposition of pollen grains in lake sediments, *Limnol. Oceanogr.*, 18, 44, 1973.
99. **Charlton, M. N.,** Downflux of sediment, organic matter and phosphorus in the Niagara River area of Lake Ontario, *J. Great Lakes Res.*, 9, 201, 1983.
100. **Bloesch, J. and Sturm, M.,** Settling flux and sinking velocities of particulate phosphorus (PP) and particulate organic carbon (POC) in Lake Zug, Switzerland, in *Sediments and Water Interactions*, Sly, P. G., Ed., Springer-Verlag, New York, 1986, 481.
101. **Stabel, H. H.,** Settling velocity and residence time of particles in Lake Constance, *Schweiz. Z. Hydrol.*, 49, 284, 1987.
102. **Burns, N. M. and Pashley, A. E.,** *In situ* measurement of the settling velocity profile of particulate organic carbon in Lake Ontario, *J. Fish. Res. Board Can.*, 31, 291, 1974.
103. **Bienfang, P. K.,** A new phytoplankton sinking rate method suitable for field use, *Deep Sea Res.*, 26A, 719, 1979.
104. **Burns, N. M. and Rosa, F.,** *In situ* measurements of the settling velocity of organic carbon particles and 10 species of phytoplankton, *Limnol. Oceanogr.*, 25, 855, 1980.
105. **Bloesch, J.,** Sedimentation und Phosphorhaushalt im Vierwaldstättersee (Horwer Bucht) und im Rotsee, *Schweiz. Z. Hydrol.*, 36, 71, 1974.
106. **Rathke, D. E., Bloesch, J., Burns, N. M., and Rosa, F.,** Settling fluxes in Lake Erie (Canada) measured by traps and settling chambers, *Verh. Int. Verein. Limnol.*, 21, 383, 1981.
107. **Tessenow, U.,** Untersuchungen über den Kieselsäurehaushalt der Binnengewässer, *Arch. Hydrobiol. Suppl.*, 32, 1, 1966.
108. **Bienfang, P., Laws, E., and Johnson, W.,** Phytoplankton sinking rate determination: technical and theoretical aspects, an improved methodology, *J. Exp. Mar. Biol. Ecol.*, 30, 283, 1977.
109. **Walsby, A. E. and Reynolds, C. S.,** Sinking and floating, in *The Physiological Ecology of Phytoplankton*, Morris, I., Ed., Blackwell Scientific, Oxford, 1980, 371.
110. U.S. GOFS Working Group, Sediment Trap Technology and Sampling, Report No. 10, U.S. Global Ocean Flux Study, August 1989.
111. **Harbison, G. R. and Gilmer, R. W.,** Effect of animal behaviour on sediment trap collections: implications for the calculation of aragonite fluxes, *Deep Sea Res.*, 33, 1017, 1986.
112. **Lee, C., Wakeham, S., and Hedges, J. S.,** The measurement of oceanic particle flux — are "swimmers" a problem?, *Oceanogr. Mag.*, 1, 34, 1988.
113. **Gundersen, K. and Wassmann, P.,** Use of chloroform in sediment traps: caution advised, *Mar. Ecol. Prog. Ser.*, 64, 197, 1990.
114. **Gardner, W. D., Hinga, K. R., and Marra, J.,** Observations on the degradation of biogenic material in the deep ocean with implications on accuracy of sediment trap fluxes, *J. Mar. Res.*, 41, 195, 1983.
115. **Knauer, G. A., Karl, D. M., Martin, J. H., and Hunter, C. N.,** *In situ* effects of selected preservatives on total carbon, nitrogen and metals collected in sediment traps, *J. Mar. Res.*, 42, 445, 1984.
116. **Karl, D.M. and Knauer, G.A.,** Swimmers: A recapitulation of the problem and potential solution, *Oceanography*, 32—35, 1989.
117. **Spiess, M.,** POC-, PP- und PN-Fraktionierung in Seen mit verschiedenem Trophiegrad, M.Sc. thesis, ETH, Zürich, 1985.
118. **Dörrstein, D.,** Sedimentation im Piburger See (Ötztal, Tirol), Ph.D. thesis, University of Innsbruck, 1977.
119. **Bloesch, J.,** unpublished data, 1983 and 1984.
120. **Wassmann, P.,** personal communication, 1990.

121. **Gardner, W. S., Eadie, B. J., Chandler, J. F., Parrish, C. C., and Malczyk, J. M.,** Mass flux and "nutritional composition" of settling epilimnetic particles in Lake Michigan, *Can. J. Fish. Aquat. Sci.*, 46, 1118, 1989.
122. **Lawacz, W.,** The characteristics of sinking materials and the formation of bottom deposits in a eutrophic lake, *Mitt. Int. Ver. Theor. Angew. Limnol.*, 17, 319, 1969.
123. **Ahlgren, I.,** Changes in Lake Norrviken after sewage diversion, *Verh. Int. Verein. Limnol.*, 18, 355, 1972.
124. **Ferrante, J. G. and Parker, J. I.,** Transport of diatom frustules by copepod fecal pellets to the sediments of Lake Michigan, *Limnol. Oceanogr.*, 22, 92, 1977.
125. **Ulen, B.,** Seston and sediment in Lake Norrviken. I. Seston composition and sedimentation, *Schweiz. Z. Hydrol.*, 40, 262, 1978.
126. **Gowing, M. M. and Silver, M. W.,** Origins and microenvironments of bacteria mediating fecal pellet decomposition in the sea, *Mar. Biol.*, 73, 7, 1983.
127. **Knauer, G. A. and Martin, J. H.,** Phosphorous-cadmium cycling in northeast Pacific waters, *J. Mar. Res.*, 39, 65, 1981.
128. **Rigler, F. H., MacCallum, M. E., and Roff, J. C.,** Production of zooplankton in Char lake, *J. Fish. Res. Board Can.*, 31, 637, 1974.
129. **Lastein, E.,** Recent sedimentation and resuspension of organic matter in eutrophic Lake Esrom, Denmark, *Oikos*, 27, 44, 1976.
130. **Hendrikson, R.,** Auf- und Abbauprozesse partikulärer organischer Substanz anhand von Seston- und Sinkstoffanalysen, Ph.D. thesis, University of Kiel, 1975.
131. **Smetacek, V., von Bröckel, K., Zeitzschel, B., and Zenk, W.,** Sedimentation of particulate matter during a phytoplankton spring bloom in relation to the hydrographical regime, *Mar. Biol.*, 47, 211, 1978.
132. **Zeitzschel, B., Diekmann, P., and Uhlmann, L.,** A new multisample sediment trap, *Mar. Biol.*, 45, 285, 1978.
133. **Wefer, G., Suess, E., Balzer, W., Liebezeit, G., Müller, P. J., Ungerer, C. A., and Zenk, W.,** Fluxes of biogenic components from sediment trap deployment in circumpolar waters of the Drake Passage, *Nature*, 299, 145, 1982.
134. **Eadie, B. J.,** Sediment trap studies in Lake Michigan: resuspension and chemical fluxes in the southern basin, *J. Great Lakes Res.*, 10, 307, 1984.
135. **Honjo, S.,** Sedimentation of materials in the Sargasso Sea at a 5376 m deep station, *J. Mar. Res.*, 36, 469, 1978.
136. **Spencer, D. W., Brewer, P. G., Fleer, A., Honjo, S., Krishnaswami, S., and Nozaki, Y.,** Chemical fluxes from a sediment trap experiment in the deep Sargasso Sea, *J. Mar. Res.*, 36, 493, 1978.
137. **Lee, C. and Cronin, C.,** The vertical flux of particulate organic nitrogen in the sea: decomposition of amino acids in the Peru upwelling area and the equatorial Atlantic, *J. Mar. Res.*, 40, 227, 1982.
138. **Honjo, S., Manganini, S. J., and Cole, J. J.,** Sedimentation of biogenic matter in the deep ocean, *Deep Sea Res.*, 29, 609, 1982.
139. **Lee, C., Wakeham, S. G., and Farrington, J. W.,** Variations in the composition of particulate organic matter in a time-series sediment trap, *Mar. Chem.*, 13, 181, 1983.
140. **Prahl, F. G. and Carpenter, R.,** The role of zooplankton fecal pellets in the sedimentation of polycyclic aromatic hydrocarbons in Dabob Bay, Washington, *Geochim. Cosmochim. Acta*, 43, 1959, 1979.
141. **Fisch, M.,** Experimentelle Untersuchungen über die Anwendung von Bakteriziden zur Verhinderung der künstlichen Mineralisation in Sedimentfallen, M.Sc. thesis, ETH, Zürich, 1985.
142. **Matsuyama, M.,** Organic substances in sediment and settling matter during spring in a meromictic lake Shigetsu, *J. Oceanogr. Soc. Jpn*, 29, 53, 1973.
143. **Bloesch, J.,** unpublished data, 1969.
144. **Fischer, K. and Cobler, R.,** Antibiotic poisoning of sediment trap, *Eos*, 60, 851, 1979.
145. **Dymond, J., Fischer, K., Cobler, R., Gardner, W., Richardson, M. J., Berger, W., Soutar, A., and Dunbar, R.,** A sediment trap intercomparison study in the Santa Barbara Basin, *Earth Planet. Sci. Lett.*, 53, 409, 1981.
146. **Ohle, W.,** Measurement and comparative values of the short circuit metabolism of lakes by POC relationship of primary production of phytoplankton and settling matter, *Arch. Hydrobiol. Beih.*, 19, 163, 1984.
147. **Kirchner, W. B.,** An evaluation of sediment trap methodology, *Limnol. Oceanogr.*, 20, 657, 1975.
148. **Schwoerbel, J.,** *Methoden der Hydrobiologie, Süsswasserbiologie*, 2nd ed., G. Fischer-Verlag, Stuttgart, 1980.

Chapter 6
SEDIMENT PRESERVATION, PROCESSING, AND STORAGE

Alena Mudroch and Richard A. Bourbonniere

I. INTRODUCTION

The objective of any sediment sampling program is to deliver to the laboratory samples as representative of the original material as possible. After sediments have been collected using grab samplers or corers at the field site, the resultant material (the sample) has to be handled and processed properly. There should be little or no disturbance due to field handling before transporting to the laboratory for physical, chemical, and biological testing. Proper sediment handling and preparation must be considered a very important operation in the overall procedures used for testing and analyses. The best analytical methods and procedures can fail and yield incorrect data if samples are improperly handled and prepared. Materials such as rocks, soils, or sediments are difficult to prepare for analyses without changing their composition in some respect. For example, the addition of foreign material (contamination) or the loss or change of certain components of the sample through evaporation, oxidation, etc. can compromise its integrity.

The objective of proper sample preparation is to minimize these undesirable effects by selecting procedures which create the least contamination or alteration of the sediment sample. Moreover, different tests and analyses carried out on bottom and suspended sediments may require different sample handling and preparation. The literature on this topic is not extensive in terms of studies that are specifically intended to define proper procedures for handling of sediment samples. Many workers in the field have had to devise or adopt protocols best suited to the needs of their studies. Often sample handling methods have been discussed only briefly in scientific publications. Workers in other fields, dealing with biological specimens, soils, and rocks, have published results of studies about problems comparable to those in sediment studies. Where their observations and examples were deemed applicable to sediment preparation, they are included in this chapter. This chapter covers methods and procedures adopted for sample handling, selection and cleaning of containers, transport, drying, screening, grinding, homogenization, and storage of sediment samples.

II. CONTAINERS FOR SEDIMENT SAMPLES

Containers and implements should be carefully selected for sediment handling in the field and laboratory with due consideration for which investigations and analyses are to be carried out. Container and implement materials can be significant sources of contamination. Containers should neither contaminate the sample nor promote loss of constituents of interest through

1-56670-027-2/94/$0.00+$.50

wall effects. For example, plastics contain plasticizers that can become contaminants in the determination of organic compounds. Metal containers, spoons, or other equipment may contaminate samples which will be analyzed for metal contaminants.

In considering sampling, handling, and long-term storage of a variety of environmental specimens (sediments included) for all types of components, many at trace levels, Luepke[1] summarized the recommendations of those studies that were conducted. Workers should take care to:

- Minimize interaction between samples and containers and implements
- Minimize interaction between samples and external environment
- Test any materials which contact the samples
- Treat sample containers with the same precautions as samples
- Wash sample containers and implements with appropriate cleaners
- Run an appropriate analytical blank to which to refer every sample

Some common types and sizes of containers used for storage of sediment samples are shown in Figure 1. Wide-mouth screw-capped clear and amber glass bottles in sizes from 30 ml to 4 l, with appropriate Teflon-lined caps and wide-mouth screw-capped polyethylene bottles of similar size with appropriate polypropylene screw closures, are recommended containers and can be obtained from any scientific supply company. The prime disadvantage of glass containers or any glass instruments is that they easily break in the field, during shipment, or particularly when the sample is frozen.

Polyethylene, Teflon, or glass implements are usually best for sediment samples which are to be analyzed for inorganic components. Plastic bags of various sizes, made of polyethylene, polypropylene, or other suitable plastic, can be used for storing wet or dry sediment samples. Sediments for biological testing can be collected, transported, and stored in plastic or glass containers.[2,3]

Containers should be appropriately cleaned in the laboratory and, after cleaning, sealed properly to avoid, or at least minimize, contamination with components of ambient air during storage and at the sampling site. A minimum cleaning procedure should involve washing the interior of the container with hot water and laboratory grade soap, then hot water followed by distilled or deionized water rinses. The initial washing may be followed either by acid washing with dilute nitric acid solution and a pure water rinse for analyses of inorganic compounds, or by rinses with solvents (e.g., methyl alcohol and dichloromethane) and drying for organic analyses. Baking at 550°C has often been considered adequate for cleaning borosilicate glass bottles for sediment samples intended for organic analyses. Large volume samples, for organic analyses, can be stored in covered, galvanized garbage pails or other metal containers cleaned with detergents and organic solvents before use. Plastic garbage pails can similarly be used for large volume sediment samples for inorganic analyses, after detergent, water, and acid washing.

The general statements made above reflect the experiences of many workers who took what they perceived to be the necessary precautions, but generally did not perform detailed studies to prove the adequacy of such precautions except by running field and procedural blanks.

In various studies[4-9] and reviews,[10-12] problems of metal contamination of water samples, not sediments, due to either container material or the addition

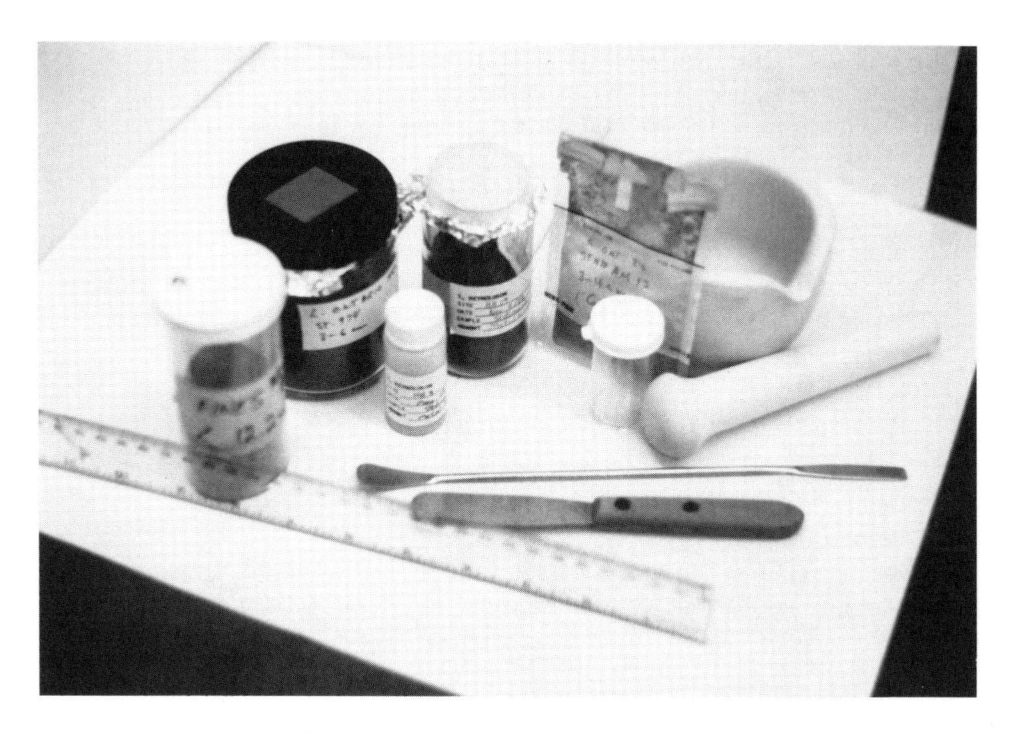

FIGURE 1. Different containers and implements for sediment samples.

of preservation agents, were investigated. It should be noted that concentrations of elements of concern (Zn, Pb, Cd, Al, Cu) in water are in the ng/g range, whereas concentrations of the same elements in the sediments are usually in the µg/g range. Consequently, contamination of water samples is by far a more serious problem than contamination of sediment samples by the material of containers.

A. EXAMPLE CLEANING PROCEDURES FOR CONTAINERS TO HOLD SEDIMENTS DESTINED FOR DETERMINATION OF ORGANIC CONTAMINANTS

Containers used for sediments to be analyzed for organic components should be carefully chosen — plastic should be avoided wherever possible. Glass, porcelain, stainless steel, Teflon, or Teflon-coated instruments should be used in handling sediment samples. Wide-mouth amber or clear glass jars and bottles with aluminum foil or Teflon-lined caps are the best containers, but certain compounds (e.g., phenols) can adsorb to these surfaces.[1]

At Environment Canada's National Water Research Institute (NWRI), in Burlington, Ontario, containers intended for the collection of sediments to be analyzed for organic compounds are first cleaned with detergent and rinsed well with water. Following water rinsing, containers are rinsed twice with methyl alcohol which can remove residual water and dissolve away polar contaminants. Following methyl alcohol are two rinses with dichloromethane in which nonpolar contaminants are soluble. All containers and implements which are cleaned in this way are immediately placed in an oven and dried thoroughly at 60°C. Aluminum foil, for lining the caps of wide-mouth jars, and Teflon-lined screw caps for narrow-mouth containers are also rinsed with methyl alcohol and dichloromethane and air-dried in a fume hood or other well-ventilated area.

The Water Quality National Laboratory of Environment Canada (WQNL), also in Burlington, uses the following procedure for cleaning glass containers which will contain sediment and water samples for organic analyses.[13]

- Wash with high-pressure tap water jet
- Wash once with chromic acid and water
- Wash once with soap water
- Rinse three times with organic-free water
- Rinse two times with washing acetone, followed by one rinse with special-grade acetone
- Rinse twice with pesticide-grade hexane
- Dry (uncapped) containers in a hot air oven at 360°C for at least 6 h

B. EXAMPLE CLEANING PROCEDURES FOR CONTAINERS TO HOLD SEDIMENTS DESTINED FOR DETERMINATION OF INORGANIC CONSTITUENTS

According to Durst,[14] materials used for the production of glass and plastic containers can contain detectable amounts of metals which are also present in sediments. Most metals can be leached from the inside surface of containers by washing with dilute acids,[1] but in some cases chelating agents may be necessary to achieve a clean surface. A detailed cleaning procedure was proposed by Durst.[14] Plastic containers made of Teflon and polyethylene are best for storage of sediments collected for the determination of inorganic constituents, including metals. Further, Teflon can also be used for samples collected for the determination of organic contaminants, but its high cost can be an important disadvantage. Moody and Lindstrom[6] studied the leaching of trace metals from a variety of plastic containers and found that containers can be considerable sources of error if not cleaned properly. They concluded that Teflon and polyethylene containers are the least contaminating if cleaned properly. Their recommended procedure for cleaning plastic containers involves nine steps:

1. Fill container with reagent-grade HCl (1:1)
2. Allow to stand 1 week at room temperature, or at 80°C for Teflon containers
3. Empty and rinse with distilled water
4. Fill with reagent-grade HNO_3 (1:1)
5. Allow to stand as in Step 2
6. Empty and rinse with distilled water
7. Fill with the purest available water
8. Allow to stand several weeks or until needed, changing the water periodically to ensure continued cleaning.
9. Rinse with purest distilled water and allow to dry in a particle- and fume-free environment (another alternative is to fill the containers with purest distilled water until sample collection)

Similar procedures are recommended by the WQNL,[13] with the exception that in addition to hydrochloric and nitric acids, sulfuric and chromic acids are used in some cases.

Containers should be sealed properly to avoid contact with the atmosphere. This minimizes contamination with components of ambient air at the sampling site, and decreases the likelihood of oxidation of the sample. It is always good practice to carry extra containers on a sampling expedition, exposing them to the same conditions as those which are actually used for samples, thus serving as field blanks. Containers should be carefully labeled with indelible ink pens. Labels should contain the following information: sediment use, date and time of collection, name of collector, and site/sample identification.

III. SAMPLE HANDLING, PROCESSING, AND STORAGE

Samples which consist mainly of fine-grained sediments have relatively uniform particle size distribution, typically particle sizes <63 μm. However, many samples can contain mixtures of fine- and coarse-grained particles. Frequently, one has to decide whether the coarse material should be sieved out and discarded since it usually contains relatively low concentrations of contaminants. Alternatively, the entire sample could be ground to a suitable particle size yielding a "bulk" sediment sample for analyses. Neither of these procedures can be recommended over the other, since there is still an ongoing discussion as to which technique properly represents the character of the sediment; choice depends mainly on the study objective(s). Irrespective of which procedure is used, a detailed description of sample preparation should be included in reporting the results of chemical analyses.

Sediments, particularly the topmost 10 cm, typically contain large and variable amounts of water (up to 95%). To permit comparison of the data, sediments are dried and analyses are carried out on dry material, or a subsample is taken for drying to determine the water content while the analyses are performed on the wet sediment. In either case, results of analyses are usually presented on a "dry weight" basis. Further, some analytical methods, such as X-ray fluorescence spectrometry, require dry material ground to a certain particle size.

In this section, the types of operations that could be required by workers in the field in handling their sediment samples are described. They are separated into three cetegories of operations: those that must be carried out in the field, handling operations that are suggested for wet sediments, and those suggested for dry sediments. Not all of these handling operations are required for all samples. Depending upon the study objectives, workers would choose those operations that are necessary to properly handle their samples. As an example, consider a study which involves the analysis of dredged harbor sediments. In order to decide how to properly dispose of dredged material, they must be analyzed for potential inorganic and organic contaminants, and bioassay tests are required by the local regulatory agency. The necessary operations that should be applied to these samples will be pointed out in each section to follow.

A. SAMPLE HANDLING AND MEASUREMENTS CARRIED OUT IN THE FIELD

The operations required for samples that should be carried out in the field are given diagrammatically in Figure 2. Not all of these handling operations will be required for all studies, but when they are, they should be performed immediately. In the dredging example, Eh (redox potential) and pH, cation exchange capacity and oxygen-free subsampling are not normally required. The worker, in this instance, will need only to describe the sample properly (Section III.A.2) and ensure proper mixing and subsampling (Section III.A.5).

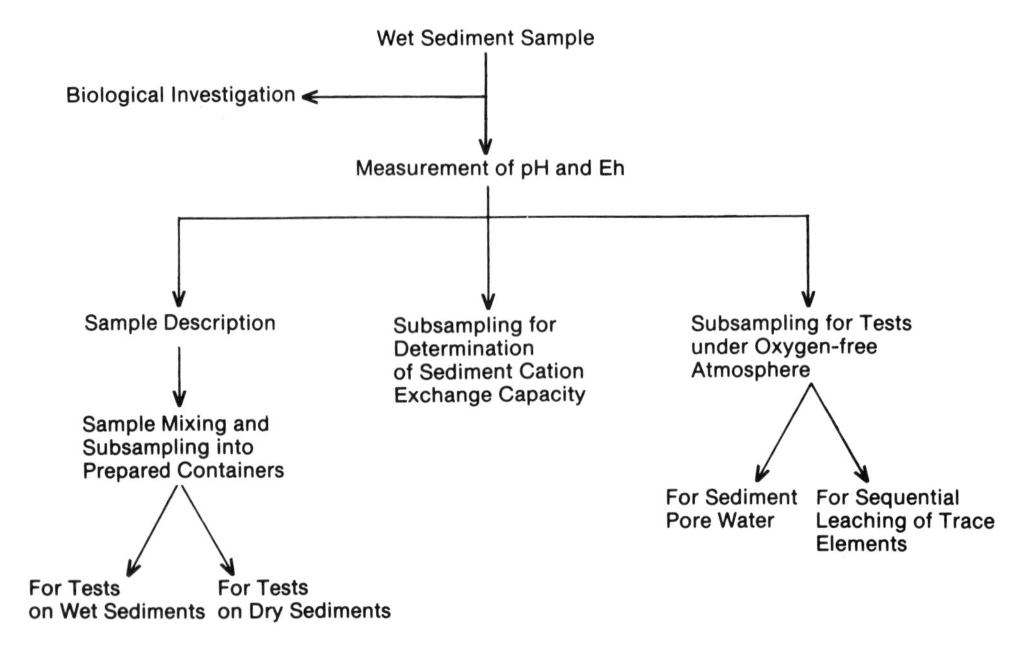

FIGURE. 2. Sample handling in the field.

1. Measurements of pH and Eh

Studies published by different scientists on the meaning and problems with measuring and interpretation of sediment pH and Eh[15-22] should be consulted prior to measuring these parameters in sediment samples. Quantitative interpretation of Eh measurements in natural aqueous systems is difficult because of problems associated with the technique of measurement, the performance of the inert metal electrode, and the thermodynamic behavior of the environment.[19] Critiques regarding the use of Eh for determining the redox level of marine sediments have pointed out that the absence of a uniform measurement procedure (choice and preparation of electrodes, their retention time in the sediment, etc.) leads to different results.[23]

Sediment pH and Eh should be measured in the field immediately upon retrieval of an undisturbed sediment sample to avoid the effect of changes of sediment chemistry on measured values. Several different methods are described in detail in the scientific literature. The reader is advised to consider the following case studies when planning such measurements, and to seek out the original literature if necessary.

For the Eh measurement, Whitfield[19] used two platinum electrodes immersed side-by-side in the sediment sample to ensure that reproducible potentials were being measured. The values of Eh measured by these two electrodes differed by 10 to 30 mV due to the surface properties of the platinum and poisoning of the platinum surface by irreversible attack at different rates at the two electrode surfaces. To measure simultaneously Eh, pH, and sulfide activity (pS^{-2}),[19] a total of five electrodes combined into a compound electrode were used which enabled all working electrodes to be introduced in a single probe into the sediment. When this probe was used in conjunction with a remote junction reference electrode, stable and reproducible results were obtained, even on highly reduced samples. The Eh of each sample was monitored until a steady value (drift less than 1 mV/min) was observed. This value was recorded and the Eh measured by the second electrode was registered. The sulfide activity and pH were measured in turn, and the Eh registered by the first electrode was checked again. If the results differed by more than 30 mV, a further 5 min were allowed for equilibration. The temperature was recorded before the sample was discarded. The whole cycle of readings was completed within about 10 min.

It was found that cleaning of the platinum electrodes by rubbing with a fine abrasive cloth was more efficient than rinsing with different chemicals and distilled water. The potential of the Eh electrode/calomel cell was measured occasionally in a Zobell solution[24] (0.003 *M* potassium ferricyanide and 0.003 *M* potassium ferrocyanide in 0.1 *M* potassium chloride) to check the performance of the liquid junction. The potential of the cell was adjusted to the standard hydrogen scale by adding 250 mV to the measured value to enable comparison with other data.

A portable Metrohm pH and Eh meter and combination glass and platinum electrodes were used for measurements of pH and Eh in surficial sediments and sediment cores collected from the Great Lakes.[25,26] Surface sediment samples were collected by a Shipek grab sampler. Upon retrieval, the sampling bucket, with the collected sediment, was placed on a special stand to keep the sediment surface horizontal. The Eh electrodes were calibrated using the Zobell solution[24] and marked vertically at 0.5-cm intervals. They were then pushed into the sediment to the desired depth. The pH electrodes were similarly marked and were calibrated using two buffer solutions of a known pH value (4.0 and 7.0). For measuring pH and Eh in sediment cores, the electrodes were inserted into the sediment after the core was placed on an extruder, uncapped, and water from the top of the core siphoned off. For measuring the pH and Eh of the top 1-cm sediment layer, the electrodes were pushed 0.5 cm into the sediment. In Figure 3, the pH and Eh electrodes inserted into the top of the sediment core were mounted on the equipment ("frying pan") designed for subsampling sediment cores and described in Chapter 4. The pH reading was taken about 1 min after inserting the electrode into the sediment. However, it took approximately 10 min to stabilize the Eh value on the meter. After measuring of the first subsample, the electrodes were removed from the sediment, cleaned with distilled water, and dried. The sediment layer, for which the measurements were made, was subsampled into a prepared container and the electrodes were inserted into the next layer of the sediment core. After every five measurements, each electrode was recalibrated as before.

Bagander and Niemisto[27] measured the Eh in marine sediment cores with a specially designed electrode-inserting attachment, adaptable to a subsample slide used in the sectioning of the cores. This attachment allowed electrode measurements with minimal air contamination and disturbance of subsamples. Two different sets of electrodes were used: a platinum wire fused into the end of a glass tube and a Ag/AgCl reference electrode was one set; an Orion combination electrode was the other.

2. Sample Description

All of the samples retrieved should be described after measuring pH and Eh. It will likely be necessary to split the sample with a spatula or knife exposing the inner sediment in the sampling bucket. Such a disturbed sample is not suitable for measuring pH and Eh. Observations should be recorded on texture, color, odor, presence of biota, foreign matter, etc. described under sediment logging in Chapter 4.

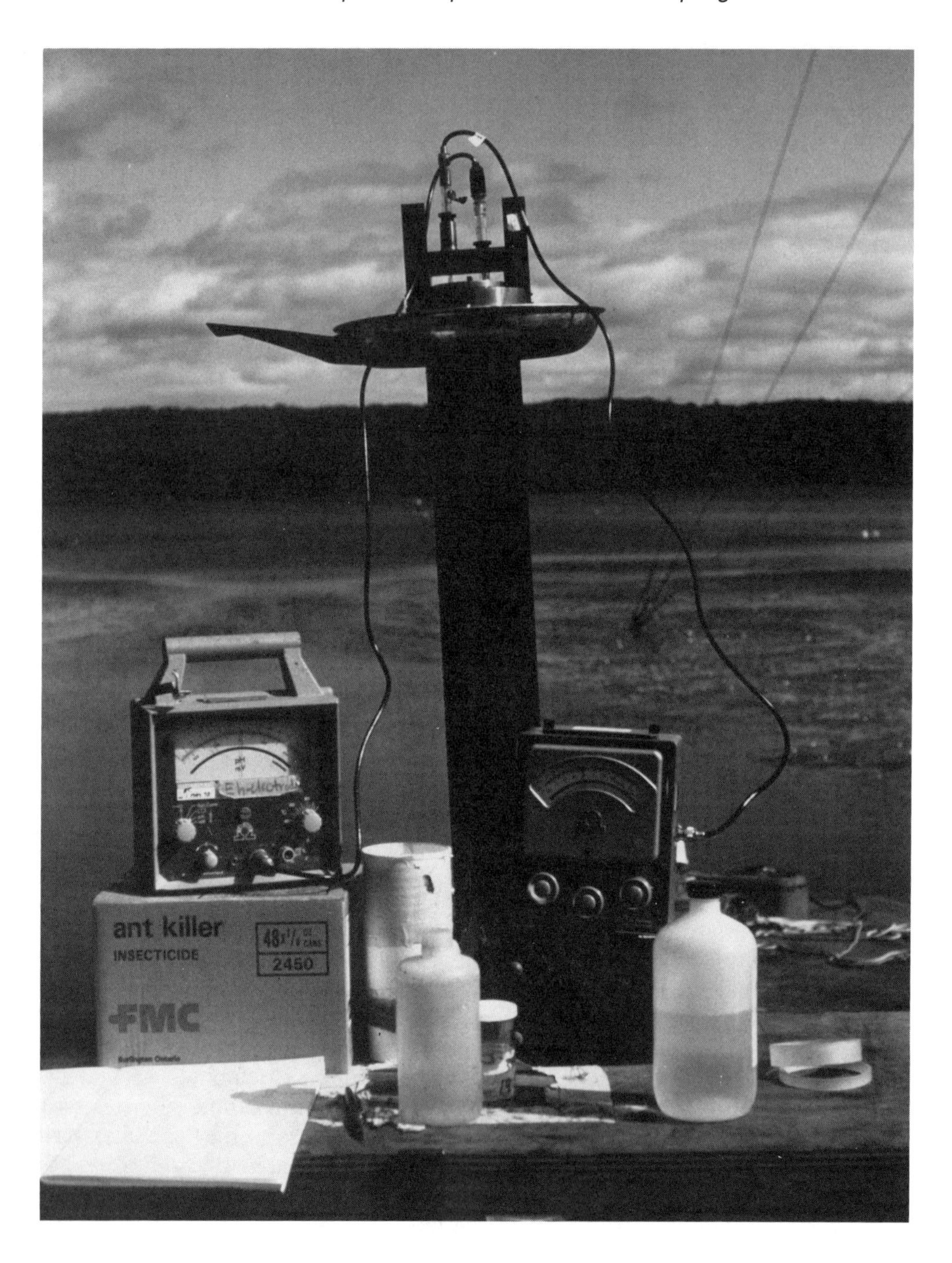

FIGURE 3. Measuring pH and Eh in the field.

3. Subsampling for Determination of Cation Exchange Capacity

Cation exchange capacity (CEC) should be determined on wet, untreated bulk sediments immediately after sample collection. Adams et al.[28] studied the effects of freeze-drying, size fractionation, organic matter removal with 30% H_2O_2, and colloidal iron removal with citrate-dithionate on changes of sediment CEC, and found from 71% increase to 40% reduction of CEC by these different treatments.

4. Subsampling Under Oxygen-Free Atmosphere

Chemical species of trace elements and their association with different sediment components have been investigated in environmental studies and in geochemical exploration.[29-31] The concentrations of different chemical forms of trace elements associated with sediment components can be used to predict bioavailability, particularly metals, in sediments.[32]

Analyses of interstitial water are used for the development of thermodynamic models to determine the partitioning of sediment associated major and trace elements in sediment/water systems.[33] Results of these investigations can be influenced by the degree to which the integrity of the sediment sample is preserved between the time of collection and chemical analyses. Rapin et al.[34] tested the effects of sample handling and storage on the results of a sequential extraction procedure for determining the partitioning of trace metals in freshwater sediments. Their study showed that drying, freeze-drying or oven-drying, should be avoided, and that freezing or short-term wet storage (at 1 to 2°C) is an acceptable preservation technique. Of nine metals tested (cadmium, cobalt, chromium, copper, nickel, lead, zinc, iron, and manganese), copper, iron, and zinc were particularly sensitive to sample handling.

The maintenance of oxygen-free conditions during extraction of anoxic sediments was of critical importance. Consequently, the preservation of sediment constituents from the effects of the air and biological activity is the fundamental requirement. Except for the surface 1- to 3-cm layer, bottom sediments are usually anoxic and become rapidly oxidized by air. Suspended sediments collected from the water column are usually oxidized.

From the results discussed above, the handling and preparation of bottom sediments collected for the determination of chemical forms of trace elements, collection of interstitial water, or measurement of pH and Eh, must be carried out under an oxygen-free atmosphere. Bulk sediment samples collected by a grab sampler should be transferred immediately upon retrieval into storage containers. Head space should be avoided by filling containers to the top and sealed air-tight. Sediment cores must be extruded in a controlled atmosphere glove box or bag filled with an inert gas such as nitrogen. Glove boxes are available from companies selling laboratory equipment. A simple piston extruder similar to that described in Chapter 4 can be used for subsampling sediment cores in a glove box. The plastic liner with the recovered sediment core has to be fitted through an opening cut in the bottom of the glove box. Before commencing sample handling and core subsampling, the air in the glove box must be replaced by a constant, controlled volume of inert gas supplied from a cylinder (see Figure 4).

The sediment core is subsampled and individual sections are transferred into desired containers and tightly closed in the glove box. When pH and Eh measurements are required, they should be carried out in the glove box prior to or during subsampling by inserting the electrodes into individual sections before they are cut from the core and placed into containers. For collection of interstitial water, sediment is often transferred into centrifuge tubes. After centrifugation they are returned to the glove box to remove the interstitial water and for further sample handling. If a sequential extraction procedure is conducted for the determination of chemical forms of trace elements, all manipulations and extraction steps should be carried out in a glove box. This includes deaeration, withdrawal and/or addition of extraction solutions, and sealing the extracts.

FIGURE 4. Glove box for handling of sediment samples under oxygen-free atmosphere.

Filipek and Owen[35] used a sequential extraction procedure on different particle size fractions on sediments from Lake Michigan. The grain size separation was achieved by wet sieving a well-mixed subsample of sediment (20 g) in a glove box under nitrogen atmosphere to avoid changes in chemical forms of the metals in the sediment.

Generally, sample handling and preparation for different sequential extraction procedures should always be carried out following instructions described in scientific literature or established protocols.

5. Sample Mixing and Subsampling into Prepared Containers

Before dividing bulk surface sediment samples collected by a grab sampler into subsamples for different analyses, they must be manually mixed in sufficiently large containers in the field. An intensive manual mixing of wet sediment is usually sufficient to homogenize the sample. Sediment can be transferred to a precleaned glass or stainless steel bowl and throughly homogenized by stirring with a glass or stainless steel spatula until textural and color homogeneity are achieved.[36] Sediment sieved through a 1-mm sieve was mixed until it appeared to be homogeneous prior to bioavailability experiments.[37] Breteler et al.[38] homogenized sediments collected for bioassays by combining fine grab samples in a Teflon-coated stainless steel pan, and through stirring with a Teflon-coated stainless steel spoon. Homogenized sediment was then added to half-gallon polypropylene bottles for chemical analyses and bioassays. For more information on wet homogenization refer to Section III.C.4. If all analyses are to be performed on dry sediment, the bulk wet sediment sample should not be subsampled in the field, but should be dried as a bulk sample and further processed in the dry state (see Section III.C.4). Containers and implements used for homogenization and subsampling should be chosen with the same considerations as for the sample containers described in Section II.

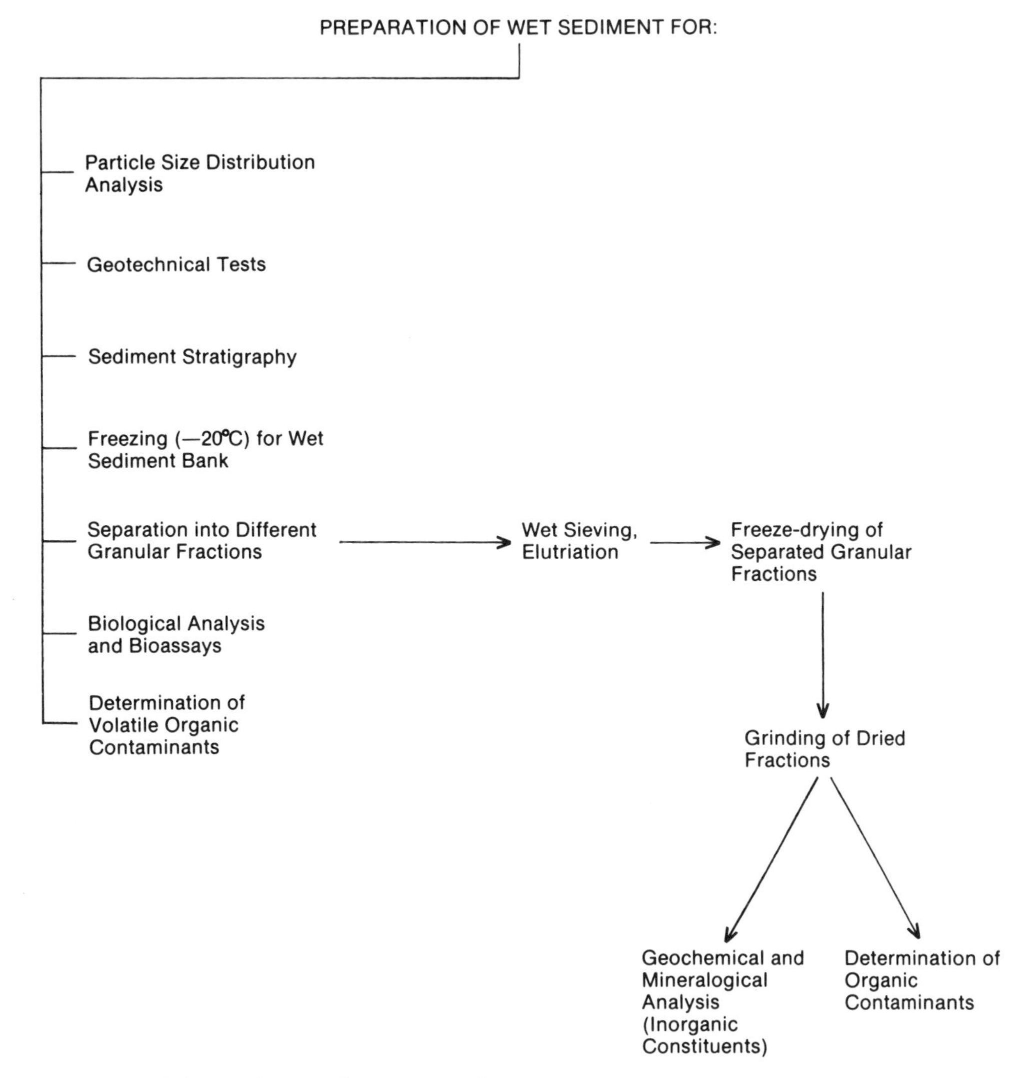

FIGURE 5. Handling samples for tests and analyses on wet sediments.

B. HANDLING SAMPLES FOR TESTS AND ANALYSES ON WET SEDIMENTS

Figure 5 outlines the possible operations that could be required for samples if tests are to be carried out on wet sediments. For the hypothetical dredging study that we are conducting as an example, we are not interested in particle size distribution or stratigraphy. It is important however to consider freezing of subsamples that will be analyzed for organic contaminants (Section III.B.3) and sieving out coarse particles while the sediment is still wet so that the contaminant-enriched fine fractions are analyzed (Section III.B.4). Subsamples for bioassays require special handling distinct from that which is appropriate for organic and inorganic contaminants (Section III.B.5).

1. Handling of Samples for Determination of Particle Size Distribution

Particle size distribution analyses should be carried out on wet sediments. Samples for this should not be frozen but stored at 4°C. Tightly sealed plastic bags, glass jars, or other containers can be used to store samples prior to

FIGURE 6. Storage of cores for sediment stratigraphy.

particle size analyses. Sediments with a high iron content should be stored in air-tight containers to avoid precipitation of iron oxides on particle surfaces, and should be analyzed as soon as possible after collection. Drying, freezing, and thawing of the sediment can cause aggregation of particles and should be avoided. Size analyses of fine-grained sediments should be carried out only on completely dispersed samples which have been treated for the removal of organic matter, carbonate, and iron coatings.

2. Handling of Samples for Geotechnical Tests and Sediment Stratigraphy

Sediment samples collected for stratigraphical or geotechnical studies can be stored at 4°C in a humidity-controlled room, without any large changes in sediment properties for several months. Long cores, such as those collected by piston coring, can be cut into lengths suitable for storage, sectioned longitudinally, described, labeled, wrapped to preserve their original consistency, and stored in a refrigerated room (see Figure 6).

3. Freezing of Wet Sediment

Freezing has long been an acceptable preservation method for sediments collected for the determination of organic and inorganic constituents. It has been widely used for sediment and biological samples.[13,39] Luepke[40] reported that rapid and deep-freezing can best maintain sample integrity and thus enable investigation for concentrations of contaminants. The lower the temperature of deep-freezing the better: a temperature of –80°C is the suggested maximum. Gills and Rook[41] found little variation in concentrations of selected trace elements on tissue samples that were stored for 1 year at –80°C compared to the fresh samples.

Freezing of sediments requires an expense in purchasing and maintaining a sufficient capacity at the required temperature. Generally, the lower the temperature the higher the cost incurred. The size of the sample set(s) to be preserved and the length of time for which they must be stored enter into the cost estimate. Also, the possibility of catastrophic freezer failure and the necessity of warning and/or backup systems must be considered. Good freezer control is required to minimize temperature fluctuations which could cause the migration of water within the container.

4. Separation into Different Size Fractions for Subsequent Analyses

Sediments are often used to monitor metal pollution in natural waters. Determination of sediment grain size, which may strongly affect the results of such investigations, is often neglected. Examples of different approaches to the separation of sediments into various particle size fractions and methods used for the separation are given below.

Several correction procedures are available for minimizing the grain size effect on concentrations of metals in sediments.[42] Studies on the correlation between metals and sediment particle size fractions[43,44] suggest that fine-grained sediments usually contain greater concentrations of metals, and that the main portion of many metals is incorporated in the silt and clay size fractions (<63-μm particle size). One of several methods to investigate the significance of the granular composition to the concentration of metals (or other contaminants) in sediments is based on the isolation of individual granular fractions and determination of metal concentrations in each fraction. Ackermann et al.[45] reviewed several studies and suggested that the choice of the <60- or <63-μm size division was based on the traditional definition of silt and sand boundary. A major advantage in using the <63-μm size fraction proposed by Forstner and Salomons[46] is the greater concentrations of trace elements associated with silt/clay particles. In addition, many studies have been carried out using this sediment fraction, so results can be compared. Also, the silt/clay sediment size fraction is close to that of material carried in suspension.

Wet sieving has been used to separate different particle size fractions in sediments. It should be noted that wet sieving involves resuspension of sediments in water and may change their original size distribution. Furthermore, because the water used for wet sieving usually does not have the *in situ* ionic composition, resuspension may even break particles that originally were agglomerated.[28] This effect may be reduced by using water collected at the sediment sampling site.

Wilberg and Hunter[47] used water from the river from which bottom sediments were collected for wet sieving in their study. Three 1-l volumes of water were passed through a continuous flow centrifuge to remove suspended matter prior to wet sieving of the sediment through six stainless steel sieves (2000, 1000, 420, 250, 125, and 63 μm). The remaining 3-l suspension of particles <63 μm was fractionated by sedimentation-decantation and successive centrifugation using a continuous-flow centrifuge. The silt/clay suspension was not treated prior to separation in order to retain the natural aggregate state of the particles. Wall et al.[48] suggested that during separation of the silt/clay fraction from the bulk sediment using chemical dispersants, mechanical stirring, and subsequent sieving, considerable quantities of smaller particles will adhere to larger ones. Similar problems can be encountered during sieving of dry sediments or wet sieving due to improper use of the ultrasonic treatment.[42]

FIGURE 7. Cyclosizer for separation of sediments into different size fractions.

Ackermann et al.[45] used plastic sieves in a beaker placed in an ultrasonic bath with 70 to 100 ml of distilled water to separate the < 60- and < 20-μm particle size fractions. The grain size distribution varied only slightly (1 to 4%) according to the duration of the sieving process. With several samples treated at the same time, the separation of a sediment sample into three fractions (< 20, 20 to 60, 60 to 200 μm) required approximately 15 to 20 min/sample.

Another system for separating particles into size ranges is based on the principle of elutriation. A wet sediment sample is separated into specific size fractions by a process which depends upon the forces present in a moving fluid. The Cyclosizer is a commercially available instrument[49] which separates particles, according to their relative size and density, in a series of hydraulic cyclones where centrifugal force produces the elutriating action (Figure 7). This unit is capable of separating silt-sized particles into six standard fractions: > 44, $33 \leq 44$, $23 \leq 33$, $15 \leq 23$, $11 \leq 15$, and < 11 μm. These size ranges are nominal and depend upon standard conditions of water flow rate, water temperature, particle density, and time of determination. Deviations from standard conditions will result in deviations in the size cutoffs. The utility of this system for silt-sized particles has been demonstrated by Mudroch and Duncan[50] on sediments from the Niagara River, and for Lake Erie sediments.[51] The cyclosizer enables large amounts of a sample in a batch system to be separated and relatively clean for subsequent chemical analysis. This method extends to silt-sized particles and the separation ability that was once only practical for coarser samples by wet sieving.

Umlauf and Bierl[52] used an elutriation procedure to study the partitioning of several organic pollutants to sediments and suspended solids in a river. The elutriation involved suspending the sediment sample in an upward flow of dispersing medium (0.01 *M* sodium pyrophosphate). Particles that settled at velocities greater than the upward velocity of the medium remained in the separation chamber, and those settling at lower velocities were carried upward to the next separation chamber. The upward flow can be calibrated by the calculation of the settling velocity using Stokes' law to achieve separation into required particle size fractions.

Cranston[53] separated sediment samples, which were previously dried at 40°C, into size fractions by sieving through a series of sieves (> 1000, 500 < 1000, 250 < 500, 125 < 250, 63 < 125, and < 63 μm) in a study of mercury distribution in estuarine sediments. The < 63-μm size fraction was further subdivided into 16 < 63- and <16-μm fractions using an Atterburg sedimentation column. The 16 < 63-μm fraction was collected from the bottom of the tube, and the finest fraction collected by centrifuging the decanted suspension from the sedimentation column. For this study, dry sediment samples were lightly pulverized using an agate mortar and pestle and soaked in water for about 3 h before wet sieving and elutriation.

5. Handling of Samples for Biological Analyses and Bioassays

Samples collected for investigation of benthic organisms are usually processed in the field by wet sieving through different size sieves. Mixing and sieving are often required before testing sediment for toxicity; the fraction less than 500 μm is usually retained.[54,55] Removal of large debris and pebbles facilitates the homogenizing of samples which are to be subsampled for multiple biological and/or chemical tests. Counting of benthic organisms is also easier on samples cleaned of debris and pebbles. If, for any reason, the samples cannot be processed in the field, they should be stored at 4°C in the dark and processed in the laboratory as soon as possible (preferably within 48 h). Biological tests should be conducted within 2 to 7 days.[38,56-59] Storage of sediments collected for bioassays is described in Section IV.

C. HANDLING OF SAMPLES FOR TESTS AND ANALYSES ON DRY SEDIMENTS

Handling operations for dry sediment are shown in Figure 8. To continue with our hypothetical study of dredged sediments, the operations required from this section are those that should be applied to the subsamples slated for analyses of inorganic contaminants. For example, these samples must be dried by an appropriate method such as freeze-drying (Section III.C.1.c). Once dry, the samples should be ground to a fine particle size (Section III.C.3) if certain determinations are to be performed, e.g., X-ray fluorescence. The ground samples must be homogenized before subsampling (Section III.C.4). Note that although these samples are screened in the wet condition to remove the coarse fraction, and no further dry sieving is required, the information given on wet sieving (Section III.C.2.a) and types of sieves (Section III.C.2.c) could be helpful.

The operations required for the preparation of dry sediment samples such as drying, sieving, grinding, mixing, and homogenization are described in detail below.

1. Drying

Three types of drying are commonly used to prepare solid samples prior to analysis: air-drying, oven-drying, and freeze-drying.

a. Air-Drying

Although time consuming, air-drying is commonly used in soil science[60] and in sedimentology.[61] It is used only rarely for the preparation of sediments for pollution studies. Air-drying may generate undesirable changes in sediment

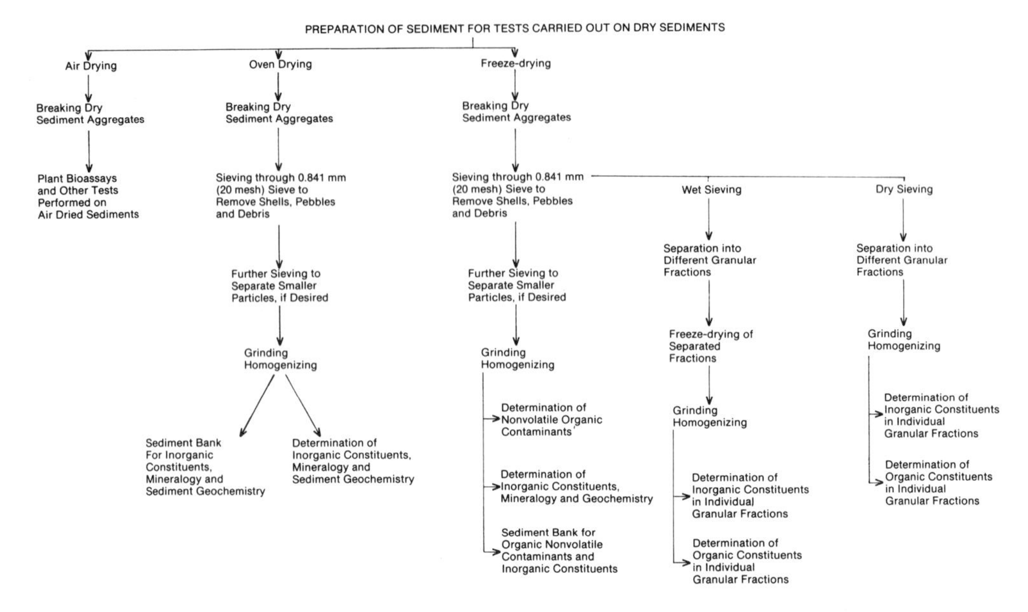

FIGURE 8. Handling samples for tests and analyses on dry sediments.

properties. For example, changes in metal availability and complexation were shown for samples that were air-dried.[62] In some cases, air-drying has been used to avoid losses of components, such as mercury, which are volatile at temperatures above 50 to 60°C.

The difficulty is to achieve thorough drying to constant weight as specified by the ASTM method D421-58.[63] MacKnight (personal communication) has tested the air-drying of marine sediments. Organic matter-rich, fine-grained sediments typical of harbors were found to take 3 to 5 days of air-drying to achieve constant weight. A typical drying period of 24 h was found to remove only 40 to 60% of the water content. Slow air-drying was attributed to hygroscopic salts and organic material in the samples. Drying of sediments from areas adjacent to pulp mills was found to be even more difficult due to wood particles and other organic material in the samples.

Sediments collected for plant bioassays[64-66] are usually partially air-dried at a temperature of about 20 to 40°C and relative humidity of 20 to 60%. Depending on the quantity of the material, air-drying is carried out in a fume hood (small samples), air-drying cabinet with air circulation, or in sheltered ventilated rooms (large samples). However, because of the possibility of air contamination of sediment samples by dust, air-drying is not recommended for accurate determination of inorganic elements and organic compounds in sediments. Similarly, because of biological activity during the drying period, air-drying is not recommended for sediment samples to be used for biological tests. For chemical analyses where preservation is required as well as drying, this method is not suitable because microbial degradation, oxidation, and other processes, which can alter the sample, are not halted.

b. Oven-Drying

Oven-drying of sediments is usually carried out on samples collected for the determination of inorganic components, such as major and trace elements. However, oven-drying is not suitable for grain size determination, since wet fine-grained sediments become hard-to-break aggregates.[61] Oven-drying is not acceptable for sediments which contain any volatile or oxidizable components, whether they be organic or inorganic,[1] and may contribute to the alteration of even nonvolatile organics. For instance, unsaturation can be created by dehydration of aliphatic alcohols, and heat, in the presence of clay catalysts, can promote pyrolytic reactions.

Geological materials in general, and sediments in particular, on heating at 100 to 110°C in an oven, release most of their hygroscopic water, which is water held by surface forces such as adsorption and capillarity. The amount of hygroscopic water is related to the physical properties and mineralogical composition of sediments. Sediment interstitial water is also evaporated during the drying procedure. ASTM method D22-63T[67] covers the laboratory determination of the moisture content of soil at a temperature of 110 ± 5°C. This is similar to the method used in geology[68] and can be used for the determination of moisture in sediments. Lower heating temperatures (less than 60°C) are essential when preparing sediment for the determination of volatile trace elements, such as mercury.[40] The effects of different drying temperatures on the determination of mercury in sediments were reviewed by de Groot and Zschuppe.[43] They concluded that preferably mercury should be determined in sediments that have either been air-dried or oven-dried at 40°C. An alternative drying procedure is freeze-drying for these analyses (see below).

Generally, containers used for drying should be made of material resistant to corrosion and not subject to change in weight or disintegration on repeated heating and cooling. Crucibles, dishes, and trays made of aluminum, nickel, glass, and porcelain are recommended for drying sediments in ovens. The selection of material also depends on subsequent analyses with much the same considerations that apply to the containers described above. Laboratory drying ovens are available from companies supplying laboratory equipment in a variety of models with capacity from 20 to 1600 l with a temperature range from 40 to 250°C, air circulation, and accurate control of temperature.

c. Freeze-Drying

In the freeze-drying process water in the frozen or solid state is sublimated and is removed from the material as a vapor. Freeze-drying (also called lyophilization) can be used for drying sediments collected for the determination of most organic pollutants as well as for analyses of inorganic components, such as the major and trace elements. Certain organic components are more susceptible to volatilization. For example, loss of lighter chlorobenzenes (di- and trichlorobenzene) was observed during testing of the effects of freeze-drying on integrity of sediment samples from the Great Lakes.[69] Some inorganic constituents, such as mercury and iodine, can also be lost.[70,71] LaFleur[72] was able to retain 95 to 100% of mercury as methyl- or phenylmercury salts, and inorganic mercury when freeze-drying biological tissues. de Groot and Zschuppe[43] reviewed reports on losses of mercury by freeze-drying of sediments. They found that no detectable loss of mercury following freeze-drying was reported by some authors, but one investigator reported that results from freeze-dried samples were 23% lower than those from air-dried samples.

The principal advantages of freeze-drying for sediments are

- Low temperatures avoid chemical changes in labile components
- Loss of volatile constituents, including certain organic compounds, is minimized[69]

- Most particles of dried sediments remain dispersed
- Aggregation of the particles is minimized
- Sterility is maintained
- Oxidation of various minerals or organic compounds is minimized or eliminated

The price of freeze-drying equipment together with special bottles and vials as well as the maintenance cost are significantly higher than the price and maintenance cost of drying ovens. At National Water Research Institute Environment Canada, the procedures for freeze-drying can vary depending on which units are available and the type of analyses to be performed on freeze-dried samples. Several different commercially available freeze-driers are operated according to recommendations which appear in the manufacturers' manuals.

Prefreezing of samples is common when freeze-drying is anticipated. This is done mainly to avoid "bumping" or spattering of the sample when evacuating the freeze-drier chamber, and has the added advantage of offering intermittent preservation. LaFleur[72] compared room temperature storage, prefreezing at –20°C, and at –196°C (liquid nitrogen), prior to freeze-drying muscle and liver tissue. No difference was found among these methods for retention of methylmercury chloride. Alternatively, unfrozen sediment samples can be loaded directly into the freeze-drier in suitable containers. Many commercial units such as that in Figure 9 have the capability of a freezing cycle prior to the drying cycle.

The essential steps in freeze-drying of sediments in most commercial freeze-driers are

- Replace the caps on vials, jars, or bottles containing samples with filter paper and special tops. If samples are freeze-dried in plastic bags, open the bags and cover the top with a filter paper or a tissue which would not contaminate the sample. Secure the filter paper or the tissue on the top of the containers by masking tape or rubber bands. The filter paper allows water to escape while retaining the sediment particles during freeze-drying and particularly during releasing of the vacuum in the drying chamber. Special freeze-dry flasks are commercially available from companies supplying freeze-drying equipment. These flasks are attached to the valves mounted on the ports outside the drying chamber by different adaptors. Samples collected in small containers can be placed in the special freeze-dry flask. Filter papers are also available which fit into the top of these flasks to prevent loss or contamination of freeze-dried samples.
- Freeze the samples for approximately 18 h at –20°C, directly in the freeze-drier on trays in the drying chamber. Omit this step when using prefrozen samples.
- Turn on the condenser on the freeze-drier and cool to –40 to –50°C. When cold enough apply vacuum using the high vacuum pump required by most units. The normal operating range is 0.010 to 0.050 torr depending upon the surface area available for sublimation of water. In some units the trays in the drying chamber can be heated to speed up the process. Such heating should be turned on only after maximum available vacuum is reached.
- The time required for complete drying in freeze-driers ranges between 24 h and 14 days, depending on sample quantity, surface area, sediment type,

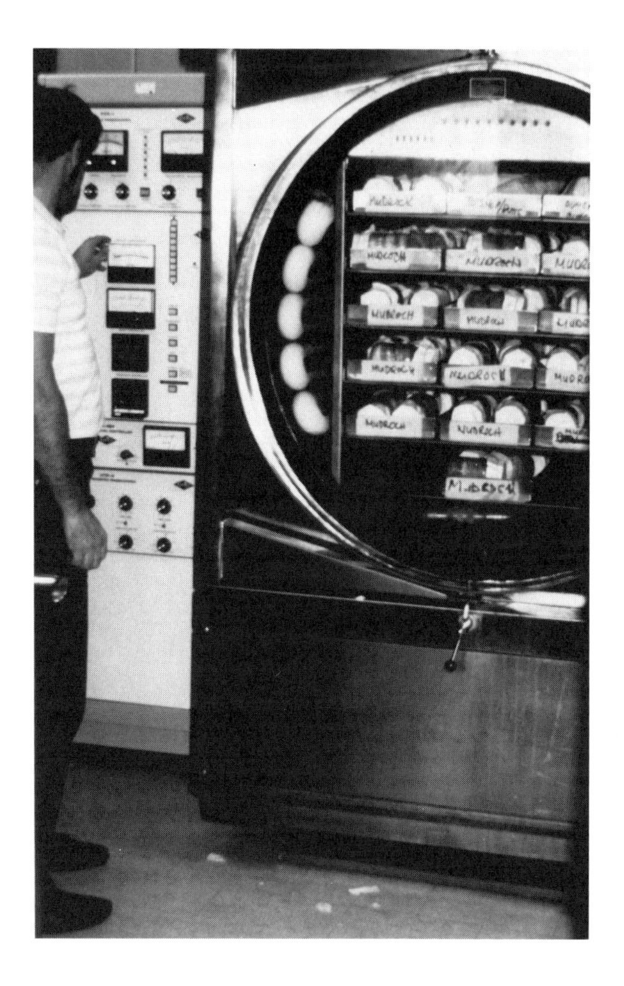

FIGURE 9. Freeze-dryer.

and water content. For example, a 100-g fine-grained sediment sample needs about 3 days, and a 500-g sample of similar consistency about 7 days of freeze-drying. The capacity of a freeze-drying unit is usually expressed as the mass of water that can be frozen onto the condenser. The rate of drying levels off as ice builds up on the condenser. A good rule of thumb is to load the unit to about 80% of the condenser capacity.

- Check the condition of the samples after the estimated drying time:

  - Slowly release vacuum and remove containers from the drying chamber.
  - Remove special tops and filter paper.
  - With a spatula or knife examine the entire sample to see if it is dry, paying special attention to the center portion of the sample, which usually needs the longest time to dry.
  - If the samples are at ambient temperature and a cold spot can be felt, especially on the bottom of the container, then ice remains.
  - If a portion of the samples is still wet, continue freeze-drying.

- When the samples are dry, mount the original caps and lids on the containers and store the samples for further processing.

- After the dry samples are removed from the drying chamber, check if the freeze-drier needs cleaning and defrosting, and control the level and quality of oil in the vacuum pump. The maintenance of the equipment should be carried out following the instructions in the manual supplied by the company for each unit.

2. Sieving

Sieving is an effective and economical process for dividing sediment samples into different fractions containing particles of more or less the same size, i.e., passing an aperture of certain dimensions and failing to pass through some smaller aperture.

It is common practice for soil scientists to work soil samples, which are very similar to sediments, through a 2-mm round hole sieve (10/20 mesh) using a rubber stopper or a rubber pestle. For routine soil testing, the fraction coarser than 2 mm is discarded. In research studies, the material retained on the 2-mm round hole sieve is examined, described, and then either discarded or, if of interest, preserved, dried, weighed, and analyzed.

If sieving of sediment samples is to be carried out, due consideration should be given to the analyses and tests planned, as discussed in the introduction to this chapter. There are three possible scenarios:

1. The whole bulk sample is ground with no sieving, or alternate grinding and sieving in a disk mill, and analyses are carried out on an homogeneous subsample.
2. The whole sediment sample is passed through a sieve of the desired mesh size, e.g., 2 mm, 63 μm (10, 250 mesh). The oversize portion is discarded while the portion which passes is used for analyses.
3. The whole sediment sample is separated into specific size fractions by dry or wet sieving and elutriation. The size fractions which result are analyzed individually in their original state or ground to an appropriate grain size for analyses. The composition of the sediment can vary considerably for different particle sizes; therefore, thorough mixing of the material after sieving is essential.

Sediments are usually classified simply as gravel (>2 mm), sand (63 μm < 2 mm), silt (2 < 63 μm) and clay (<2 μm). These definitions are based on arbitrary cuts of median diameter between clay and silt, silt and sand, and sand and gravel; there are several different classification systems.[61] Particle size can be expressed on a millimeter scale (mm), micrometer scale (μm), or phi scale (ϕ). The mathematical definition of the latter is:

$$\text{phi}(\phi) = -\log_2 (d)$$

where d = particle size diameter in mm.

Classification of particles of a different size together with designated sieves for separating is given in Table 1.

a. Wet Sieving

Wet sieving is particularly useful for the processing of fine-grained sediments. Wet sieving of a small quantity of sediment is generally carried out manually and requires one or more sieves and two types of containers, such as buckets, dishes, beakers, bowls, and jars. When a dry sediment is to be wet sieved, the sample is placed in a container, usually a metal or plastic bucket or large dish, clean water is added to cover the sample, and a minimum of 2 h is allowed for soaking. A portion, which can also be weighed, of soaked sample is then placed on the standard sieve with the appropriate opening, i.e., 0.5 mm (35 mesh). The sieve is placed in a dish filled with water and gently swirled so that the particles smaller than the selected sieve size are washed through the sieve into the dish. In a method described by ASTM[73] all material in the container is transferred to the sieve and washed with the smallest volume of running water. To facilitate passing through the sieve a nylon brush can be used. Particles retained in the sieve (the coarse fraction) are examined and, if they are of interest to the study, retained and redried. The sediment washed through the sieve (the fines) is redried and weighed. After use, all sieves should be thoroughly cleaned under running water, with special care to remove the material caught in the screen.

b. Dry Sieving

Dry sieving is used for the separation of coarse samples into several size fractions, and quite often to separate coarse from fine fractions. However, when sediments contain very fine particles, particularly clays, fine sieves become clogged, thus impeding proper sieving of the sediment. In such cases wet sieving is employed. A combination of dry and wet sieving is often effective and is recommended: dry sieving for coarser particles and wet sieving for particles passing easily through the finest sieve. Material smaller than the finest sieve, usually 44 and 37 μm (325/400 mesh), is determined by difference, although these fines can be recovered and weighed.

Hand sieving of dry material is common. The sieve is placed on a pan, an appropriate quantity of the sediment is placed on the sieve, the sieve is covered with a lid, and it is shaken with a rotating intermittent tapping action until separation is complete. After use, all sieves should be cleaned either with a soft paint brush or under running water and air- or oven-dried. Care must be taken to avoid contamination by the material of the sieve. If trace metals are to be determined, plastic woven sieves should be used. After use the plastic sieve is either discarded or cleaned and reused.

c. Types of Sieves

U.S. standard sieves consist of a set of fine and coarse 20-cm- (8-in.) diameter sieves made in accordance with specifications outlined in ASTM E-11-61,[74] approved U.S. standard Z23.1, AASHO M92, and federal specifications RR-S-3368. The U.S. standard sieve designations (in mm and μm) correspond to test sieve aperture values recommended by the International Standards Organization. The U.S. series alternate sieve designations (by number) are the approximate number of openings per linear inch.

Sieves are circular frames, made of stainless steel or brass, the standard size being 20 cm in diameter, either 5 or 2.5 cm high, with a wire cloth carefully soldered to the frame. Sieve covers and receivers (catch pans) made of brass or stainless steel are essential parts of any sieving equipment. For special purposes, both large and small diameter frames, covers, and catch pans are commercially available.

TABLE 1
Wentworth Size Classes, Grain Size Scale and Sieve Numbers

Wentworth size class	Phi	Metric		U.S. standard sieves			
		mm	µm	mm	µm	No.	in.
Cobble gravel	–8	256					
	–6	64		64			2 1/2
Pebble gravel	–5	32		32			1 1/4
				25			1
				19			3/4
	–4	16		16			5/8
				12.5			1/2
				9.5			3/8
	–3	8		8			5/16
				6.3			1/4
				5.6		3 1/2	
				4.75		4	
	–2	4		4		5	
Granule gravel				3.35		6	
				2.80		7	
	–1	2		2		10	
Very coarse sand				1.70		12	
				1.40		14	
				1.18		16	
Coarse sand	0	1		1		18	
					850	20	
					710	25	
					600	30	
	1	0.50			500	35	
Medium sand					425	40	
					355	45	
					300	50	
	2	0.25			250	60	
Fine sand					212	70	
					180	80	
					150	100	
	3	0.125			125	120	
Very fine sand					106	140	
					90	170	
					75	200	
	4	0.063			63	230	
Coarse silt					53	270	
					45	325	
					38	400	
	5		31				
Medium silt	6		15.6				
Fine silt	7		7.8				
Very fine silt	8		3.9				
Clay	9		2.0				
	10		0.98				
	11		0.49				
	12		0.24				
	13		0.12				
	14		0.06				

Available screening surfaces are woven wire cloth, plastic woven polymer filter screens, punched plate, and bar screens. However, only the first two screen surfaces are suitable for laboratory sediment handling. The opening, wire diameter, and open area are to be carefully considered when selecting a screening machine and sieves. Woven wire cloth has by far the greatest selection as to screen openings, from 100 mm to 37 μm (4 in. to 400 mesh), wire diameter (6.3 mm to 37 μm), and percentage of open area. Woven wire screens are made mainly of stainless steel and brass, but can be made from other metals and alloys when required.

Synthetic polymer woven screens are available in four different materials: polyethylene, polypropylene, nylon, and polytetrafluoroethylene (Teflon or PTFE), with a range of sieve openings from 1 μm to 1.24 mm. They are ideal for screening solutions or suspensions containing particles to be analyzed for trace elements. For small volumes of sediment, a polyethylene microsieve set 51 mm I.D. × 29 mm supplied with all components needed to build a stack of four sieves is handy.

A number of laboratory electromagnetic and mechanical sieve shakers are on the market that can automatically carry out dry or wet sieving with accuracy and reliability (see Figure 10). They are designed to hold and handle from 1 to 13 standard 200-mm-diameter sieves with different openings. Any laboratory planning to purchase a sieve shaker should consider these factors: ease of operation, noise level, space requirement, number of sieves shaken at one time, and suitability for dry and wet sieving.

Screening for industrial utilization of sediments is outside the scope of this book and the reader is referred to reference books: *Chemical Engineers' Handbook*[75] and *Handbook of Mineral Dressing.*[76]

3. Grinding

Geological samples — rocks from about 3 to 15 cm containing minerals of different hardness — are reduced, in three steps, to powder prior to analysis: crushing, pulverizing (180 to 150 μm or 80/100 mesh), and fine grinding (150 to 38 μm or 100/325 mesh). In contrast, sediment samples commonly require only fine grinding. The required final size is usually between 149 and 44 μm (100/325 mesh). Grinding of geological samples and sediments is a batch operation. In the process of reducing bulk geological and sediment samples to powder, frequent sieving of the ground material during the course of grinding removes the finer fractions and speeds up the grinding process, and also ensures that the bulk of the powder will be of the desired size. No material is discarded.

In general, hard materials (e.g., quartz), coarse particles, and fast motion are conducive to wear or abrasion in mills. This can cause contamination of the samples, particularly with trace metals. To obtain true concentrations the sediment samples are split. One split is ground in an agate or ceramic dish, and the other is ground in a metal dish in a disk mill. Both splits are then analyzed for trace elements of interest, the results compared, and the lower values selected as true concentrations. However, if samples are to be used for organic analyses, the metals present in the chamber and disks do not pose any contamination problem.

a. Grinding Equipment

The choice of equipment generally depends on the quantity of sediment to be ground, hardness of the particular mineral particles, and contamination considerations. Since the sediments are collected wet and if the presence of water is not objectionable, wet grinding can be applied with advantage. In fine dry pulverizing or disintegration steps, surface forces come into action to

FIGURE 10. Mechanical shakers for sediment sieving.

cause flocculation. An intermediate product in size between 425 and 150 μm (40/100 mesh) is made by removing fines and screening the coarse material. Grinding of any sample generates grains and particles of different sizes. Moreover, under the same conditions, hard minerals (e.g., quartz, garnets, amphiboles) disintegrate less than soft minerals and rocks (e.g., talc, clays, limestone).

Alternate grinding and sieving is an efficient method to obtain particle uniformity. For example, in the treatment of freeze-dried Great Lakes sediments for archive storage, grinding and sieving were alternated until more than 90% of the sample passed through a sieve opening of 250 μm (60 mesh).[69] However, for analyses requiring the grains as small as possible (e.g., X-ray fluorescence analyses) the stored sediment will be subjected to another finer grinding. Sediment reference materials from the National Research Council of Canada (NRCC)[77] are sieved and/or ground to pass through a 125-μm (120 mesh) sieve.

b. Mortars and Pestles

A mortar and pestle is an important and indispensable tool in the preparation of sediment samples for testing and analyses. Mortars and pestles of suitable size are commonly used manually, or operated mechanically, to grind small samples to the desired particle size. Manual grinding is time consuming and, consequently, relatively costly. If a homogeneous sample is required, alternate grinding and sieving using small sieves about 50 mm in diameter is essential. The abrasion of the mortars and pestles is greater in mechanical grinding than in manual grinding. Users should take the abrasion into account and assess the potential for contamination from the equipment, particularly with respect to the accuracy of elemental analyses.

Mortars and pestles, in sizes from 35 to 200 mm, are made of various materials: agate, alumina, steel, glass, and porcelain. Agate is a waxy variety of cryptocrystalline quartz with submicroscopic pores. Aluminum oxide mortars and pestles are more resistant to abrasion and less porous than agate. Diamonite mortars are made of finely powdered synthetic sapphire molded under high pressure and sintered at high temperature. Mineral sapphire is blue corundum, nearly pure aluminum oxide with minor amounts of iron and titanium. Porcelain and glass mortars and pestles are considered unsuitable for grinding sediments to be analyzed for trace elements because of contamination. Porcelain mortars and pestles, as well as rubber tipped pestle and mortars, can be used for breaking up aggregates formed by air- or oven-drying sediments containing clays prior to sieving or mechanical grinding. Alternating grinding and sieving is usually effective for an adequate disaggregation of the sample and thorough homogenization.

c. Grinders

There are several commercially available grinders able to reduce sediment samples to 100-μm (150 mesh) and smaller grains. Ball and disk mills are very effective for disintegration, but there is less control over the final particle size. The disk mill is a high speed disintegration apparatus that breaks agglomerates of various minerals and produces a blend of particles in sizes between 212 and 75 μm (70/200 mesh). For example, a container lined with tungsten carbide or agate, a ring, a solid grinding stone made of the same material, and a cover are used in the swing mill, produced by Siebtechnik, Germany (Figure 11). Shaterbox, a vibratory disk mill, such as that produced by Spex Industries, Michigan, is a similar machine, commonly used in geological and environmental laboratories for fine grinding of various materials, including sediments.

Hammer type mills are designed for capacities from 2 to several thousand kg/h. Small hammer type mills are suitable for the reduction of small quantities of fragments which are <6 mm in size to various degrees of fineness. The mill features shaft-carrying swing hammers, pivoting on a disk which rotates at high speed. Gravity and suction feeds material to the mill through a spout in the cover. The pulverized particles are forced through a grate into a collection receptacle. The required grain size is achieved by changing the screens.

Ball and pebble mills have a steel- or stone-lined cylindrical steel shell, rotating on a horizontal axis. They contain a charge of steel balls or stone pebbles and the sediment to be reduced. Size reduction is effected by the tumbling of the balls or pebbles on the material between them. The size reduction can be obtained from 3.5 mm (10 mesh) to 45 μm (325 mesh). The laboratory ball mill is a mechanically rotated steel or ceramic container filled with ceramic balls and the sediment sample to about one third of its volume. It produces a fine powder and mixes thoroughly, but cleaning is inconvenient.

Large quantities of sediment (up to 300 l wet sediment) collected for large scale tests are usually air-dried and ground using special grinders, such as the Kelly Duplex grinder (Duplex Mill and Manufacturing Company, Springfield, OH). This equipment was used for grinding air-dried sediments to pass a 2-mm screen intended for plant bioassays.[64]

4. Mixing and Homogenization

The final goal in processing sediment samples is to produce an homogeneous sample that would yield precise results in replicate determinations of inorganic

FIGURE 11. Grinder with a grinding dish.

or organic components. Aside from the actual representativeness of the method used for the original sample collection, the degree of success attained in sample homogenization and splitting is largely responsible for the variability in analytical results. Grant and Pelton[78] treat the subject of homogeneity and sampling of solids in a theoretical and statistical manner and the reader is referred to that paper for methods for the prediction of homogeneity from theoretical considerations. The discussion here will cover the practical aspects and illustrate the success of homogenization with examples.

When one considers the causes of inhomogeneity, the unit which determines segregation becomes very important. The goal in creating an homogeneous sample is to make this unit as small as possible. As an example, a soil or sediment sample which is poorly sorted may contain relatively large sand grains which are made up of dense minerals mixed with fine clay particles. If this were a dry sample and it was mixed without any treatment, the finer grains would tend to collect in the lower portions of the mixing container. If this were a wet sample the opposite could occur; the heavier and larger grains would tend to settle first leaving the finer particles enriched in the upper portions of the container. The unit of segregation in this example is the large size and greater density of the sand grains. Treatment, which would be appropriate in this case, might be to even off the grain size distribution of the sample by grinding.

Mixing of untreated or ground sediment is required for homogenization and can be carried out by a variety of simple operations:

- Coning and quartering
- Turning the sample over and over with a spatula
- Rolling the sediment sample spread on a sheet of paper, plastic, foil, or cloth (see example below)

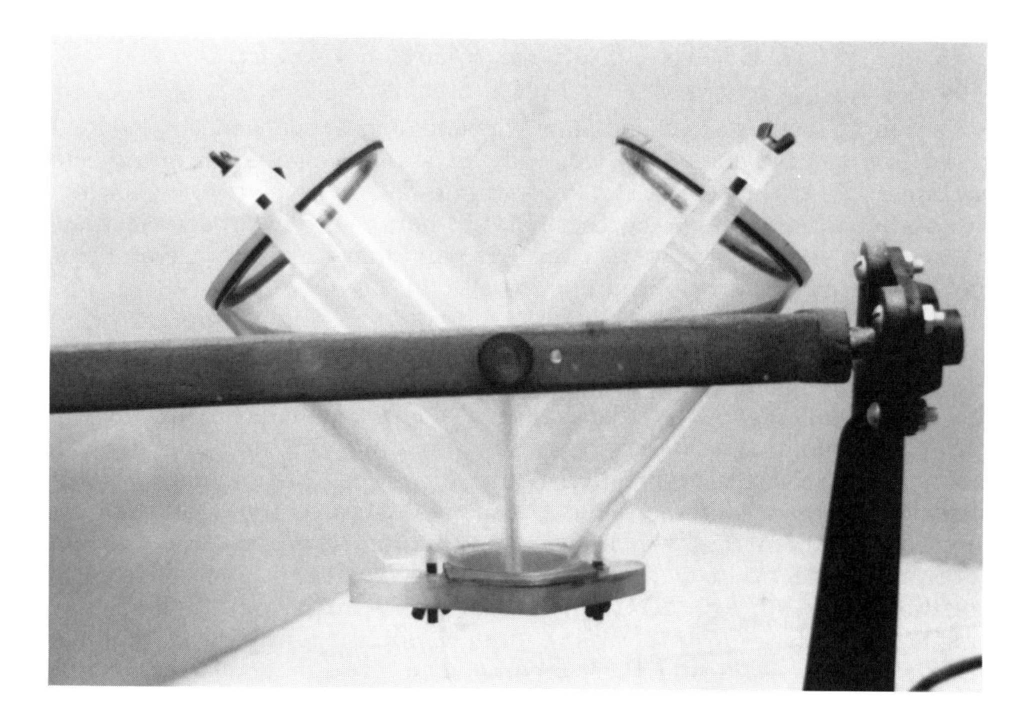

FIGURE 12. Sediment mixer.

- Using mechanical rotating mixers consisting of boxes of various shapes, e.g., V-shaped, cones, rectangles, cylinders, or cubes on a diagonal (see Figure 12)
- Using a Jones riffle splitter for mixing and splitting finely ground material (the sample is poured through the splitter and divided in half; halves are alternately resplit until the desired homogeneity is obtained)

The following method is modified from one used by the Geological Survey of Canada for the preparation of geological samples for analyses. It is also useful for the mixing of sediment samples weighing from 50 g to several kilograms:

- Place the sediment sample, generally of uniform particle size, on a sheet of glazed paper, plastic, or rubber mat.
- Roll the material from one corner of the sheet to the opposite corner by raising one corner and causing the material to tumble over upon itself; repeat the process by raising each corner in succession until thorough mixing of the sample has been achieved (usually between five and ten times for each corner).
- If the original quantity of the sediment sample needs to be reduced to a particular quantity, such as 5 to 20 g, use the coning and quartering technique. Place the material in a conical shape in the center of the sheet, spread it out into a circular cake, divide it into four quarters, and remove opposite quarters; repeat mixing, coning, and quartering and remove opposite quarters. A quarter tool can be used for separating the cone into four quarters. Continue, until the desired quantity is attained.
- Transfer the final sample to a labeled plastic or glass vial, seal it with a tight cap, and store it until needed.

IV. STORAGE OF SEDIMENT SAMPLES

The storage conditions of sediments depend on practical aspects and limits as well as the previous knowledge of the stability of known or expected contaminants in the sediments. Preservation and storage are two aspects of sample handling which go together hand-in-hand. For our dredging example the sections on preservation (Section IV.A) and temporary storage and shipment (Section IV.B) are appropriate.

A. PRESERVATION

Preservation techniques are usually intended to retard microbial degradation, oxidation, and/or loss of volatile components. Methods are limited to pH control, poisoning, drying, refrigeration, freezing, and isolation from the atmosphere. No single preservation method is applicable to all constituents, so often it is necessary to take replicate samples, or subsamples, and preserve them by different methods when a variety of parameters are required. Selection of the most appropriate methods should be based on the purpose of the study and the components to be determined.[79] The requirement for multiple samples, or splits of single samples increases handling and the time between sampling and preservation — both of these should be kept to a minimum.

Two parameters are important to preserve the integrity of samples: temperature and storage time. Temperature is an important factor that can variously affect the collected sediment samples from the time of recovery of the sample through handling and processing to the final analyses. Sediment samples intended for analyses or experiments after air- or oven-drying can be stored in containers, cans, plastic bags, etc. at ambient or room temperature. However, sediments collected for the determination of organic contaminants as well as mercury should be stored in a refrigerator (about 4°C). The higher the temperature, the higher is the risk of losses or changes of volatile compounds. Timing of collection, shipment to the laboratory, and analytical work is very important and should be discussed thoroughly during project planning.

B. TEMPORARY SAMPLE STORAGE AND SHIPMENT

Shipment of collected samples has to be planned, specified, and defined prior to any sediment sampling. It should be a general procedure to ship the containers with the samples, in coolers filled with ice cubes, as soon as possible for processing, preservation, and analyses. Coolers are available in various sizes with volumes ranging from 5 to 60 l and can be obtained from various sources, such as scientific supply companies, hardware or department stores, etc. The personnel in charge of the sediment sampling project has to decide, in advance, about the number and size of the required coolers with respect to the number of samples to be collected and the number and size of the containers to be used, as well as the availability of ice cubes, blocks, or dry ice in the field. Storage of samples in coolers is useful for insulation during sampling in winter when freezing of the samples has to be avoided (Figure 13).

Because of operational considerations, it may not be possible to store samples in the field in the same way that they will be ultimately stored. A temporary storage method may have to be adopted. One should strive to compromise the samples as little as possible and implement the best storage conditions as soon as possible. For instance, Bourbonniere et al.[69] stored samples on shipboard

FIGURE 13. Collecting samples into portable cooler.

frozen in solvent-cleaned glass trays at –20°C immediately after sampling from a box core and maintained them at that temperature until freeze-dried. In this example, the temporary storage method (freezing) was also a necessary prerequisite for the ultimate storage method (freeze-drying).

In cases where samples must be frozen, dry ice is a good choice. When the receiving laboratory is not ready to process the samples, these must be stored in refrigerators or freezers. In such cases, the holding time is exceeded with unknown consequences.

The ASTM[80] defines holding time as "the period of time during which a water sample can be stored after collection and preservation without significantly affecting the accuracy of analyses." In the lack of any guidelines and scientific data dealing with the integrity of sediment samples between collection and analyses or other testing, this definition of the holding time for water samples can be applied to define the holding time for sediment samples.

Samples collected for bioassays should be refrigerated and transported or stored in coolers filled with ice. In cases where these samples are also to be used for chemical analyses, they should be collected and stored in the appropriate containers, as discussed previously in this chapter. While freezing is considered appropriate for sediment samples collected for chemical analyses, it is not recommended for toxicity tests[81] because it can affect the toxicity of sediments.

Biological tests should be conducted as soon as possible, and the 2- to 7-day time frame has been recommended.[38,56-59] However, Tatem[82] showed that some contaminated sediments can be stored at 4°C for up to 12 months without significant changes in toxicity. If, for any reason, sediments collected for bioassays have to be stored for a long time, an easy, short duration toxicity assay should be carried out periodically to determine to what degree change has occurred.

In some studies, sterilization was used to inhibit biological activity in collected sediments. This was done by autoclaving[3,83] additions of antibiotics[57,84] or the addition of chemical inhibitors such as formaline or sodium azide.[3,85]

If the vessel used for sediment collection has a freezer and/or refrigerator, the sediment samples are placed in appropriate containers which are immediately transferred to the freezer or a refrigerator. The transportation of samples from the freezers and refrigerators on the vessel to a permanent storage room or laboratory is usually in coolers filled with ice. After transporting the samples from the field, storage in refrigerators (usually at 4°C) and/or freezers (usually at –20°C) is essential to preserve the integrity of the collected sediments. Refrigerators and freezers are available from various sources, such as scientific supply companies and department stores. General purpose laboratory or kitchen upright freezers designed for the storage of pharmaceuticals and chemicals as well as food, which requires a freezing temperature in the range –12 to –30°C, equipped with adjustable epoxy-coated steel shelves, are available in various sizes with volumes from 150 to 750 l. General purpose upright refrigerators with a temperature range 0 to 14°C, usually set at 4°C by the manufacturer, are available in various sizes with firm or adjustable shelves. For a small laboratory, a freezer/refrigerator can be suitable storage equipment for two sets of sediment samples requiring different storage temperatures, namely 4 and –20°C. Laboratories involved in a large number of analyses of sediments and/or other materials may have an enclosed, separate room for storing samples at room temperature, and walk-in refrigerators and freezers, such as the walk-in refrigerator in Figure 14.

Typically, after being analyzed or tested, the samples in the original, clean containers are either immediately discarded or are kept until the analyst or the customer is satisfied with the results. Simultaneously, the laboratory decides the fate of the containers, either to clean them for further use or, if they are heavily contaminated or disposable, discard them.

C. ARCHIVAL STORAGE

A special case is the permanent storage of sediment samples in sediment archives from which they are issued only when required for different projects.

A number of environmental specimen banks were established to store frozen materials at –40, –80, and –196°C, particularly the U.S. National Environmental Specimen Bank,[86,87] the environmental specimen bank program of the Federal Republic of Germany,[88,89] the Canadian Wildlife Service National Specimen Bank,[90,91] and the Great Lakes Biological Tissue Archive.[92] The Great Lakes Sediment Bank, stored at NWRI,[69] contains sediment samples collected with a box corer from a vessel, placed in solvent cleaned glass trays, frozen on shipboard at –20°C, and freeze-dried in the laboratory. Drying times were kept as short as possible. Dry samples were stored at room temperature.

FIGURE 14. Sediment cores stored in walk-in refrigerator.

V. EXAMPLES

This section contains descriptions of techniques which have been actually used by different workers for handling and preparation of sediment samples collected for various purposes.

A. PREPARATION OF MARINE SEDIMENT REFERENCE SAMPLES HS-1, HS-2, AND CS-1 BY NATIONAL RESEARCH COUNCIL, CANADA (NRCC)[77]

Sediments were collected using a 0.25-m^2 Van Veen grab sampler, stored wet, and unprocessed in sealed 185-l polypropylene barrels. The material was freeze-dried and later tumbled in a modified cement mixer to break the clumps formed during the drying. Materials retained on the screen were discarded. HS-1 and HS-2 were sieved through a 100 mesh stainless steel screen. CS-1 was sufficiently fine that sieving was not necessary. The sieved sediments were returned to the cement mixer and homogenized. Tests for homogeneity were carried out while blending proceeded. The sediments were packed into quart-sized solvent-rinsed steel cans and numbered sequentially in order of packing. A series of cans were selected at random for final tests of homogeneity and quantitative determination of the contaminant of interest, in this case PCBs.

B. SAMPLE PREPARATION AND HOMOGENEITY TEST OF LARGE QUANTITIES OF WET AND DRY SEDIMENT REFERENCE MATERIALS

Chau and Lee[93] described the preparation of large quantities of wet and dried lake sediments to be used as analytical reference materials for long-term PCB quality control studies. The procedures for the preparation are described as informative examples.

1. Wet Sediment Reference Material

The bulk sediment used to prepare the wet sediment reference material was collected in western Lake Ontario and stored at 4°C in covered galvanized garbage pails cleaned with detergent and organic solvents before use. Because the sediment was free of larger rock pieces, no sieving was necessary. About 12 kg of water-saturated sediment was transferred to a cutter/mixer. After adding 3 l of pure water and 150 g of concentrated HCl to the sediment, the material was blended at full speed for 12 min to obtain a smooth slurry. To prevent settling, the sediment operations had to be done quickly and carefully. The slurry was poured into 1-l beakers and from these beakers about 15-ml slurry subsamples were immediately poured into 30-ml bottles with plastic snap caps. The bottles were numbered sequentially according to the pouring sequence. The tightly capped sample bottles were stored in the freezer at –20°C ready for the determination of PCB.

2. Dry Sediment Reference Material

The sediment (about 450 kg) for the dry reference material was collected in Hamilton Bay, Lake Ontario, and was stored at 4°C in covered galvanized garbage pails cleaned with detergent and organic solvents before use. It was frozen in the same pails at –20°C for 1 week. Numerous 5-mm diameter holes were drilled on the sides of the pails and the sediment was allowed to thaw for 3 days. During this time, excess water drained from the perforations. The partially dried sediment was transferred in 20-kg lots to a commercial freeze-drying chamber and dried at a reduced pressure and elevated temperature.

After drying, the sediment was in the form of small aggregates and weighed 205 kg. The aggregated sediment was crushed in a Denver roller and then passed through a 125-μm (120 mesh) vibrating screen. The oversized fraction was set aside. The remainder was passed through a 45-μm (325 mesh) vibrating screen and the fraction passing through was collected. The oversize fraction was ground in a ball mill for 1.5 h and the material was passed through the 45-μm screen again. The oversize fraction was rejected at this stage. The combined 45- μm sediment fractions were tumbled in one lot for approximately 8 h in a 570-l conical steel shell blender. At the end of tumbling, six 50-g samples were removed from the top, middle, and bottom level of the bulk sediment for homogeneity testing. After the homogeneity of the bulk sample was established, the sediment was blended for another 5 h before manual bottling. To avoid settling, the sediment was blended for 1 h before the commencement of bottling 25-g sediment samples into 100-ml brown bottles equipped with plastic screw caps. The shiny side of an 8 × 8 cm aluminum foil liner, prewashed with petroleum ether or ethyl ether, was inserted under the cap before making a tight seal of the bottle to avoid moisture absorption. Homogeneity testing was carried out during bottling. All bottles were placed in cardboard boxes, sealed with thick plastic bags, and stored at –20°C ready for use.

Cheam and Chau[94] described an homogeneity testing protocol which they used to describe the success in homogenizing certified sediment reference materials from Great Lakes sources. The technique utilizes a two-way analysis of variance to simultaneously test the effects of variation between subsamples and within subsamples. They adopted a significance level of 95% for the F statistic to demonstrate that mean values for between and within

TABLE 2
As, Se, and Hg Data from Homogenized Bulk Sediment (Reference Materials Prepared from Great Lakes Sites)[a]

Sample set	N	Mean conc.[b]	Standard deviation	RSD[c]
Arsenic				
WQB-1	48	22.71	±0.53	±2.33
WQB-3	30	19.1	±0.80	±4.19
Selenium				
WQB-1	48	1.07	±0.05	±4.67
WQB-3	30	1.25	±0.04	±3.20
Mercury				
WQB-1	30	1.08	±0.05	±4.63
WQB-1	48	1.08	±0.03	±2.78
WQB-3	30	2.95	±0.11	±3.73

[a] WQB-1 sediment data from Reference 30; WQB-3 sediment data from Reference 32.
[b] Concentrations in μg/g.
[c] Relative standard deviation expressed as % of mean.

subsamples are not different, thus the bulk sample is considered homogeneous. The procedure has been demonstrated for determinations of As, Se, and Hg.[95]

The success of homogenization in the 75- to 45-μm (200/325 mesh) range is indicated by the data in Table 2. These data for As, Se, and Hg are from freeze-dried Great Lakes sediment reference materials described by Cheam and Chau[94] and Cheam et al.[95] The authors have shown that these samples are homogeneous for these elements when tested at the 95% significance level according to a two-way analysis of variance. A relative standard deviation (RSD) criterion of ±5% of the mean was chosen as an homogeneity criterion for single parameter-single method testing. The WQB-3 sediment was shown to be homogeneous for Mn, Fe, Co, Ni, and Zn, as well as for As, Se, and Hg.[95]

C. PREPARATION OF IN-HOUSE STANDARD SEDIMENT SAMPLES

Because of the high price of commercially available standard reference materials, it is advantageous for laboratories involved in performing many analyses of sediments to prepare an in-house set of reference materials for different analyses. A sediment sample, collected at a selected location in a freshwater or marine system, usually in quantities of 20 to 100 l volume, and properly prepared, including sieving, drying, grinding, and mixing, is appropriate. The material should be placed in air-tight glass jars with screw caps. A sample is assigned a name and a label is mounted on the jar. The prime requirement is that the sample be analyzed many times, and if possible, by different analysts using various methods of determination, e.g., X-ray fluorescence spectrometry and atomic absorption spectrometry for the determination of concentrations of major and trace elements. The results of these analyses can be used for the calculation of a consensus reference value for each parameter measured.

As part of a comprehensive quality control scheme, the in-house standard can then be run with every batch of analyses. However, it is also recommended

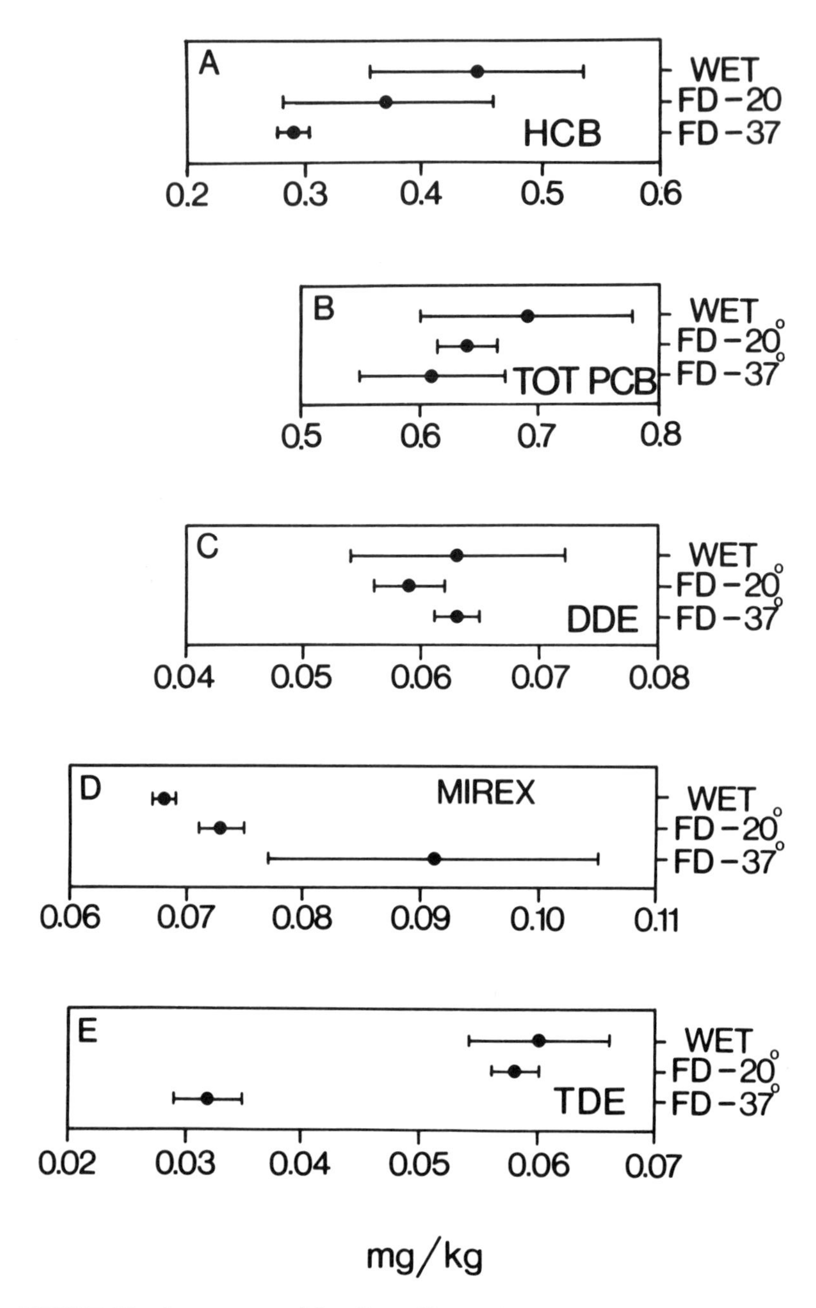

FIGURE 15. Comparison of the effect of freeze-drying on the concentration of five organochlorines. Values plotted are means of triplicate determinations ± standard deviations.

that commercial standard reference materials be used on at least a weekly basis as an aid to monitoring analytical quality assurance.

D. SELECTION OF THE BEST PROCEDURE FOR PRESERVATION AND STORAGE OF SEDIMENT SAMPLES TO BE ANALYZED FOR ORGANIC CONTAMINANTS

In preparation for establishing the Great Lakes Sediment Bank at NWRI, comparative studies were carried out to propose the best procedure for processing and storage of sediments to be analyzed for organic pollutants.[69] Bottom sediments were collected at two sites in Lake Ontario and subjected to three different preparation and storage treatments.

They were analyzed in triplicate for five organochlorine substances shortly after collection, and then after 2 and 6 months. Samples designated as "WET" were stored for 6 months at –20°C and analyzed after collection, and then after six months. Samples labeled "FD20" were freezed-dried during the first week after collection, with a drying temperature of 20°C, and subsequently stored dry at room temperature for 6 months, and analyzed after 2 and 6 months. Samples designated "FD37" were stored frozen in a freezer at a temperature of –20°C for 6 months and freeze-dried just prior to analyses with an initial chamber temperature of 37°C, until half of the water was removed, followed by an ambient chamber temperature until dry. The differences in the results of analyses carried out at various time intervals were insignificant. The results of different treatments were also insignificant, as shown in Figure 15. All ranges of concentrations as represented by standard deviations about the mean overlapped, regardless of the treatment.

ACKNOWLEDGMENTS

We would like to thank Mr. O. Mudroch for his invaluable help with the preparation of this chapter. Mr. R. D. Coker provided the photographs of the equipment in this chapter. We also thank Ms. Dianne Crabtree and Ms. Virginia Vader for their help in preparing the manuscript.

REFERENCES

1. **Luepke, N.-P.**, Final report on workshop, in *Monitoring Environmental Materials and Specimen Banking*, Luepke, N.-P., Ed., Martinus Nijhoff, The Hague, 1979, 1.
2. **Giesy, J.P., Graney, R.L., Newsted, J.L., Rosiu, C.J. Benda, A., Kreis, R.G., and Hovarth, F.J.**, Comparison of three sediment bioassay methods using Detroit River sediments, *Environ. Toxicol. Chem.*, 7, 483, 1988.
3. **Jafvert, C.T. and Wolfe, N.L.**, Degradation of selected halogenated ethanes in anoxic sediment water systems, *Environ. Toxicol. Chem.*, 6, 827, 1987.
4. **Robertson, D.E.**, Role of contamination in trace element analysis of seawater, *Anal. Chem.*, 40, 1067, 1968.
5. **Robertson, D.E.**, The adsorption of trace elements in seawater on various container surfaces, *Anal. Chim. Acta*, 42, 533, 1968a.
6. **Moody, J.R. and Lindstrom, R.M.**, Selection and cleaning of plastic containers for storage of trace element samples, *Anal. Chem.*, 49, 2264, 1977.
7. **Carr, R.A. and Wilkniss, P.E.**, Mercury: short-term storage of natural waters, *Environ. Sci. Technol.*, 7, 62, 1973.
8. **Karin, R.W., Bunon, J.A., and Fasching, J.L.**, Removal of trace metal impurities from polyethylene by nitric acid, *Anal. Chem.*, 47, 2296, 1975.
9. **Calabrese, E.J., Tuthill, R.W., Seiger, T.L., and Klar, J.M.**, Lead and cadmium contamination during acid preservation of water samples, *Bull. Environ. Contam. Toxicol.*, 23, 1979.
10. **Erickson, P.**, Marine trace metal sampling and storage, Marine Analytical Chemistry Stds. Prog. Rep. #1, National Research Council Canada, 1977, 49.
11. **Green, D.R.**, Sampling seawater for traced hydrocarbon determination, Marine Analytical Chemistry Stds. Prog. Rept. #2, National Research Council Canada, 1977, 73.
12. **Batley, G.E. and Gardner, D.**, Sampling and storage of natural waters for trace metal analysis, *Water Res.*, 11, 745, 1977.

13. **Water Quality National Laboratory**, Protocol on preservation, container type, and bottle preparation, Water Quality National Laboratory, Environment Canada, Burlington, Ontario, 1985.
14. **Durst, R.A.**, Container materials for the preservation of trace substances in environmental specimens, in *Monitoring Environmental Materials and Specimen Banking*, Luepke, N.-P., Ed., Martinus Nijhoff, The Hague, 1979, 263.
15. **Mortimer, C.H.**, The exchange of dissolved substances between mud and water in lakes, *J. Ecol.*, 29, 280, 1941.
16. **Mortimer, C.H.**, The exchange of dissolved substances between mud and water in lakes, *J. Ecol.*, 30, 147, 1942.
17. **Berner, R.A.**, Electrode studies of hydrogen sulfide in marine sediments, *Geochem. Cosmochem. Acta*, 27, 563, 1963.
18. **Morris, J.C. and Stumm, W.**, Redox equilibria and measurements in the aquatic environment, *Adv. Chem. Ser.*, 67, 270, 1967.
19. **Whitfield, M.**, Eh as an operational parameter in estuarine studies, *Limnol. Oceanogr.*, 14, 547, 1969.
20. **Tinsley, I.J.**, *Chemical Concepts in Pollutant Behaviour*, Wiley-Interscience, New York, 1979, 92.
21. **Berner, R.A.**, A new geochemical classification of sedimentary environments, *J. Sediment. Petrol.*, 51, 359, 1981.
22. **Bates, R.G.**, The modern meaning of pH, *CRC Crit. Rev. Anal. Chem.*, 10, 247, 1981.
23. **Rozanov, A.G.**, Pacific sediments from Japan to Mexico: some redox characteristics, in *The Dynamic Environment of the Ocean Floor*, Fanning, K.A. and Manheim, F.T., Eds., Lexington Books, D. C. Heath, Lexington, MA, Toronto, 1982, 239.
24. **Zobell, C.E.**, Studies on the redox potential of marine sediments, *Bull. Am. Assoc. Pet. Geol.*, 30, 1946.
25. **Kemp, A.L.W. and Lewis, C.F.M.**, A preliminary investigation of chlorophyll degradation products in the sediments of Lakes Erie and Ontario, in *Proc. 11th Conf. Great Lakes Res.*, International Association of Great Lakes Research, Ann Arbor, MI, 1968, 206.
26. **Kemp, A.L.W., Savile, H.A., Gray, C.B., and Mudrochova, A.**, A simple corer and a method for sampling the mud-water interface, *Limnol. Oceanogr.*, 16, 689, 1971.
27. **Bagander, L.E. and Niemisto, L.**, An evaluation of the use of redox measurements for characterizing recent sediments, *Estuarine Coastal Mar. Sci.*, 6, 127, 1976.
28. **Adams, D.D., Darby, D.A., and Young, R.J.**, Selected analytical techniques for characterizing the metal chemistry and geology of fine-grained sediments and interstitial water, in *Contaminants and Sediments*, Vol. 2, Baker, R.A., Ed., Ann Arbor Science Publishers, Ann Arbor, MI, 1980, 3.
29. **Gupta, S.K. and Chen, K.Y.**, Partitioning of trace metals in selective chemical fractions of nearshore sediments, *Environ. Lett.*, 10, 129, 1975.
30. **Tessier, A., Campbell, P.G.C., and Bisson, M.**, Sequential extraction procedure for the speciation of particulate trace metals, *Anal. Chem.*, 51, 844, 1979.
31. **Tessier, A., Campbell, P.G.C., and Bisson, M.**, Particulate trace metal speciation in stream sediments and relationships with grain size: implications for geochemical exploration, *J. Geochem. Explor.*, 16, 77, 1982.
32. **Jenne, E.A. and Luoma, S.N.**, Forms of trace elements in soils, sediments, and associated waters: an overview of their determination and biological availability, in *Biological Implications of Metals in the Environment*, Wildung, R.R. and Drucker, H., Eds., U.S. Energy Res. Develop. Admin. Sym. Ser. 42, 1977, 110.
33. **Carignan, R., Rapin, F., and Tessier, A.**, Sediment porewater sampling for metal analysis: a comparison of techniques, *Geochim. Cosmochim. Acta*, 49, 2493, 1985.
34. **Rapin, F., Tessier, A., Campbell, P.G.C., and Carignan, R.**, Potential artifacts in the determination of metal partitioning in sediments by a sequential extraction procedure, *Environ. Sci. Technol.*, 20, 836, 1986.
35. **Filipek, L.H. and Owen, R.M.**, Geochemical associations and grain-size partitioning of heavy metals in lacustrine sediments, *Chem. Geol.*, 26, 105, 1979.
36. **Chapman, P.**, Marine sediment toxicity tests, in *Chemical and Biological Characterization of Municipal Sludges, Sediments, Dredge Spoils, and Drilling Muds*, Lichtenberg, J.J., Winter, J.A., Weber, C.I., and Fradkin, L., Eds., ASTM Special Technical Publication (STP) 976, American Society for Testing and Materials, Philadelphia, 988, 391.

37. **Landrum, P.F., Faust, W.R., and Eadie, B.J.**, Bioavailbability and toxicity of a mixture of sediment-associated chlorinated hydrocarbons to the amphipod *Pontoporeia hoy*, in *Aquatic Toxicology and Hazard Assessment: 12th Volume*, Cowgill, U.M. and Williams, L.R., Eds., American Society for Testing and Materials, Philadelphia, 989, 315.
38. **Breteler, R.J., Scott, K.J. and Shepherd, S.P.**, Application of a new sediment toxicity test using the marine amphipod *Ampelisea abdita* to San Francisco Bay sediments, in *Aquatic Toxicology and Hazard Assessment: 12th Volume*, Cowgill, U.M. and Williams, L.R., Eds., American Society for Testing and Materials, Philadelphia, 989, 304.
39. **Environment Canada**, Analytical Methods Manual, Inland Waters Dir., Water Quality Branch, Ottawa, Ontario, 1979.
40. **Luepke, N.-P.**, State-of-the-art of biological specimen banking in the Federal Republic of Germany, in *Monitoring Environmental Materials and Specimen Banking*, Luepke, N.-P., Ed., Martinus Nijhoff, The Hague, 1979, 403.
41. **Gills, T.E. and Rook, H.L.**, Specimen bank research at the National Bureau of Standards to insure proper scientific protocols for the sampling, storage and analysis of environmental materials, in *Monitoring Environmental Materials and Specimen Banking*, Luepke, N.-P., Ed., Martinus Nijhoff, The Hague, 1979, 263.
42. **Ackermann, F.**, A procedure for correcting the grain size effects in heavy metal analyses of estuarine and coastal sediments, *Environ. Technol. Lett.*, 1, 518, 1980.
43. **de Groot, A.J. and Zschuppe, K.H.**, Contribution to the standardization of the methods of analysis for heavy metals in sediments, *Rapp. P. V. Reun. Cons. Int. Explor. Mer*, 181, 111, 1981.
44. **de Groot, A.J., Zschuppe, K.H., and Salomons, W.**, Standardization of methods for analysis of heavy metals in sediments, *Hydrobiologia*, 92, 689, 1982.
45. **Ackermann, F., Bergmann, H., and Schleichert, U.**, Monitoring of heavy metals in coastal and estuarine sediments — a question of grain size: <20 mm versus <60 mm, *Environ. Technol. Lett.*, 4, 317, 1983.
46. **Forstner, U. and Salomons, W.**, Trace metal analysis on polluted sediments. I. Assessment of sources and intensities, *Environ. Technol. Lett.*, 1, 494, 1980.
47. **Wilberg, W.G. and Hunter, J.V.**, The impact of urbanization on the distribution of heavy metals in bottom sediments of the Saddle River, *Water Res. Bull.*, 15, 790, 1979.
48. **Wall, G.J., Wilding, L.P., and Smeck, N.E.**, Physical, chemical, and mineralogical properties of fluvial unconsolidated bottom sediments in northwestern Ohio, *J. Environ. Qual.*, 7, 319, 1978.
49. **Warman International Ltd.**, Particle size analysis in the sub-sieve range, Cyclosizer Instruction Manual, Bulletin WCS/2, Warman International Ltd., Sydney, Australia, 1981.
50. **Mudroch, A. and Duncan, G.A.**, Distribution of metals in different size fractions of sediment from the Niagara River, *J. Great Lakes Res.*, 12, 117, 1986.
51. **Stone, M. and Mudroch, A.**, The effect of particle size, chemistry and mineralogy of river sediments on phosphate adsorption, *Environ. Technol. Lett.*, 10, 501, 1989.
52. **Umlauf, G. and Bierl, R.**, Distribution of organic micropollutants in different size fractions of sediments and suspended solid particles of the River Rotmain, *Z. Wasser Abwasser Forsch.*, 20, 203, 1987.
53. **Cranston, R.E.**, Accumulation and distribution of total mercury in estuarine sediments, *Estuarine Coastal Mar. Sci.*, 4, 695, 1976.
54. **Landrum, P.F. and Poore, R.**, Toxicokinetics of selected xenobiotics in *Hexagenia limbata*, *J. Great Lakes Res.*, 14, 427, 1988.
55. **Swindoll, C.M. and Applehaus, F.M.**, Factors influencing the accumulation of sediment-sorbed hexachlorobiphenyl by midge larvae, *Bull. Environ. Contam. Toxicol.*, *39*, 1055, 1987.
56. **Klump, J.V., Krezoski, J.R., Smith, M.E., and Kaster, J.L.**, Dual tracer studies of the assimilation of an organic contaminant from sediments by deposit feeding oligochaetes, *Can. J. Fish. Aquat. Sci.*, 44, 1574, 1987.
57. **Burton, G.A. and Stemmer, B.L.**, Factors affecting effluent and sediment toxicity using cladoceran, algae and microbioal indicator assays, *Abstr. Annu. Meeting Soc. Environ. Toxicol.*, Pensacola, FL, 1987.
58. **Swartz, R.C., DeBen, W.A., Jones, J.K., Lamberson, J.O., and Cole, F.A.**, Phoxocephalid amphipod bioassay for marine sediment toxicity, in *Aquatic Toxicology and Hazard Assessment*, 7th Symp., ASTM STP 854, Cardwell, R.D., Purdy, R., and Bahner, R.C., Eds., American Society for Testing and Materials, Philadelphia, 1985, 284.

59. **Anderson, J., Birge, W., Gentile, J., Lake, J., Rodgers, J., Jr., and Swartz, R.**, Biological effects, bioaccumulation, and ecotoxicology of sediment-associated chemicals, in *Fate and Effects of Sediment Bound Chemicals in Aquatic Systems*, Dickson, K., Maki, A., and Brungs, W., Eds., Pergamon Press, Elmsford, NY, 1987, 267.
60. **McKeague, J.A., Ed.**, *Manual on Soil Sampling and Methods of Analysis*, 2nd ed., Canadian Society of Soil Science, Ottawa, 1978.
61. **Folk, R.L.**, *Petrology of Sedimentary Rocks*, Hemphill, Austin, TX, 1974.
62. **Kersten, M. and Forstner, U.**, Cadmium associations in freshwater and marine sediment, in *Cadmium in Aquatic Environment*, Nriagu, J.O. and Sprague, J.B., Eds., John Wiley & Sons, New York, 1987, 51.
63. **ASTM**, *1969 Book of ASTM Standards*, Part 11, American Society for Testing and Materials, Philadelphia, 1969.
64. **Folsom, B.L., Jr., Lee, C.R., and Preston, K.M.**, Plant bioassay of materials from the Blue River dredging project, Miscellaneous Paper EL-81-6, U.S. Army Engineer Waterways Experiment Station, Vicksburg, MS, 1981, 24.
65. **Van Driel, W., Smilde, K.W., and van Luit, B.**, Comparison of the heavy metal uptake of *Cyperus esculentus* and of agronomic plants grown on contaminated dutch sediments, Miscellaneous Paper EL-81-6, U.S. Army Engineer Waterways Experiment Station, Vicksburg, MS, Miscellaneous Paper EL-81-6, 1985, 67.
66. **Mudroch, A. and Painter, S.**, Comparison of growth and metal uptake of *Cyperus esculentus*, *Typha latifolia* and *Phragmites communis* grown in contaminated sediments from two Great Lakes harbours, Contribution No. 87-70, National Water Research Institute, Environment Canada, Burlington, Ontario, 1987.
67. **ASTM**, *Procedures for Testing Soils*, 4th ed., American Society for Testing and Materials, Philadelphia, 1964.
68. **Maxwell, J.A.**, *Rock and Mineral Analysis*, Interscience, New York, 1968, 584.
69. **Bourbonniere, R.A., VanSickle, B.L., and Mayer, T.**, The Great Lakes sediment bank. I. Report 86-151, National Water Research Institute, Environment Canada, Burlington, Ontario, 1986.
70. **Pillay, K.K.S., Thomas, C.C., Jr., Sondel, J.A., and Hyche, C.M.**, Determination of mercury in biological and environmental samples by neutron activation analysis, *Anal. Chem.*, 43, 1419, 1971.
71. **Harrison, S.H. and LaFleur, P.D.**, Evaluation of lyophilization for the preconcentration of natural water samples prior to neutron activation analysis, *Anal. Chem.*, 47, 1685, 1975.
72. **LaFleur, P.D.**, Retention of mercury when freeze-drying biological materials, *Anal. Chem.*, 45, 1534, 1973.
73. **ASTM Annual Book of Standards**, Section 14, General Methods and Instrumentation, American Society for Testing and Materials, Philadelphia, 1986, 5.
74. **ASTM**, *1964 Book of ASTM Standards*, Parts 9 and 30, American Society for Testing and Materials, Philadelphia, 1969.
75. **Perry, J.H.**, *Chemical Engineers' Handbook*, McGraw-Hill, New York, 1963.
76. **Taggart, A.F.**, *Handbook of Mineral Dressing*, John Wiley & Sons, New York, 1945.
77. **National Research Council, Canada**, Marine reference materials and standards, Circular, National Research Council of Canada, Halifax, Nova Scotia, 1988.
78. **Grant, C.L. and Pelton, P.A.**, Role of homogeneity in powder sampling, in *Sampling, Standards and Homogeneity*, ASTM Spec. Tech. Publ. 540, American Society for Testing and Materials, Philadelphia, 1973, 16.
79. **Plumb, R.H., Jr.**, Procedure for the handling and chemical analysis of sediment water samples, Tech. Rep. EPA/CE-81-1, U.S. Environmental Protection Agency and U.S. Army Corps of Engineers, Vicksburg, MS, 1981.
80. **ASTM**, *Standard Practice for Estimation of Holding Time for Water Samples Containing Organic Constituents*, American Society for Testing and Materials, ASTM D4515-85, Philadelphia, 1987.
81. **Malueg, K.W., Schuytema, G.S., and Krawczyk, D.F.**, Effects of sample storage on a copper spiked freshwater sediment, *Environ. Toxicol. Chem.*, 5, 248, 1986.
82. **Tatem, H.E.**, Use of *Daphnia magna* and *Mysidopis almyra* to assess sediment toxicity, in *Water Quality '88*, Seminar Proc., Willey, R.G., Ed., U.S. Army Corp of Engineers Common Water Quality, Washington, D.C., 1988.
83. **Clark, J.R., Patrick, J.M., Jr., Moore, J.C., and Lores, E.M.**, Waterborne and sediment-source toxicities of six organic chemicals to grass shrimp (*Palaemonetes pugio*) and amphious (*Branchiostoma Caribaeum*), *Arch. Environ. Contam. Toxicol.*, 16, 401, 1987.

84. **Danso, S.K.A., Habte, M., and Alexander, M.**, Estimating the density of individual bacterial populations introduced into natural ecosystems, *Can. J. Microbiol.*, 19, 1450, 1973.
85. **Wolfe, N.L., Kitchens, B.E., Macalady, D.L., and Grundl, T.J.**, Physical and chemical factors that influence the anaerobic degradation of methyl parathion in sediment systems, *Environ. Toxicol. Chem.*, 5, 1019, 1986.
86. **Wise, S.A. and Zeisler, R.**, The pilot environmental specimen bank program, *Environ. Sci. Technol.*, 18, 302A, 1984.
87. **Wise, S.A., Fitzpatrick, K.A., Harrison, S.H., and Zeisler, R.**, Operation of the U.S. pilot national environmental specimen bank program, in *Environmental Specimen Banking and Monitoring as Related to Banking*, Lewis, R.A., Stein, N., and Lewis, C.W., Eds., Martinus Nijhoff, Boston, 1984, 108.
88. **Kayser, D., Boehringer, U.R., and Schmidt-Bleek, F.**, The environmental specimen banking project of the Federal Republic of Germany, *Environ. Monit. Assess.*, 1, 241, 1982.
89. **Stoeppler, M., Backhaus, F., Schladot, J.-D., and Nurnberg, H.W.**, Concept and operational experiences of the pilot environmental specimen bank program in the Federal Republic of Germany, in *Environmental Specimen Banking and Monitoring as Related to Banking*, Lewis, R.A., Stein, N., and Lewis, C.W., Eds., Martinus Nijhoff, Boston, 1984, 45.
90. **Elliot, J.E.**, Collecting and archiving wildlife specimens in Canada, in *Environmental Specimen Banking and Monitoring as Related to Banking*, Lewis, R.A., Stein, N., and Lewis, C.W., Eds., Martinus Nijhoff, Boston, 1984, 45.
91. **Elliot, J.E.**, Specimen banking in support of monitoring for toxic contaminants in Canadian wildlife, in *International Review of Environmental Specimen Banking*, Wise, S.A. and Zeisler, R., Eds., NBS Spec. Publ. 706, U.S. Department of Commerce, Washington, D.C., 1985, 4.
92. **Hyatt, W.H., Fitzsimons, J.D., Keir, M.J., and Whittle, D.M.**, Biological tissue archive studies, Canadian Tech. Rep. Fish. Aquat. Sci. 1497, Fisheries and Oceans Canada, Burlington, Ontario, 1986.
93. **Chau, A.S.Y. and Lee, H.**, Analytical reference materials. III. Preparation and homogeneity test of large quantities of wet and dry sediment reference materials for long-term polychlorinated biphenyl quality control studies, *J. Assoc. Off. Anal. Chem.*, 63, 947, 1980.
94. **Cheam, V. and Chau, A.S.Y.**, Analytical reference materials. IV. Development and certification of the first Great Lakes sediment reference material for arsenic, selenium and mercury, *Analyst*, 109, 775, 1984.
95. **Cheam, V., Aspila, K.I., and Chau, A.S.Y.**, Analytical reference materials. VIII. Development and certification of a new Great Lakes sediment reference material for eight trace metals, *Sci. Tot. Environ.*, 87/88, 517, 1989.

Chapter 7
SEDIMENT PORE WATER SAMPLING

Donald D. Adams

I. INTRODUCTION

Coarse-grained surface sediments, such as sand and mixtures of silt and sand, contain typically about 30 to 40% water. Fine-grained surface sediments, such as silty clays found in lake depositional areas, contain about 90 to 95% water. Surface fine-grained sediments with high organic matter, found typically in focusing centers of smaller, eutrophic lakes, may contain up to 99% water. Some of this water is held by surface forces such as adsorption and capillarity and is bound to the crystalline lattice of minerals in the sediments. The content of this water is related to the physical properties and mineralogical composition of the bottom deposits. The rest of the water filling the space between sediment particles is called pore water or interstitial water.

Studies on the physical and chemical nature of sediment pore water have centered, in most cases, on the identification of equilibrium reactions between minerals and water and early diagenetic mineralogical and chemical changes within the sediments. Recently, processes and impacts of the transport of chemical contaminants, dissolved in the pore water, across the sediment-water interface into overlying water have been the central themes. In most of these studies the collection of pore water containing original *in situ* chemical forms of elements and compounds was of common concern. With the exception of the uppermost (top 1 to 3 cm) bottom sediments, most fine-grained deposits are anoxic but become rapidly oxidized upon exposure to air. The oxidation of sediments brings about rapid changes of redox-sensitive chemical species dissolved in the sediment pore water. During pore water collection and sample processing the maintenance of an oxygen-free atmosphere is critical to measuring some of the original chemical constituents. For example, Krom and Berner[1] found that the adsorption of phosphate by anoxic marine sediment is greatly increased when the system is aerobic, probably due to reaction with iron oxyhydroxides. In their extensive reviews Manheim[2] and Kriukov and Manheim[3] summarized some fundamental problems which must be considered when conducting pore water studies. These included techniques for extraction of sediment gases and labile sulfur components as well as a discussion of changes in sediment pore water composition resulting from pressure and thermal effects during sample processing. Such changes have to be considered during pore water sampling and processing. In addition, Froelich et al.[4] and Malcolm et al.,[5] and references cited therein, should be consulted concerning processing techniques and methodology centered on the avoidance of producing oxidative artifacts during sample handling. Therefore, it is imperative that *in situ* conditions be simulated as closely as possible.

Sampling of sediment pore water is tedious, and requires special equipment and some experience and knowledge concerning the possible effects and

1-56670-027-2/94/$0.00+$.50

potential chemical artifacts resulting from incorrect procedures. Considering these difficulties, pore water sampling and analyses should be carried out by professionals and only if these data are required for the interpretation of sediment-water exhange, sediment pore water quality, or other necessary protocols.

II. PORE WATER SAMPLING — GENERAL TECHNOLOGY

Pore water sampling methods can be generally divided into two groups: (1) separation of pore water from sediments collected by different coring devices or grab samplers, and (2) recovery of pore water from various *in situ* sampling devices. The advantage of pore water separation from a recovered sediment sample is the capability of measuring and interpreting particle-pore water relationships by analyzing both solution and solid phases at the same location. However, there are many factors which have to be considered to avoid changes in pore water chemical constituents and creation of chemical artifacts by improper sediment sampling and pore water separation techniques. The utilization of an *in situ* pore water sampling device would potentially eliminate possible errors and chemical artifacts produced during the collection and processing of sediments for contained pore water. One major disadvantage, however, is that divers are required for deployment of many of the *in situ* pore water sampling devices, thus restricting this technique to water depths suitable for scuba (self-contained underwater breathing apparatus).

Squeezing and centrifugation are two of the major techniques which were used for the extraction of pore water from recovered sediments. However, several other systems, such as vacuum filtration and leaching, were reported in the literature. The removal of pore water from recovered sediments requires special preparation of the sediments prior to extraction. Typically, the sediment collected by a coring device is sectioned, for example, into 1- to 2-cm-thick slices which are further used for pore water recovery. To obtain a pore water sample approximating the natural sedimentary environment, exposure of the sediment and sampled pore water to the atmosphere must be completely avoided by employing oxygen-free conditions during the entire processing period. This can be accomplished using an inert gas such as argon, nitrogen, and in some cases helium or carbon dioxide.

A few authors described the handling and subsampling of sediment cores collected for the separation of pore water. For example, Matisoff et al.[6] had scuba divers collect sediment cores with plastic liners normally used with a Benthos gravity corer. The collected core was capped and held in a vertical position while the length was measured, sediment appearance described, and overlying water removed by siphoning. This technique was also employed by Adams et al.[7] for a Lake Erie pore water sampling program. To minimize oxidation of the sediment, a plastic bag was immediately taped over the end of the core liner and nitrogen gas was continuously flushed into the liner during sediment processing. While maintaining a vertical position, the top of the core liner was inserted into a nitrogen gas-filled glove bag and the sediment extruded and subsampled into desired sections. A rubber plunger inserted into the bottom of the core liner was used to force the sediments upward into the glove bag with hydraulic pressure. The subsamples were spooned into nylon squeezers.

Generally, glove bags or glove boxes should be used for the subsampling of sediments collected for pore water extraction. The cores should be extruded while in a vertical position to avoid mixing of the soft, surficial sediments which possess a high water content. Subsampling should be carried out as soon as possible after recovery of the core and under an inert atmosphere. As an example, Emerson[8] reported changes in pore water phosphate concentrations between cores processed under nitrogen and in the open. He also evaluated storage time and adjustment of sample pH during storage. Equipment required for core subsampling and storage, including labeled containers for the sediments, should be organized and readily available inside the glove bag. If centrifugation is used for the separation of pore water, centrifuge tubes or bottles need to be purged with inert gas prior to adding sediments in the glove bag. The bottles should be filled with sediments (preferably to the lip to exclude a headspace) and fitted with gas-tight lids to avoid air contact before removal from the inert gas atmosphere. The glove bag or box should be evacuated and refilled with inert gas numerous times before commencement of the sediment subsampling and each time after reopening the bag. During sediment subsampling the inert gas should continually flow into the glove bag to maintain a slightly positive internal pressure (to minimize oxygen invasion around arm sleeves, etc.). It is also recommended to purge the sampling containers, squeezers or centrifuge tubes with an inert gas to remove residual atmospheric oxygen just before and after sediments are added. Even though nitrogen is most commonly used, other inert gases such as argon, helium, and carbon dioxide were employed for glove bag operations.[4,9-13] Because oxygen diffuses through most plastics, it is also advisable to initiate centrifugation as soon as possible or store the containers under oxygen-free conditions. An oxygen meter for monitoring air contamination in the glove box during sediment subsampling and processing is also strongly suggested. After centrifugation, pore water should be collected and filtered within a glove box or bag under oxygen-free conditions. After filtration, an inert gas environment is not necessary for long-term storage.

Last, as described in previous chapters, shortening of the core sediments during sampling would also have an effect on the pore water profile. This is discussed in Lebel et al.[14] where they suggested undisturbed box coring rather than gravity coring to preserve pore water chemical gradients. Emerson et al.[15] used a piston to avoid core shortening when subsampling a box core while Schimmelmann et al.[16] applied a slight vacuum to insure that the sediment interface inside and outside of their acrylic tubes remained at identical levels.

A. PORE WATER SEPARATION FROM RECOVERED SEDIMENTS

1. Centrifugation

Earlier techniques for removing pore water by centrifugation have often suffered from lack of speed to efficiently separate pore water from sediments. Moreover, the long centrifugation time tended to heat the sediments. High-speed refrigerated centrifuges are presently capable of rapid and efficient separation of pore water from sediments at desired temperatures. Capping of centrifuge tubes in the glove box under an inert gas atmosphere prevents oxidation and evaporation of samples. The use of syringes (with in-line filters) for withdrawing the supernatant pore water after centrifugation permits accurate and rapid transfer into containers for storage and analyses. As an example, Engler et al.[17] placed sediments into oxygen-free polycarbonate centrifuge tubes in a glove bag followed by centrifugation in a refrigerated centrifuge (4°C) at 9000 rpm for 5 min. Approximately 40% of the total sediment water was recovered. Extracted pore water was vacuum filtered under nitrogen through rinsed 0.45-μm membrane filters, immediately acidified to pH 1 and stored in rinsed plastic bottles. Centrifugation at different speeds from 7000 to 19,000 rpm was tested by Adams et al.[18] who found little changes in pore water Ca, Fe, Mn, and Zn, but a doubling of phosphate. For processing large volumes of pore water essential for studying sediment accumulation and mixing rates, Anderson et al.[19] collected about 20 cm of surface sediments with an Ekman grab sampler. The sample was quickly transferred into six 250-ml plastic centrifuge bottles, filled to the top to exclude air and closed tightly. After centrifugation the supernatant was filtered as quickly as possible through a glass fiber filter and then through a 0.45-μm-pore-size Millipore filter. Total processing time to extract 1 l was about 6 h. The pore water was acidified with HCl prior to further processing. Regardless, chemical artifacts from oxidation and temperature variations, as described by Fanning and Pilson,[20] Bray et al.,[21] Troup et al.,[22] Emerson,[8] and Lyons et al.,[23] are normally present as a result of sample processing by centrifugation and filtration, especially if an inert gas glove bag is not employed. This should be recognized as one of the major problems.

Saager et al.[24] reported a simple device for simultaneous centrifugation and filtering of coarse, sandy sediments at 1500 *g* for 5 min on board ship. Approximately 80% of the original pore water is collected through 0.45-μm Nuclepore filters; recovery was better than squeezing (25 to 30% recovery). A similar type of system was designed by Bauer et al.[25] for sampling pore water in a hydrocarbon seep area in the southern California Bight.

A comparison between centrifugation and *in situ* dialysis, which is described later, was reported by Carignan et al.[26] for measurements of pore water dissolved Ca, Cd, Co, Cr, Cu, Fe, Mg, Mn, Ni, Zn, and organic carbon. Sediment centrifugation at 5000 rpm followed by filtration (0.45-μm membrane) was equivalent to *in situ* dialysis for Co, Cr, Fe, Mn, and Ni, but gave higher and more variable concentrations for Cu, Zn, and organic carbon. Concentrations similar to those obtained by *in situ* dialysis were found when the centrifugation speed was increased to 11,000 rpm and 0.2- or 0.03-μm membranes were used to filter the pore water supernatant. This procedure gave values similar to *in situ* dialysis for Cd, Co, Cr, Ni, and organic carbon. The authors also described a procedure, generally considered a reasonable method for the tested chemicals except Cu, to remove pore water from sediments by centrifugation. Subsampling of sediment cores was carried out in a glove box under a nitrogen atmosphere with the oxygen partial pressure continuously monitored and kept below 10^{-3} atm. Aliquots from each sediment section were transferred with a plastic spatula into polycarbonate centrifuge tubes, which were tightly closed while in the glove box. Tubes were then removed from the glove box and centrifuged at 5000 rpm for 20 min. After centrifugation the tubes were returned to the glove box and the supernatant pore water was collected into plastic syringes and passed through 0.45-μm HA Millipore membrane filters using a Nuclepore in-line plastic filter holder. In a modified procedure, to remove colloidal particulates in the supernatant pore water, the centrifugation speed was increased to 11,000 rpm and the supernatant was filtered sequentially using Nuclepore 0.2- and 0.03-μm filters.[26] Since low concentrations of trace metals were expected, all materials, including membranes, were previously soaked for 12 to 18 h in

1% HNO_3 and rinsed with deionized water. In each case, filtrates were collected in clean polystyrene vials, acidified with Ultrex 1 *N* HNO_3 and stored at 4°C. It is important not to freeze the pore water samples,[18] and in fact Rutledge and Fleeger[27] reported that freezing of whole cores could disrupt the vertical profile of a core through sediment convective flows (fast freezing tended to cause the greatest disruption) and thus make further sectioning for analysis useless. Such a technique, however, was used by Vanderborght and Billen[28] and Vanderborght et al.[29] for preserving cores for later analysis.

2. Squeezing

A commonly employed procedure for pore water removal is sediment squeezing, followed by filtration, using either low-pressure mechanical squeezers,[30,31] low-pressure gas squeezers,[32,33] or high-pressure mechanical squeezers.[34,35] The devices have usually been constructed from stainless steel. Presley et al.[33] pointed out that stainless steel squeezers are satisfactory for the determination of anions, alkali metals, or alkaline earths in pore waters but are not suitable for measurements of iron and certain transition elements because of potential contamination from the stainless steel. As an example, Brown et al.[36] used the Presley et al.[33] squeezer to study pore water organic constituents in Saanich Inlet, Canada. Squeezers have been designed to maintain the original conditions of the squeezed sediments, such as temperature and redox potential, and to minimize the preparation and handling of sediments for pore water extraction.

Proper preparation of sediments before the actual squeezing plays an important role in maintaining the integrity of pore water chemistry. For obtaining pore water from consolidated and unconsolidated sediments, Manheim[35] developed a mechanical squeezer based on an earlier design from Kriukov[37] and Kriukov and Komorova.[34] The entire system consisted of a standard laboratory press and a filter unit containing a stainless steel screen, perforated steel plate, steel filter holder, and filters in addition to other essential parts. Approximately 2 to 15 g of wet sediment was transferred into the cylinder through the top followed by rubber Neoprene and Teflon disks. The whole unit was placed in a press and pore water was removed, at approximately 200 to 600 kg/cm^2, through a bottom filter into a disposable syringe mounted at the bottom of the device. The extraction time was generally 3 min. After squeezing, the sediment sample could be dried, to determine the total water content of the original sample, and then used for further studies. The squeezing pressure did not appreciably affect the composition of the extracted pore water. According to Manheim,[35] the squeezer would have to be modified for the separation of pore water from poorly compressible but permeable materials such as coarse sands. In a more recent publication Manheim[38] used only disposable plastic syringes fitted with screen discs and filter paper circles to extract small quantities (1 to 5 ml) of pore water from unconsolidated sediments containing greater than 30% bulk water. A wooden screw frame or large "C" clamp was used to control pressure to the syringe. The equipment was suitable for field operations with pressures of 40 kg/cm^2 and 20 kg/cm^2 for 20- and 50-ml syringes, respectively. Another syringe system using hand pressure to extract 1 to 5 ml of pore water, for measuring sulfide, iron and manganese by differential pulse polarography, was described by Davison et al.[39]

Presley et al.[33] developed a stainless steel low-pressure gas-operated squeezer, completely lined with Teflon, for the study of transition elements in pore waters from ocean sediments. The squeezer could handle a 300-g sample of sediment which was packed firmly into the apparatus. Nitrogen gas was used to generate a pressure up to 70 kg/cm^2 for a period of about 10 s through a pressure-reducing valve. The squeezer was placed in a suitable cage for safety purposes. The pore water exited through a perforated Teflon filter plate, covered with a piece of hardened filter paper, at the bottom of the squeezer. In experiments with a 5-m core of marine sediments, 40% of the pore water (which contained 45% original water content) was removed by squeezing at a pressure of 70 kg/cm^2. The separated pore water was subsequently filtered through a 0.4-μm Millipore filter. Siever[30] described a portable mechanical squeezer (filter press) suitable for shipboard work. Parts of the apparatus were made of stainless steel with a bronze piston and thrust bearing, and neoprene O-rings and gaskets. A perforated stainless steel plate provided the necessary support for a stainless steel screen with filter paper. The dimensions of the squeezer were designed to take a 7.5-cm diameter core, a common dimension for piston corers. Maximum pressure of 20 atm was applied to the squeezer. Fifteen minutes were usually sufficient to extract 20 to 30 ml from a 100-g sample of slightly compacted clay containing 60 to 70% water. The method was not suitable for the quantitative extraction of pore water, but it was useful to obtain a representative sample for chemical analysis. Further modifications were made to carry our the squeezing operation under an inert atmosphere to allow for measurements of redox potential in the extracted pore water.

Howes et al.[40] developed a low-pressure (0.1 atm) gas squeezer for removing pore water from marsh sediments. Holes were predrilled into polycarbonate core tubes used to collect and store the sediments. The holes were covered with watertight plastic tape. The sediment surface was adjusted in the laboratory with a bottom piston so that samples were withdrawn at 2-cm intervals by application of argon or nitrogen pressure to the core headspace. Pore water (100 μl) was collected from the center of the core with glass micropipettes which fit snugly against the 2-mm holes. Jahnke[41] modified this sampler by tapping each hole to accommodate nylon screws and O-rings and added top and bottom pistons for pressure application. After sediment collection, the nylon screws were removed and a male-to-female luer fitting, connected to a disposable filter and plastic syringe, was connected to the hole. Either the top or bottom pistons, or up to 300 to 400 kPa of gas pressure, were used to pressurize the core and expel pore water.

For separation of pore water with minimal air contact Kalil and Goldhaber[31] designed a low-pressure mechanical squeezer utilizing, as the inside surface of the squeezer, the original plastic core liner in which the sediments were collected. The apparatus was designed for core liners of various diameters. It consisted of two identical Plexiglas plungers with O-rings, machined to tightly fit a specific core liner. A recessed perforated Teflon disc held two sheets of filter paper above the bottom plunger. A 7- to 10-cm section of recovered sediments, with core liner, were cut with a hacksaw. Two filter papers were placed against the sediment surface, followed by the Plexiglas plunger/Teflon screen assembly. Outgassing and oxidation were minimized since the sediments were not extruded. An insulated water jacket was used to regulate the temperature during squeezing. The entire apparatus was mounted in a 3-ton press. Pressure was applied and pore water collected, after trapped air bubbles were vented, through an in-line filter holder connected to a plastic syringe. Squeezing time varied from 10 min to over 1 h depending upon sediment characteristics and the amount of pore water desired. Generally, about 50 to 100 ml of water were recovered in an hour from a variety of sediments containing various water contents. As an example, Aller[42] used the Kalil and Goldhaber[31] squeezer to study the effects of deposit feeders on pore water chemistry and the flux of redox-sensitive inorganic chemicals across the sediment-water interface.

Bender et al.[43] designed a whole-core mechanical squeezer to be used aboard ship for studies of interfacial gradients and fluxes of O_2, NO_3^-, SiO_2, and other particle-unreactive ions and molecules in sediments. However, this method was considered unsuitable for the determination of pore water profiles of trace metals and other particle-reactive chemicals. The squeezer consisted of a Lucite piston with an attached rod to which pressure was applied manually. A narrow hole and tubing through the center of the piston accommodated the flow of extracted pore water, which was filtered through a 10-μm polyester screen mounted at the bottom of the piston. Another piston, secured to the bottom of the core liner, retained the sediment. By collecting the water in about 3-ml aliquots it was possible to obtain sediment pore water profiles at millimeter depth resolutions. The calculated depth of each pore water sample was determined from the volume of squeezed water, the radius of the core liner, and the porosity of the sediment. Extraction of pore water to a sediment depth of 2 cm typically took 30 to 60 min. Calculations of SiO_2 fluxes and oxygen profiles determined by this technique were similar to those obtained by other methods.

For a study of trace mineral equilibria in Lake Erie sediments, Matisoff et al.[6] used nylon squeezers clamped to a squeezing rack. Nitrogen gas (at 3.4 atm pressure) acting against a rubber diaphragm forced sediment pore water through two circles of nylon mesh supports overlain by two 0.45-μm Whatman filters and one 0.22-μm Millipore filter. Exposure to oxygen was minimized during sediment subsampling in a N_2 filled glove bag. Ten separate pore water aliquots were collected during squeezing of each sediment sample for a variety of different analyses.

Hartmann[32] described a squeezer for recovering pore water from sediments of all grain sizes by the simultaneous application of gas and mechanical pressure. He modified a commercial filter press by adding a removable spacer ring, an O-ring, and a PVC membrane disc for transferring pressure to the enclosed sediment sample. About 50 ml of pore water was collected in 5 min from a medium-grained beach sand, leaving 5% residual moisture after sample extraction. Up to 30 ml of pore water was collected from fine-grained sediments in 1 to 3 h with most of the recovery taking place in the first half-hour. Recovered pore water was used mainly for studies of interstitial Fe and Mn in Baltic Sea sediments. This device was modified again to obtain 500 ml of pore water from 1000 ml of wet sediments collected from open ocean sediment cores.[44] Reeburgh[45] described an inexpensive nonmetallic squeezer suitable for pore water trace metal studies. The squeezer was gas-operated and had no piston or moving parts. The sediment was compressed by gas pressure acting against a rubber diaphragm with pore water forced through filters into sampling bottles. Delrin and nylon were used for squeezer construction to prevent corrosion and for easy cleaning. Up to 14 kg/cm^2 (200 psi) pressure was possible for extraction of 25 ml of pore water in 30 to 45 min from 100 g of sediment. Robbins and Gustinis[46] described a modified version of Reeburgh's squeezer designed for separation of pore water from small volumes of unconsolidated, anoxic Great Lakes fine-grained sediments. At the standard operating pressure of 6.9 atm, about 20 ml of pore water was recovered from 50 g of wet sediment (porosity 0.8 to 0.9) in 10 to 20 min. Recovered pore water volume averaged 50 to 60% of the sediment bulk water.

There are a variety of problems with squeezer techniques. At MANOP sites in the eastern equatorial Pacific, Klinkhammer[47] reported anomalously high concentrations of trace metals. Iron exhibited the highest concentrations in the first few milliliters of pore water even after careful sediment processing under helium. It was suspected that iron-enriched, colloidal-sized particles bypassed, or were not retained by, the filters in the squeezers. Emerson et al.[15] reported the first aliquots of squeezer solution had higher ammonium than measured in later portions; therefore they only used the fourth or later sequential solution for ammonium analysis. In studying sulfate reduction and pore water chemistry in a salt marsh, Hines et al.[48] found that *in situ* pore water samplers (called "sippers") consistently gave higher values for sulfide than squeezing, even when extreme caution was employed to avoid oxidation during core processing. Last, Froelich et al.[4] provided ellaborate information concerning squeezer techniques and compared the Reeburgh[45] and "USC"[31] squeezers. For example, pore water total carbon dioxide was much higher, by 1 m*M* or more, in the first aliquot when $CaCO_3$-rich sediments were squeezed through acid-washed filters. They also reported that N_2 and CO_2 values were about 7% less with the Reeburgh-type than the USC-type squeezer.

3. Other Techniques

Desiccation, vacuum filtration, leaching, and displacement techniques are the other typical procedures used for the separation of pore water from sediments. As described by Scholl,[49] desiccation procedures are only useful for semiquantitative studies. Vacuum filtration utilizes different filtering equipment, such as Buchner funnels, for the separation of pore water from sediments. The major problems with this technique are surface evaporation of the pore water during the relatively long time necessary for filtration, particularly of fine-grained material, and inefficiency in comparison with other methods. Leaching sediments to determine the composition of pore water appears to be a simple procedure; however, there were potential errors, particularly solubilization of solids and changes in sediment-water equilibria, during the leaching process.[3] Glass and Poldoski[50] employed a deoxygenated water leach (where pore water was diluted by two- to threefold) followed by vacuum filtration, both conducted under a nitrogen gas atmosphere. An immiscible liquid, such as Paraplex G-60 (a high molecular weight ester), is used for the displacement technique procedure. Extraction is typically made with a filter press whereby an immiscible liquid is poured over the sediment followed by either piston or gas pressure. The extracted pore water is suitable for quantitative analyses.[49] The choice of immiscible liquids was described by Scholl[49] and Kriukov and Manheim.[3] Kinniburgh and Miles[51] used trifluoroethane to extract up to 50% of the pore water from soils and porous rocks containing low (19 to 49%) water content. Examples of the displacement technique using gas (in the absence of expandable membranes) for extracting sediment pore water were described in the previous section under squeezing techniques.[32,33]

An additional technique, which could be listed under pore water separation from recovered sediments, is that described in van Raaphorst and Brinkman.[52] Sediments in undisturbed 5-cm diameter perspex cores were sampled through 2.5-mm polyethylene tubing containing cotton threads. With a slight underpressure about 5 ml of pore water was slowly collected in a 10-ml bottle within 2 to 3 days. Pore water phosphate did not interact with the cotton. It was considered that this slow passive technique would not disturb existing pore water gradients in the sediments.

B. *IN SITU* PORE WATER SAMPLING

It has long been recognized that the collection of pore water from recovered sediments through various techniques, such as centrifugation, pressure filtration, and gas displacement (as well as other methods involving sediment manipulation outside of the benthic environment), can result in changes in pore water concentrations of dissolved species when collection temperatures or pressures differ from *in situ* conditions.[53] An even greater problem is the potential for oxidation during sample manipulation. To avoid these errors, different samplers were designed to collect pore water *in situ* at the lake or seafloor. Since some sampling devices must be deployed by divers, water depths at the sampling station are usually the limiting factor. Most of the *in situ* devices were designed to determine concentration gradients of different pore water constituents for indirect flux calculations of dissolved pore water species between the sediment and the water column. Consequently, they were built to obtain pore water close to the sediment-water interface, for intervals within several centimeters above and below the interface. Most of these samplers, the first of which were proposed by Hesslein[54] and Mayer,[55] are based on diffusion-controlled transport. Other sampling systems utilizing *in situ* sediment pore water suction and filtering systems were first described by Sayles et al.[56]

The *in situ* pore water sampler developed by Hesslein,[54] and its modifications, have been used in many sediment pore water investigations. The sampler (called a "peeper" by Hesslein, or dialyzer) is based on the principle that, given enough time, a contained quantity of water in the sampler will diffuse and equilibrate through a dialysis membrane, or other materials such as porous Teflon, with the surrounding water and its dissolved solutes. The *in situ* equilibrator can either be removed by divers or the pore water can be collected through attached tubing while the device remains in the sediments. The use of dialysis or other membranes as the separating material allows for the discrimination of various size molecules, depending on the chosen membrane, and eliminates filtering of the sample as particulate matter is totally excluded.[54] The size, shape, and type of sampling device and membrane were modified for different studies, but the principle of the *in situ* pore water collection system has remained the same. A modified version of Hesslein's sampler is shown in Figure 1. Generally, samplers of this type are made of clear acrylic plastic (Plexiglass, Lucite, etc.). Two sheets of acrylic plastic (a 0.3-cm-thick cover sandwiched with a 1.3-cm-thick body, as given in the original design by Hesslein[54]) are held together by a series of nylon or stainless steel screws. Horizontal, elongated sampling compartments are machined, usually 1 cm apart, through the 0.3-cm cover and into the acrylic body. Different types of dialysis membranes were used with these samplers. This membrane is usually pierced with a syringe needle when sampling pore water (Figure 2). Examples are 0.2-μm-pore-size polysulfone filtration membranes,[57] 0.2-μm-pore-size polycarbonate membranes mounted in the sampler with the shiny side out to provide less surface area for attachment of microorganisms,[53] biologically inert PVC membranes with 0.45-μm pore size[58] and 3-μm Teflon.[59] A nonporous 75-μm (3-mil Teflon) membrane would allow for diffusion of gases but not ionic

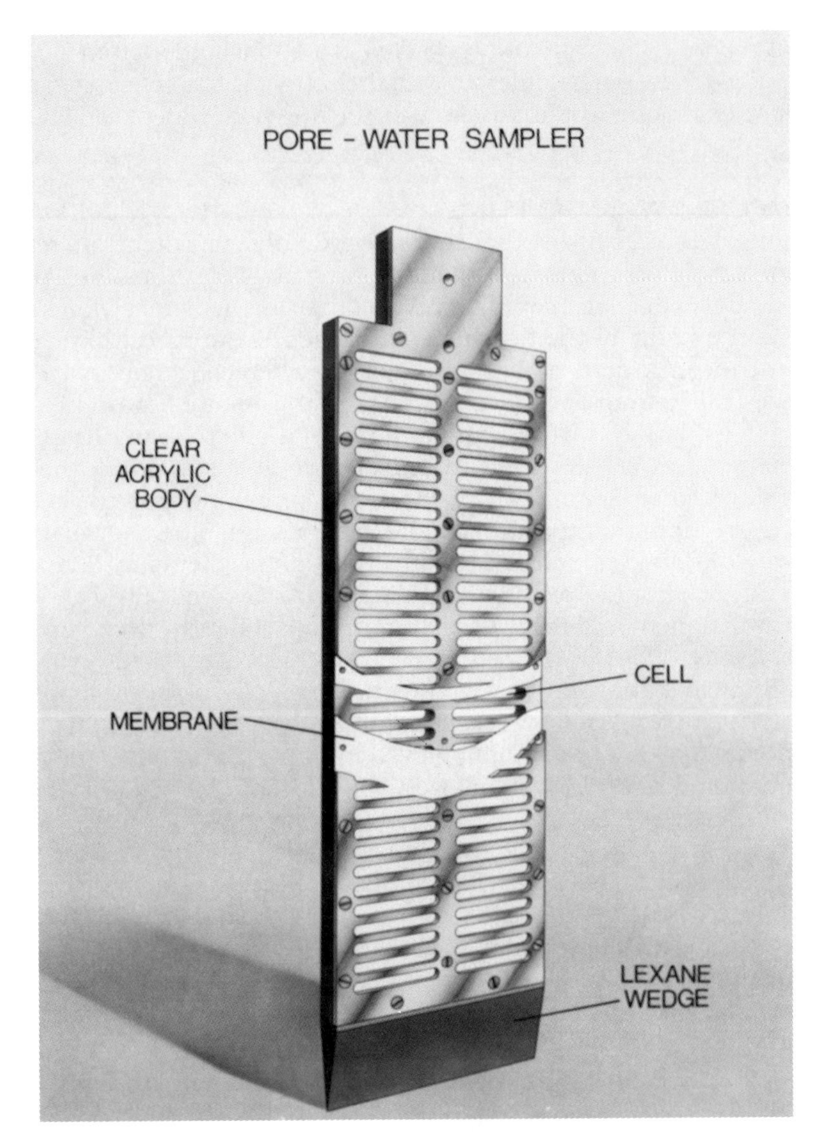

FIGURE 1. An *in situ* dialyzer sampler.

chemical species into the sampling compartments of the *in situ* equilibrator.[60] Because of microbial attack, cellulose-based dialysis membranes should not be used.[59] As an example, profiles of soluble reactive phosphorus (SRP) in sediment pore water are shown in Figures 3A to C.

To prepare the apparatus for sampling, the compartments of the peeper body are filled with distilled oxygen-free water, and a piece of dialysis membrane is placed over the filled compartments in such a way as to exclude air bubbles. The acrylic cover, containing elongated openings spaced exactly opposite the compartments of the peeper body, is placed on top of the membrane and tightly connected to the body with screws. The entire sampler is then carefully placed into an enclosed chamber containing deoxygenated distilled water, where it is allowed to equilibrate. The exact degassing procedure is somewhat different with each investigator and the type of pore water analysis. For example, Kelly et al.[61] used the sampler for measuring

FIGURE 2. Sampling for pore water through the membrane of an *in situ* dialyzer.

changes in sediment pH. To avoid changes of pH by introducing oxygen to the pore water, the sampler was placed in a N_2 atmosphere overnight. To minimize the introduction of oxygen during sampler deployment, it was placed into anoxic sediments and left for a 6- to 7-day period of equilibration.

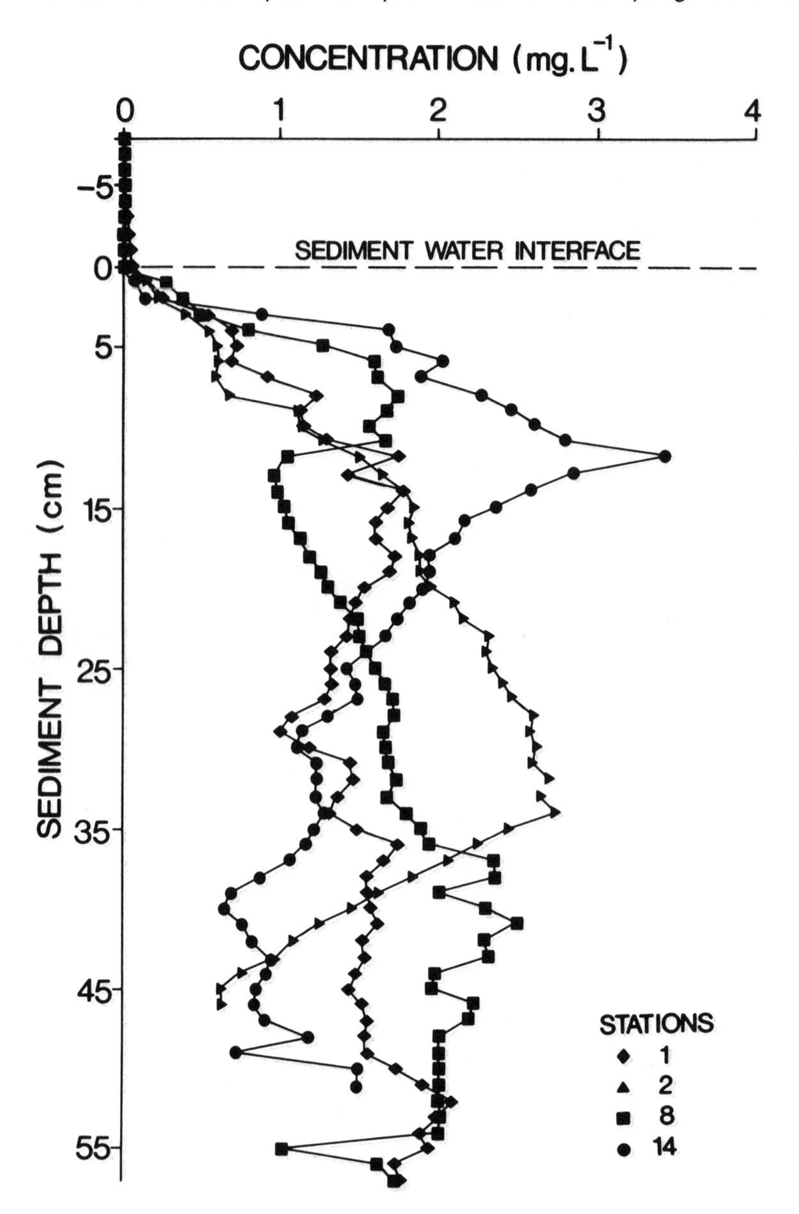

A

FIGURE 3. Examples of soluble reactive phosphorus (SRP) in sediment pore water: (A) profiles at four different stations, (B) six sampling replications, and (C) temporal variability of SRP (mg/l) at the same station (courtesy of F. Rosa).

In studies evaluating the partitioning of Zn between sediment pore water and the overlying water column, Carignan and Tessier[62] and Tessier et al.[57] deaerated equilibration samplers in Plexiglass/stainless steel containers, filled with distilled or demineralized water, by bubbling with nitrogen gas for 24 h.

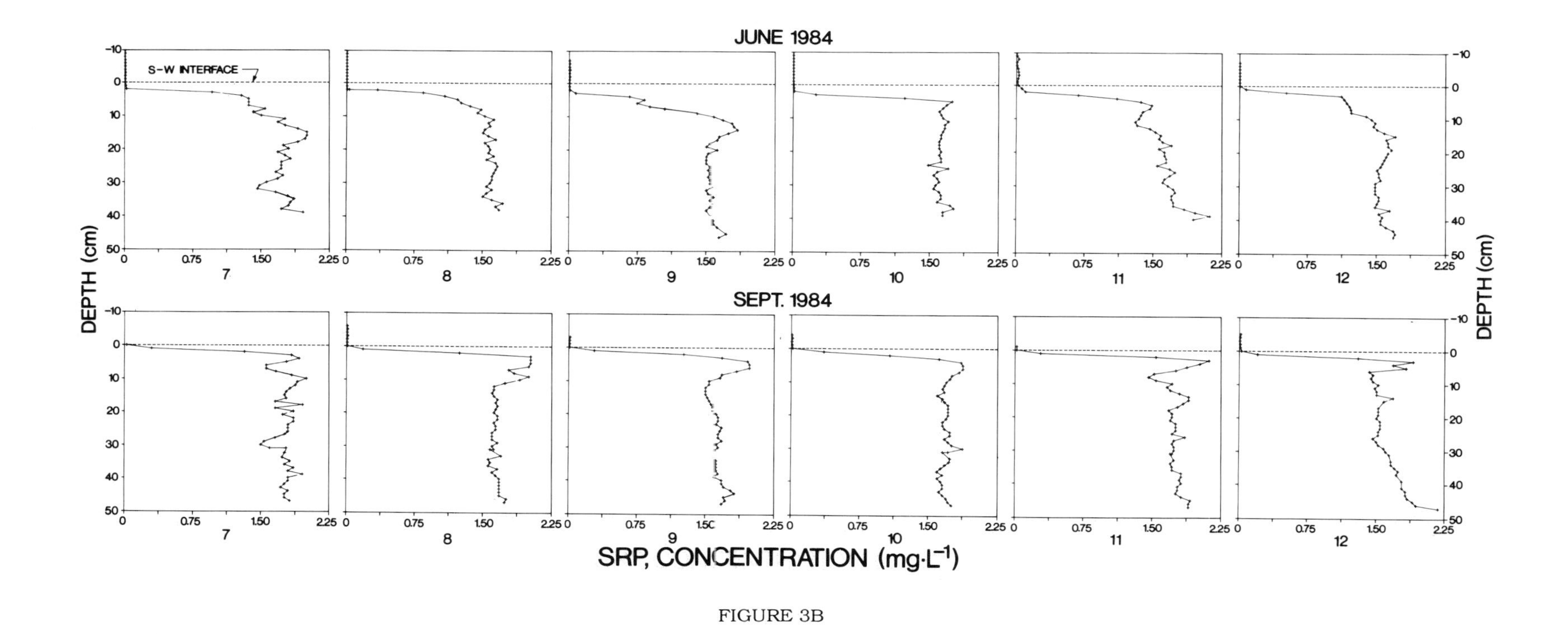
JUNE 1984
S-W INTERFACE
DEPTH (cm)
7
8
9
10
11
12
SEPT. 1984
SRP, CONCENTRATION (mg·L^{-1})

FIGURE 3B

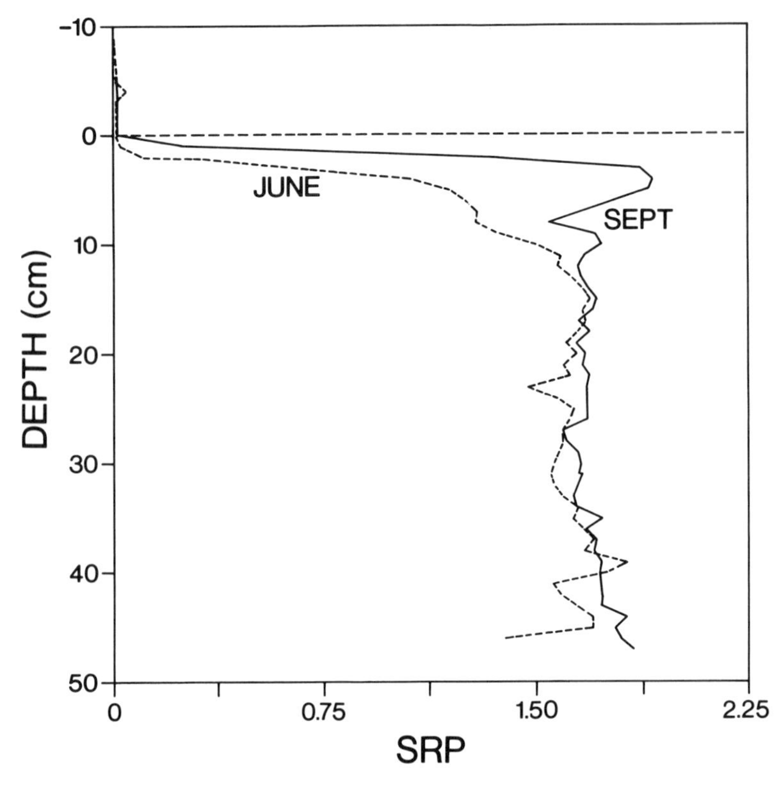

FIGURE 3C

Simon et al.[53] designed an *in situ* pore water sampler similar to that of Hesslein,[54] except it contained two cover sheets located flush to each side. These were made of 3-mm-thick acrylic plastic, each cut to match the elongated compartments in the body of the sampler. After covering the sampler with a membrane and attaching the first cover sheet, the sampler was immersed into a N_2-purged solution of sodium chloride having a concentration of not more than one-half the highest concentration of NaCl anticipated at the sampling site. The diluted salt solution minimized the osmotic loss of water from the sampler while maintaining a concentration gradient favorable for the transport of dissolved species into the sampler. A second acrylic cover sheet, containing similar elongated openings, was then slipped carefully over the membrane-covered, solution-filled cells without trapping gas bubbles and held in place with screws.

Montgomery et al.[63] developed a 5-cm diameter *in situ* porous Teflon sampler, connected to PVC pipe, to determine nutrient concentrations in anoxic pore waters of estuarine sediments. The sampler, used in muddy sands overlain by water 0.5 m deep, could be kept in place for extended periods. Several samplers were inserted to various depths in the sediment and allowed to equilibrate for 1 week to eliminate any possible oxygen contamination introduced during deployment. Pore water samples (50 to 75 ml) were withdrawn periodically using vacuum along with an inert displacement gas. This sampler was used to measure the pore water chemistry associated with diel changes in a seagrass bed[64] and the physiology of an overwash mangrove forest on the Florida eastern coast.[65] A futher refinement to this sampler allowed for *in situ* pore water collection through porous Teflon filter rings, 1 cm wide, spaced at 2-cm intervals. Each of the four chambers was separated by Plexiglass washers with 1.59-mm holes to accommodate polyethylene tubing for periodic sampling from the surface.[66] A similar pore water sampling device designed for rivers and other sandy environments was constructed by Hertkorn-Obst et al.[67] using a perforated brass outer tube and four internal chambers connected with tubing to the surface.

A sampler related to the design of Montgomery et al.[63] was developed by Howes et al.[40] to collect pore water from shallow marsh environments. This consisted of a Teflon sleeve connected to glass tubing of various lengths. The Teflon sleeve was sealed at the bottom with a glass plug. A serum stopper was attached to the top for removal of pore water. After 1 week of equilibration, pore water samples were collected with gas-tight glass syringes. The depth of pore water retrieval was determined by the length of the glass tubing. Only 5 to 10 ml were removed at any one time.

Mayer[55] developed a simple, light-weight (<0.5 kg) *in situ* dialysis bag equilibration device. This sampler was made of a perforated Lucite tube fitted with a removable cone. Internal chambers were separated by rubber washers fitted over a solid Lucite rod. The washers held Lucite spacers which fitted snugly over the inner rod. Each chamber contained a dialysis bag wrapped around the central rod. After equilibration in the sediments for 4 days (at 20 to 25°C), the sampler was retrieved and dialysis bags cut open to collect pore water. A modification of this sampler was designed by Bottomley and Bayly[68] for sampling pore water at 4-cm intervals in submerged macrophyte root zones. Ten small vials (10 to 12 ml internal volume) were stacked on top of each other; each had ports covered with dialysis membranes which were glued and secured with O-rings. Equilibrium for phosphate required 10 days in anoxic sediments.

Hopner[69] described an *in situ* diffusion device with 20 chambers placed over a distance of 33 cm. It was used for sampling pore water from aerobic sediments and, with modifications, from anoxic sedimentary environments. This device consisted of a 14-mm stainless steel rod with 20 chambers (each holding 3 ml) covered by a membrane which was mechanically protected and held in place by a stainless steel plate. The sampler had a pointed bottom and board at the top to facilitate insertion into the sediment. For pore water from anoxic sediments, the device was slightly modified by storing it within a plastic liner to prevent oxygen contamination. This liner was driven into the sediment around the sampler so that an external core filled with sediments was present during retrieval.

van Eck and Smits[70] described an improved modification (called a memocell) of the Hesslein[54] sampler for studies of nutrient fluxes across the sediment-water interface in shallow lakes. Major modifications were the use of a second solid acrylic plate and the construction of side ports for collecting pore water (Figure 4). The sampler with membrane and first acrylic sheet were placed in an equilibration chamber of nitrogen gas bubbling for deoxygenation for 48 h. While in the degassed water, the second solid acrylic plate was slid onto the top of the first acrylic sheet to protect the membrane and minimize oxygen invasion. The equilibration chamber and dialysis sampler were transported to the sampling site. After deployment in the sediments the solid plate would be removed by a diver; before retrieval the plate was reinserted once again. During pore water removal the sampler was kept at the *in situ* temperature. Pore water was withdrawn into special 10-ml plastic monovette syringes (Sarstedt,

FIGURE 4. An *in situ* dialyzer with solid cover plate and side ports for removing pore water under an inert gas atmosphere (courtesy of G.Th.M. van Eck).

Numbrecht-Rommelsdorf, FDR), complete with septa, through a sliding apparatus mounted over the side ports of the sampler. During sample recovery the entire pore water to syringe transfer system was bathed in an atmosphere of nitrogen gas. Pore water profiles at the Grote Rug freshwater reservoir in the Netherlands are shown in Figure 5. Another pore water sampler with shutters positioned over the dialysis membrane was also designed by Kepkay et al.[71] This device was used to measure *in situ* whole sediment diffusion coefficients for dissolved sulfate and total dissolved CO_2 (mainly as bicarbonate ion).

Pore water equilibration times for *in situ* sediment samplers varied consid erably depending on the investigator. The reported time of equilibration ranged from 6 to 15 days, and even as long as a month. After recovery, pore water was typically collected in specially prepared syringes and injected into washed, and often acidified, containers. Storage was usually at 4°C for various time intervals. Carignan[72] reported several potential sources of errors in the sampling methodology: the type of membrane, the material for construction of the sampling device, the chemical state of the initial degassed filling water, the general design of the device, and the selection of appropriate equilibration times. It was felt that few investigators actually demonstrated the suitability of their particular design and sampling technique. According to Carignan[72]

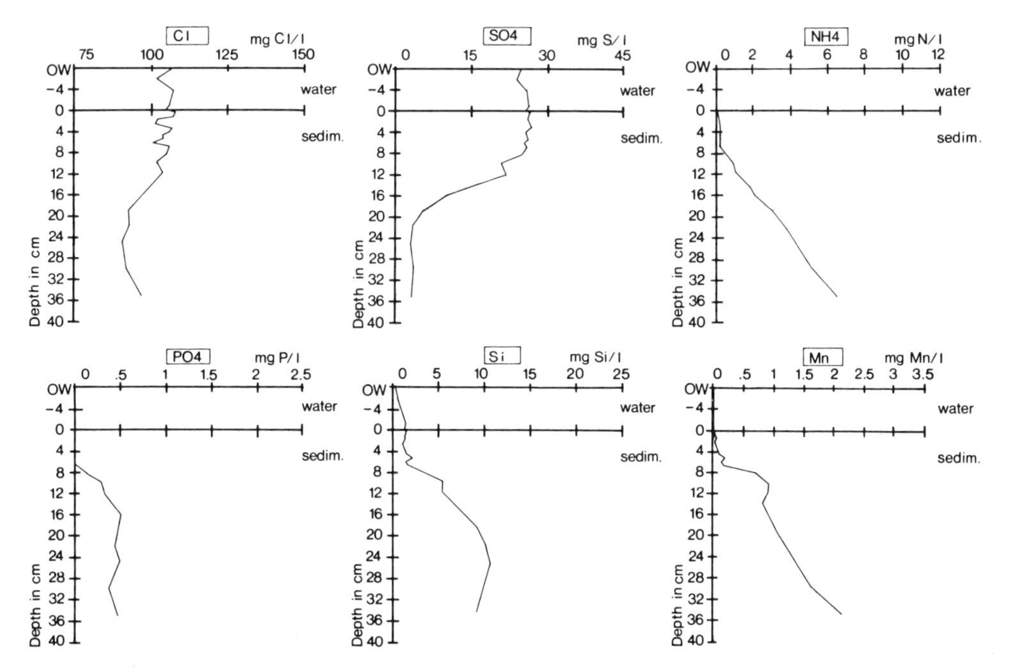

FIGURE 5. Chloride, sulfate, ammonium, orthophosphate, silicon, and manganese profiles in the pore water of the Grote Rug freshwater reservoir, the Netherlands, collected with an *in situ* dialyzer shown in Figure 4 (courtesy G.Th.M. van Eck).

pore water should be recovered from the sampler within 5 min of retrieval from the sediments. Postponement of sampling significantly reduces concentrations of dissolved reactive phosphorus and iron. For nonvolatile species, pore water was injected into acidified vials adjusted to a final pH of 2.5 to 3.5, stored at 4°C, and analyzed within 2 weeks. Pore water collected for the determination of dissolved organic carbon and methane was quickly injected into evacuated glass tubes, treated with 0.05 ml of saturated $HgCl_2$ and analyzed within 3 days. It was shown that cellulose-based dialysis membranes can seriously lead to the underestimation or overestimation of pore water solutes because of membrane breakdown. Compared to Plexiglass dialyzers, material constructed of polycarbonate gave continued problems with iron precipitation. Some precipitation of iron was also observed in Plexiglass dialyzers. It was suggested that oxygen, initially present in the plastic of the dialyzer body, diffused into the degassed water after placement into anoxic sediments. This can be prevented by storing the dialyzers in an inert gas while not in use and by minimizing atmospheric contact during preparation and field use. Furthermore, the initial presence of dissolved oxygen in the degassed water of the dialyzer compartments can significantly effect sample composition. These devices should therefore be deoxygenated before sampling, especially for elements and compounds known to be unstable in the presence of oxygen. This is the case for many pore water constituents. Carignan[72] suggested equilibration times between 3 to 20 days depending on the chemical species and the temperature and sediment composition of the particular site. In most environments, 20 days for cold (4 to 6°C) and 15 days for warm (20 to 25°C) sediments would allow for a sufficient equilibration period for most major ions and nutrients.[72]

Last, an example of a special application for a diffusion-controlled pore water sampling device is the Lucite "dipstick" containing slots casted with polyacrylamide gels. The gels were dosed with lead acetate for studying the distribution of sulfide in anoxic marine sediments.[73] Since the gels are 90% water, reaction with pore water solutes is controlled by diffusive transport and occurs rapidly for 2-mm gel "dipsticks" (calculated 90% equilibration within 30 min) in bottom sediments. A black precipitate forms because dissolved sulfide in the pore water reacts with lead in the gel. The degree of darkening of the gel was measured to determine the pore water sulfide content.

Sayles et al.[56,74] developed a different type of *in situ* sampler to quickly extract pore water from deep oceanic sediments. The sampler collected filtered (Whatman No. 3) pore water simultaneously at several sediment depths as well as from the overlying water. It consisted of 2-m long, heavy wall stainless steel tubing with a pointed tip for sediment penetration and had five to six filter-covered sampling ports. A broad base plate above the ports halted sediment penetration, and one port above this plate was positioned for sampling overlying water. The instrument functioned as a large syringe, with a spring-loaded master cylinder providing the suction. After lowering to the bottom, a full 2-m penetration triggered the system and initiated the sampling. At each of the sampling ports pore water was drawn through a filter into a Teflon tubing (1.6-mm o.d., 0.8-mm i.d.) capillary storage system. After about 30 min the sampler was retrieved. With the exception of a few sampling stations, including two at water depths greater than 1800 m, pore water (usually greater than 10 ml) was recovered at each of six depths from 5 to 200 cm. In further studies dealing with the measurement of pore water gradients and the calculation of diffusive fluxes across the sediment-water interface in the Atlantic Ocean, Sayles[75] pointed out two problems with his sampler. Overpenetration of the sampler into the sediments could result in an incorrect assigned depth, and disturbance of the probe during the sampling period (20 to 30 min) could cause leakage along the barrel and contamination of pore water with overlying bottom water. Murray and Grundmanis[76] replaced the Teflon capillary system with nylon sample loops (approximately 4-ml calibrated pressure tubing) for pore water gas analysis. A comparison of pore water collected with this sampler and two other techniques (centrifugation and squeezing) was reported by Jahnke et al.[77] for box and gravity cores collected from the MANOP site in the central Pacific Ocean and by Murray et al.[78] for squeezing of Saanich Inlet sediments.

Another tube system, designed after the Sayles et al.[56] sampler, was used for studies of time variations in pore water nutrient concentrations in the Dutch Wadden Sea.[79] A 90-cm perspex tube (30-mm o.d., 20-mm i.d.) was designed with 16 sampling ports at distances spaced close together near the sediment-water interface. Sampling ports were covered by a 100-μm nylon gauze. The sampler was used in shallow subtidal environments where pore water samples were removed at low tide (water depth of 40 to 80 cm) through stainless steel capillary tubing connected to 10-ml glass syringes located at the water surface.

Brinkman et al.[80] also developed a shallow water sampler similar to the system designed by Sayles et al.[56] Pore water is collected in a tube, either by hydrostatic pressure or vacuum, after passage through a filter. The device is only useful in waters less than 10 m deep.

Whiticar[81] collected pore water and gases using a modified *in situ* sampler designed by Barnes.[82] This sampler included replacement of the delay valve mechanism, which prevented premature filtering, with a stainless steel rupture disk punctured approximately 10 s after the sampler penetrated the sediment surface. It was deployed at 20- to 5500-m water depths and can be inserted up to a sediment depth of 12 m (usually attached as an outrigger to a core barrel[82]). The sampler would admit interstitial gases and fluids by hydrostatic pressure through 0.5-μm filters into 15-ml sample cylinders which had been previously flushed with helium and evacuated.

Last, in shallow water areas pore water was collected with scuba divers by direct insertion of syringe needles (13 gauge) into the sediments. Long needles were marked at 0.5-cm intervals and attached to plastic syringes. Approximately 1 to 5 ml of pore water was drawn into the syringe from each depth by the diver.[25]

III. SPECIAL APPLICATIONS — SEDIMENT GASES

The sampling of pore water gases represents a special technical problem. Various methodologies for sampling sediment gases are reviewed in this section. These techniques should be considered when devising a sampling program for pore water gases. In addition to the special sediment handling and processing techniques to avoid oxidation as described in previous sections, gases represent a special problem because they can exist in the sediments as both dissolved or gaseous phases. Rapid changes between these two phases can occur as well. For example, changes in pressure and temperature from *in situ* conditions during sample collection and warming of sediments before core processing can initiate, at times, spontaneous bubble formation. The worst reported cases have been exploding cores and the frothing of sediments in core liners; other examples were the formation of fissures or numerous bubbles throughout the sediments, the appearance of bubbles collecting along core liners, and the possible upward movement of larger gas bubbles which destroy sedimentary laminae or structure and invalidate later gas analyses. Because of these constraints, it is recommended that professionals be knowledgeable not only in the technology of pore water sampling but that they also be cognizant of the problems and shortcomings of sediment gas sampling.

Some basic suggestions that should be adhered to in any sediment pore water gas sampling program are (1) a clear or opaque plastic core liner should be used to observe sediment bubbles and potential degassing phenomena, (2) disturbance of the sediments should be minimized to avoid loosening bubbles, (3) cores should be subsampled immediately if sediments are stored for later gas analysis, (4) if subsampling cannot be done immediately, then sediments should be cooled to 1 to 2°C to lower bacterial action and slow bubble production and possibly repressurized, (5) since oxygen will diffuse through most plastics (probably not through thick plastic core liners), sediment or pore water subsamples collected in plastic syringes should be stored in an inert atmosphere (or resubmerged into sediments to avoid oxidation), and (6) *in situ* pore water dialyzer samplers should be processed very rapidly (5 min after recovery, if a solid plate is not part of the dialyzer system). If possible, it is also advisable to protect against oxygen diffusion through the dialyzer membrane by subsampling in an inert atmosphere. Refrigerating sediment cores for cooling could be futile when the hydrostatic pressure is released during retrieval of deep sediments; this time delay could cause bubble formation and degassing. It is suggested that after collection (and perhaps after storage) a capped core be sacrificed. It should be aggitated to determine the amount of bubbles or potential for gas ebullition.

As described in the previous sections, sediment pore water gases can either be collected from recovered sediments and processed at the surface or sampled by *in situ* devices. For recovered sediments, pore water gases are usually extracted by squeezing or headspace equilibration of the wet sediments with an inert gas. Even though some investigators used centrifugation to remove pore water for gas analysis,[83-86] because of possible bubble loss this technique is not recommended. Gases dissolved in pore water collected from squeezers and *in situ* dialyzers are usually extracted for analysis by headspace equilibration or by gas stripping techniques.

A. COLLECTION FROM RECOVERED SEDIMENTS

Gas bubbles escaping from sediments were measured by Conger[87] and Gould,[88] both cited in Reeburgh,[89] and in later studies by Chau et al.,[90] Ohle,[91,92] Martens and Klump,[59] Bartlett et al.,[93] and Chanton et al.[94] With the exception of some later studies where bubble flux was compared to other transport processes, measurements of bubble composition is only of quantitative interest since alteration occurs during migration through the sediments and overlying water column.

Reeburgh and Heggie,[95] Reeburgh,[9] and Reeburgh and Alperin[96] reviewed some of the earlier sediment gas studies in the freshwater and marine environment. As with the development of any new technique, these early investigations exhibited numerous analytical problems.[89] The first definitive studies were conducted by Reeburgh[97,98] who squeezed Chesapeake Bay sediment pore water in a glove bag filled with either CO_2 or helium. The pore water was introduced into a helium gas stripper[97] and gases were analyzed following the methodology developed by Swinnerton et al.[99,100] Linnenbom et al.[101] provided a statistical evaluation of this technique. The gas stripping system was improved by von Reusmann[102] and Weiss and Craig.[103] Smith[104] reviewed the analytical techniques for sampling and measuring soil atmospheres, while methods dealing with methane in sedimentary environments were evaluated by Conrad and Schutz.[105]

Loading of squeezers always presents a problem with bubble losses during sediment transfer and potential contamination with trace atmospheric components present in the glove bag. Martens[106] and Martens and Berner[107] developed a core-to-squeezer interlock transfer system for handling 1-m gravity cores with 5-cm o.d. plastic liners. Rubber stoppers were rammed firmly into the liner against the sediment immediately after coring. The cores were iced in the field, refrigerated before analysis, and attached to the interlock system. After flushing the interlock for 10 min with either CO_2 or helium, the rubber stopper was displaced and a spring-loaded capper was deployed. The first 2 to 3 cm of sediment was cut off and discarded. The core liner was then pushed into the squeezer, and the liner was carefully backed off with the aid of a plunger leaving about 5 to 8 cm of sediment. The core was capped again, and a push-rod was used to manipulate a bottom cap on to the squeezer. This was always carried out under a continuous flow of inert gas. About 5 ml of squeezed pore water was analyzed for argon, methane, and nitrogen.[108] It is suspected that the core processing time would be considerable.

In other studies, sediments for pore water gases were collected by subcoring the sediments through predrilled holes in the plastic liner (which were previously covered with plastic tape). Tipless disposable plastic syringes (distal end sawn off) were used for subcoring. This technique is usually employed to collect sediments for microbial incubation experiments[109-112] but can also be used for transferring sediments to other containers for gas analysis.[113] For example, Adams et al.,[114] Adams and Fendinger,[115] and Adams and van Eck[116] dispensed sediments directly into glass or plastic syringes for headspace analysis of the pore water gases following a procedure outlined in Fendinger and Adams.[12] A similar technique was also used by Novelli et al.[117] to measure n*M* concentrations of hydrogen gas in marine sediments. Other investigators have reported sectioning of sediments and injecting them directly into jars of various sizes for pore water headspace analysis of different chemical components in relatively low concentrations: radon,[118,119] low molecular weight hydrocarbons,[13,120] as well as for methane,[121] which is usually in high enough concentrations to employ more precise techniques. Sediments might also be sectioned directly for incubation and measurements of sediment gas production.[122] Another similar technique was employed by Atkinson and Hall,[123] where marsh sediments were sampled with a 10-ml tipless disposable syringe and injected into a flask containing degassed water. After stirring, the headspace gases were sparged with helium for 30 min to remove and trap methane on a charcoal column submerged in an acetone/dry ice bath, as described in Swinnerton and Linnenbom.[124]

The subcoring technique, described in Fendinger[125] and Fendinger and Adams,[126] is one of the rapid methods of collecting sediments for pore water gas analysis. After collection, cores were immediately covered with ice (or stored upright in a refrigerator). Sediments were processed from cores placed horizontally in a helium-filled glove bag. The sediments were protected from lateral movement with a piston carefully placed against the sediment surface so as not to disturb the interface. As described above, sediments were collected in 50-ml tipless disposable plastic syringes through 2.8-cm predrilled holes in the core liner. Subsampling was conducted within an hour of collection. An 80- to 100-cm scuba-collected core took 20 to 40 min to process; 14 to 20 subcores (syringes) were obtained. Nonetheless, in many environments bubbles formed on the inside of the core liners. These bubbles remained in approximately the same horizon where they first appeared, and thus did not migrate up the core or collect under the top piston. With sediments containing numerous bubbles, 2 to 31% of the methane was lost (Adams and Naguib, in preparation) during the sediment transfer process from tipless disposable syringes to normal plastic or glass syringes used for later headspace gas analysis. It is expected that this would represent a similar problem during filling of centrifuge tubes and loading squeezers. In order to avoid bubble loss during sample handling, a special apparatus was designed and custom machined at the Max Planck Institute for Limnology, Plon, Germany (Figure 6) for use with fluid lake sediments. Small diameter (4-cm o.d.) plastic liners, normally used with a Zulich (Rheineck, Switzerland) gravity corer, were tightly fitted to this core adapter syringe sampling (CASS) system. A plastic slider, located at the top of the CASS system, contained threaded holes to hold a 25-ml serum monovette (Sarstedt, Numbrecht-Rommelsdorf, FRG) plastic syringe and a glass scintillation vial (both preweighed). This slider could be adjusted to either the open or closed position. After coring, the core was kept vertical and the CASS system (which replaced a top rubber stopper or piston) was attached directly to the top of the liner. A bottom piston was inserted and overlying water expelled with the slider in the open position. As the sediment surface appeared the slider was closed and the entire CASS system and core placed horizontally into a helium-filled glove bag.

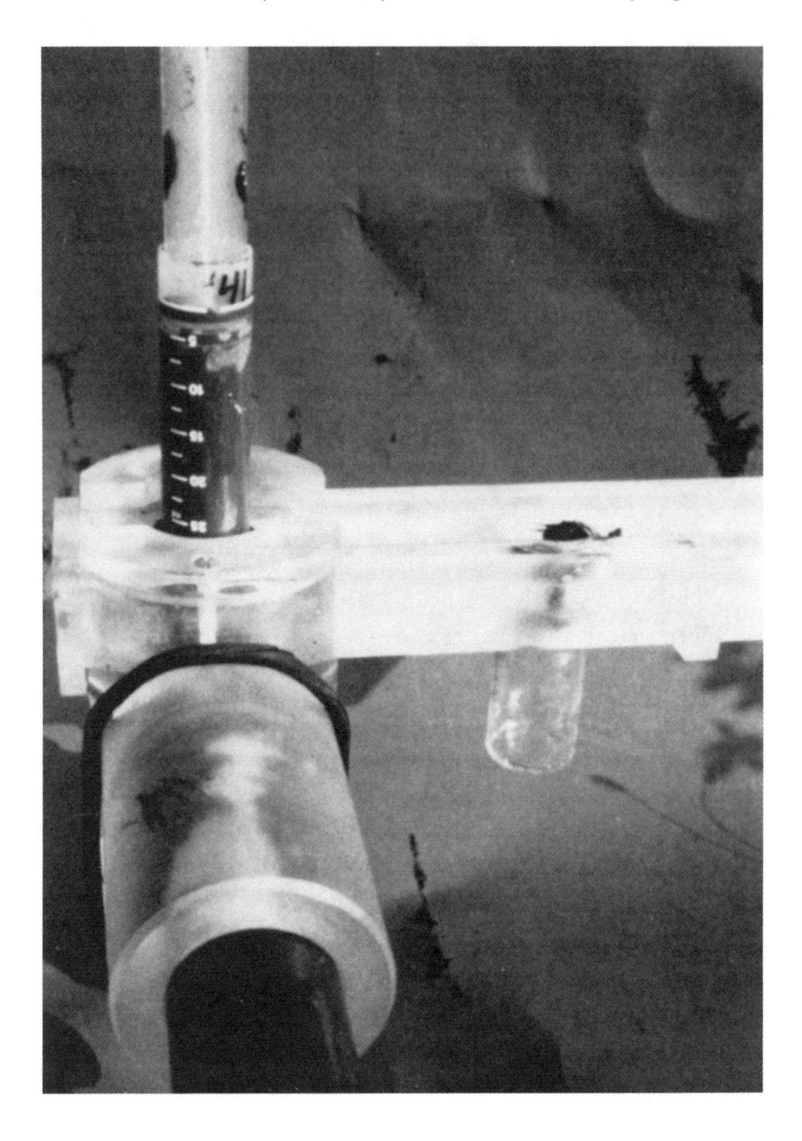

FIGURE 6. The core adapter syringe sampling (CASS) system to collect sediments for gases, including 4-cm o.d. sediment core and top plastic slider containing filled 25-ml serum monovette plastic syringe, empty scintillation vial (for determination of water content), and syringe cap (illustrated from left to right).

A second person applied pressure against the bottom piston, with a rod (marked at 1-cm intervals) extending outside the glove bag, to force sediments into the monovette syringes. Syringes and vials were attached to the slider, which was opened for extrusion of sediments (2 to 3 cm core displacement, 10 to 20 g wet sediment) into the monovette. The slider was then closed to protect the core sediment surface and the open-ended monovette syringe was rapidly moved along the slider to a vial for dispensing a small aliquot of sediment (2 to 5 g per percent water) and then for attachment of the monovette cap. This cap contained a septum (of poor quality), which was reinforced with a silicon insert, for later injection of a helium headspace for pore water gas extraction. Small bubbles collecting against the top curvature of the liner during processing usually migrated with the sediments during extrusion, so it was likely these gas bubbles were retained in the syringe at approximately the same sediment horizon. Processing a 40-cm gravity core with 10 to 12 monovette syringes would take 20 to 30 min. Because of the thickness of the slider 1.5 cm of sediment was lost (discarded) between each syringe. Syringes were then placed in a helium-filled plastic bag (double lined), stored on ice or resubmerged into cold sediments if storage was longer than 12 to 24 h. In the laboratory the syringes were cleaned, weighed, and an exact volume (5.0 ml or less) of helium headspace added. Syringes were vigorously shaken for 2 min and 500 μl of headspace was removed with a side-port needle for gas analysis.

Headspace equilibration is based on the technique developed by McAullife[127] for the analysis of hydrocarbons in water. Earlier studies were also conducted by Bassette et al.[128] for sulfides, esters, and alcohols in aqueous solutions and by Kepner et al.[129] for other volatile organic compounds. Gas extraction with an inert headspace has been adapted to sediment pore waters by Kaplan et al.[130] for N_2-gas production during denitrification, by Conrad et al.[131] for H_2 and CH_4 in sewage sludge and lake sediments, and by Naguib[132] for CH_4 and Fendinger and Adams[12] for Ar, O_2, N_2, CH_4, and CO_2 in lake sediments. A technique for measuring CO_2 by headspace equilibration is described in Stainton.[133] With this technique, a specified volume or weight of wet sediments is transferred to a sample container or syringe and a measured amount of inert gas is added. After agitation, an aliquot of the headspace is removed and injected into a gas chromatograph. As an example, Williams and Crawford[134] used the technique described in Naguib[132] for extracting methane from peatland cores. Headspace equilibration is also one of the techniques used for extracting gases from pore water recovered from *in situ* dialyzer samplers and squeezers.

The squeezer technique to remove pore water from recovered sediments for gas measurements is the most widely used system. Reeburgh[97,98] conducted the first squeezer pore water gas extractions. Gas pressure acting against a rubber diaphragm in the squeezer[45] expressed pore water into a gas stripper for analysis, as described above. Squeezers were loaded in glove bags flushed with either CO_2 or helium. This technique was used for extracting pore water gases from Long Island Sound[107,108] and Skan Bay[135] sediments. Another type of squeezer designed by Kalil and Goldhaber,[31] as described in Section II.A.2, was employed by Warford et al.[136] for Santa Barbara sediments and by Martens and Goldhaber[137] to measure changes in pore water gases during the transition from terrestrial to marine conditions in a North Carolina estuary. Luther et al.[138] used the squeezer technique to extract pore water organic sulfur compounds, some being volatile, at micromolar levels in salt marsh sediments.

B. *IN SITU* PORE WATER SAMPLING

In situ equilibration and direct sampling techniques for sediment pore water were described in Section II.B. Only special problems and methods related to gas sampling of *in situ* dialyzer systems will be addressed in this section.

Equilibrated pore water recovered from *in situ* dialyzers must be protected from invasion of atmospheric oxygen and the loss of gases through the membrane. This can be accomplished either by rapid sampling of the dialyzer (by piercing the membrane) or through various techniques to cover the membrane surface. Removal of pore water from dialyzer compartments can be conducted under an inert atmosphere in a glove bag, but this technique is extremely unwieldy, especially under field conditions. An inert gas would lower oxygen diffusion across the membrane surface but have little effect on minimizing the gas concentration gradient between pore water and the glove bag atmosphere. Use of a solid acrylic plastic plate to cover the dialyzer membrane during recovery and sample processing, as described in van Eck and Smits,[70] could avoid both of these problems. The solid plate would be inserted into the dialyzer by a scuba diver before removal from the sediments, thus tightly covering the membrane. During insertion of the cover plate a thin, firm layer of sediments collects between the membrane and plate, thus impeding oxygen invasion and minimizing any gas concentration gradient. The design of this sampler also allows for pore water removal through side ports flushed in an atmosphere of inert gas during the sampling procedure. During pore water removal, air infiltration can be avoided by packing mud along the opening between the cover plate and dialyzer body. Ice cushions would also slow gas bubble formation within the dialyzer compartments.

Two other dialyzer techniques were used to study sediment gases. Rudd and Hamilton[139] designed an open-ended Plexiglass box which was placed into the sediments. Along one side of the box holes were drilled at 0.5-cm intervals on a diagonal from top to bottom. Sediments within the box were removed and thin-walled Tygon sampling tubes (volume 0.1 ml) filled with degassed lake water were pushed into the sediments. The tubes were sealed at each end with glass rods and kept rigid with aluminum wires. The Plexiglass box was refilled and left in place for 6 days. Another technique described by Howes et al.[40] was used to collect dimethylsulfide in marsh sediment pore water using a long-term *in situ* equilibrator (described previously in Section II.B).

Pore water removed from the dialyzer chambers is usually analyzed for dissolved gases by headspace equilibration or by injection into a gas stripper. Both techniques were described in the previous section.

Sediment equilibration times for gases using the dialyzer samplers is a function of the diffusion coefficient of the particular gas,[140] the temperature, and porosity of the sediments. Equilibration of CH_4 at 5.5°C and a sediment porosity of 0.91 in lake sediments was complete in about 2 weeks,[72] while Hesslein[54] suggested a 1-week period. Kuivila and Lovley[148] reported 2 to 4 weeks for H_2 equilibration. Chanton et al.[94] used a 3-week equilibration period with a 3-mil-thick Teflon membrane; in their earlier studies with 3-µm-Teflon a 10- to 14-day equilibration was considered sufficient.[59] However, this was not the case in a study by Adams and van Eck[116] (Figure 7), where CH_4 concentrations from an *in situ* dialyzer (1 week equilibration time, sediment porosity 0.4) were much lower than measured by total core gas analysis for sediments collected nearby (within 0.5 m) by scuba divers. Wet sediments in the cores were analyzed within 1 to 2 days by headspace techniques. The two techniques (*in situ* dialyzer vs. total sediments) were comparable for surface 0- to 20-cm sediments with a 0.7 porosity. It was suspected that pore water methane would not equilibrate with water in the dialyzer compartments in the deeper, clayey layers exhibiting low porosity.

Winfrey and Zeikus[141] collected pore water from Lake Mendota to measure sulfide and methane and clarify the mechanism of sulfate inhibition of methanogenesis. Volatile reduced sulfur gases were measured by Morgan et al.[149] from peat pore water collected with *in situ* "sippers".[150] Welch et al.[142] used the dialyzer system to study methane metabolism in an arctic lake. Adams et al.[114] calculated the diffusive flux of CH_4 in Lake Erie sediments and determined the percentage of sediment oxygen demand resulting from methane oxidation at the sediment-water interface. Kipphut et al.[143] and Chanton et al.[94] used the

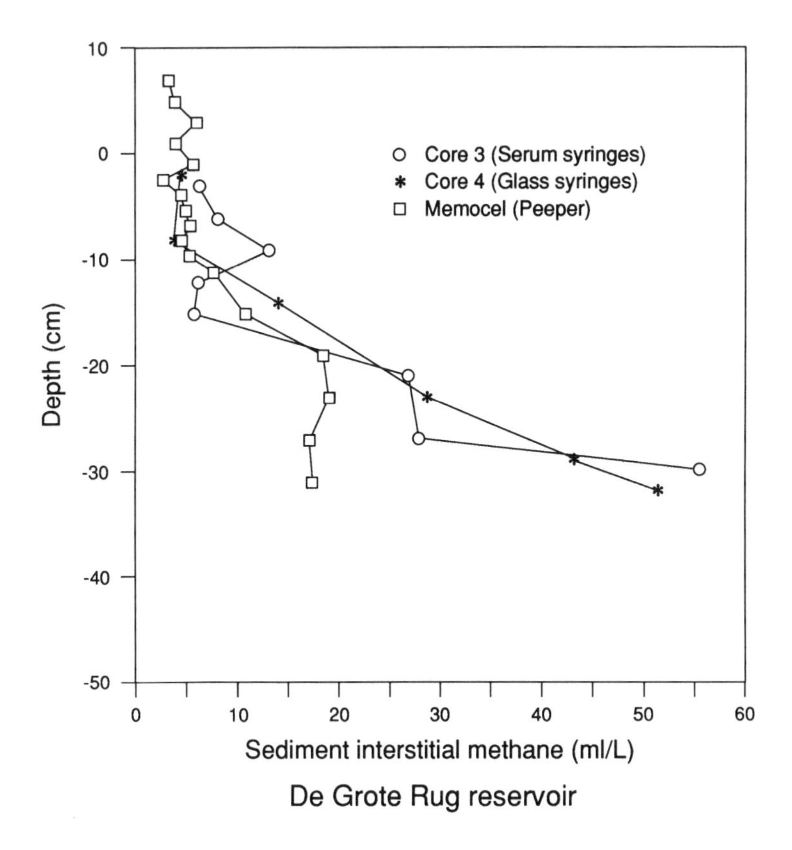

FIGURE 7. Dissolved methane (ml/l) profiles in the Grote Rug freshwater reservoir (see Figure 5 for other variables), with depth above and below (-) the sediment-water interface (0 cm). One profile was from an *in situ* dialyzer (memocel); the other two were from sediments collected through predrilled holes and stored (1 to 2 days) in either plastic (serum) or glass syringes for headspace analysis. The CASS system described above was not used. (From Adams and van Eck.[116] From Proceedings of the Third International Workshop on the Measurement of Microbial Activities in the Carbon Cycle in Aquatic Ecosystems, Cappenberg and Steenbergen, Eds., Archiv fur Hydrobiology Beiheft 31, E. Schweizerbart'sche Verlagsbuchhandlung, Stuttgart, Germany, 1988. With permission.)

dialyzer with 75-μm (3 mil) Teflon membranes to measure pore water Ar, N_2, and CH_4 in the sediments at Cape Lookout Bight, NC. Bartlett et al.[144] used the dialyzer to calculate the total amount of methane in marsh sediments and determine the balance between microbial methanogenesis and oxidation in a Virginia marsh. Sebacher et al.[145] showed that CH_4 concentrations were highest near the surface area of marsh sediments surrounding the root zones of cattails. Last, profiles of methane were reported by Bartlett et al.[93] using a dialyzer in the sediments of the central Amazon floodplain.

The *in situ* sampler designed by Sayles et al.,[56,74] described in Section II.B, was used to collect deep-sea sediment pore water for measurements of Ar, N_2, and CH_4.[81] Pore water dissolved oxygen profiles were reported by Murray and Grundmanis[76] for pelagic marine sediments in the central equatorial Pacific. They used a modified version of the Sayles sampler. This same sampler was also used by Kuivila et al.[146,147] to collect and measure pore water methane in the sediments of Lake Washington and two marine basins (Saanich and Princess Louisa Inlets) in British Columbia, Canada.

ACKNOWLEDGMENTS

The author would especially like to thank Bert van Eck for the loan of memocells for testing and sediment gas studies in Europe. He has also most generously provided figures for this publication. Professor Dr. Jurgen Overbeck and Dr. Monir Naguib of the Max Planck Institute for Limnology, Plon, Germany, are thanked for providing funding and help with improvements in sediment gas sampling techniques, and Mr. F. W. Scholler was most instrumental in designing and machining the CASS core-to-syringe sampling system. Last, I would like to acknowledge Alena Mudroch for her tremendous enthusiasm and understanding throughout the preparation of this chapter, and to F. Rosa and G. H. Adams for their critical review of the manuscript.

REFERENCES

1. **Krom, M.D. and Berner, R.A.,** The diffusion coefficients of sulfate, ammonium, and phosphate ions in anoxic marine sediments, *Limnol. Oceanogr.*, 25, 327, 1980.
2. **Manheim, F.T.,** Interstitial waters of marine sediments, in *Chemical Oceanography*, Vol. 6, Riley, J.P. and Chester, R., Eds., Academic Press, New York, 1976, 115.
3. **Kriukov, P.A. and Manheim, F.T.,** Extraction and investigative techniques for study of interstitial waters of unconsolidated sediments: a review, in *The Dynamic Environment of the Ocean Floor*, Fanning, K.A. and Manheim, F.T., Eds., Lexington Books, D.C. Heath, Lexington, MA, 1982, 3.
4. **Froelich, P.N., Klinkhammer, G.P., Bender, M.L., Luedtke, N.A., Heath, G.R., Cullen, D., Dauphin, P., Hammond, D., Hartmann, B., and Maynard, V.,** Early oxidation of organic matter in pelagic sediments of the eastern equatorial Atlantic: suboxic diagenesis, *Geochim. Cosmochim. Acta*, 43, 1075, 1979.
5. **Malcolm, S.J., Kershaw, P.J., Lovett, M.B., and Harvey, B.R.,** The interstitial water chemistry of $^{239,240}Pu$ and ^{241}Am in the sediments of the north-east Irish Sea, *Geochim. Cosmochim. Acta*, 54, 29, 1990.
6. **Matisoff, G., Lindsay, A.H., Matis, S., and Soster, F.M.,** Trace metal mineral equilibria in Lake Erie sediments, *J. Great Lakes Res.*, 6, 353, 1980.
7. **Adams, D.D., Hess, G.G., Fendinger, N.J., Deis, D.A., Wagel, D.M., Parrish, D.M., and Henry, J.,** Chemical study of the interstitital water dissolved organic matter and gases in Lake Erie, Cleveland Harbor and Hamilton Harbour bottom sediments — composition and fluxes to overlying waters, OWRT A-059-OHIO, Ohio State University, Columbus, 1984.
8. **Emerson, S.,** Early diagenesis in anaerobic lake sediments; chemical equilibria in interstitial waters, *Geochim. Cosmochim. Acta*, 40, 925, 1976.
9. **Reeburgh, W.S.,** A major sink and flux control for methane in marine sediments: anaerobic consumption, in *The Dynamic Environment of the Ocean Floor*, Fanning, K. and Manheim, F., Eds., Lexington Books, D.C. Heath, Lexington, MA, 1982, 203.
10. **McCaffrey, R.J., Myers, A.C., Davey, E., Morrison, G., Bender, M., Luedtke, N., Cullen, D., Froelich, P., and Klinkhammer, G.,** The relation between pore water chemistry and benthic fluxes of nutrients and manganese in Narragansett Bay, Rhode Island, *Limnol. Oceanogr.*, 25, 31, 1980.
11. **Giblin, A.E. and Howarth, R.W.,** Porewater evidence for a dynamic sedimentary iron cycle in salt marshes, *Limnol. Oceanogr.*, 29, 47, 1984.
12. **Fendinger, N.J. and Adams, D.D.,** A headspace equilibration technique for measurement of dissolved gases in sediment pore water, *Int. J. Environ. Anal. Chem.*, 23, 253, 1986.
13. **Oremland, R.S., Miller, L.G., and Whiticar, M.J.,** Sources and flux of natural gases from Mono Lake, California, *Geochim. Cosmochim. Acta*, 51, 2915, 1987.
14. **Lebel, J., Silverberg, N., and Sundby, B.,** Gravity core shortening and pore water chemical gradients, *Deep Sea Res.*, 29, 1365, 1982.

15. **Emerson, S., Jahnke, R., Bender, M., Froelich, P., Klinkhammer, G., Bowser, C., and Setlock, G.,** Early diagenesis in sediments from the eastern equatorial Pacific. I. Pore water nutrient and carbonate results, *Earth Planet. Sci. Lett.*, 49, 57, 1980.
16. **Schimmelmann, A., Lange, C.B., and Berger, W.H.,** Climatically controlled marker layers in Santa Barbara Basin sediments and fine-scale core-to-core correlations, *Limnol. Oceanogr.*, 35, 165, 1990.
17. **Engler, R.M., Brannon, J.M., Rose, J., and Bigham, G.,** A practical selective extraction procedure for sediment characterization, in *Chemistry of Marine Sediments*, Yen, T.F., Ed., Ann Arbor Science Publishers, MI, 1977, 163.
18. **Adams, D.D., Darby, D.A., and Young, R.J.,** Selected analytical techniques for characterizing the metal chemistry and geology of fine-grained sediments and interstitial water, in *Contaminants and Sediments*, Vol. 2., Baker, R.A., Ed., Ann Arbor Science Publishers, MI, 1980, 3.
19. **Anderson, R.F., Schiff, S.L., and Hesslein, R.H.,** Determining sediment accumulation and mixing rates using Pb-210, Cs-137, and other tracers: problems due to postdepositional mobility of coring artifacts, *Can. J. Fish. Aquat. Sci.*, 44, 231, 1987.
20. **Fanning, K.A. and Pilson, M.E.,** Interstitial silica and pH in marine sediments: some effects of sampling procedures, *Science*, 173, 1228, 1971.
21. **Bray, J., Bricker, J.T., and Troup, O.P.,** Phosphate in interstitial waters of anoxic sediments, *Earth Planet. Sci. Lett.*, 18, 1362, 1973.
22. **Troup, B.N., Bricker, O.P., and Bray, J.T.,** Oxidation effect on the analysis of iron in the interstitial water of Recent anoxic sediments, *Nature*, 249, 237, 1974.
23. **Lyons, W.B., Gaudette, H.E., and Smith, G.M.,** Pore water sampling in anoxic carbonate sediments: oxidation artefacts, *Nature*, 277, 48, 1979.
24. **Saager, P.M., Sweerts, J.-P., and Ellermeijer, H.J.,** A simple pore water sampler for coarse, sandy sediments with low porosity, *Limnol. Oceanogr.*, 35, 747, 1990.
25. **Bauer, J.E., Montagna, P.A., Spies, R.B., Prieto, M.C., and Hardin, D.,** Microbial biogeochemistry and heterotrophy in sediments of a hydrocarbon seep, *Limnol. Oceanogr.*, 33, 1493, 1988.
26. **Carignan, R., Rapin, F., and Tessier, A.,** Sediment pore water sampling for metal analysis: a comparison of techniques, *Geochim. Cosmochim. Acta*, 49, 2493, 1985.
27. **Rutledge, P.A. and Fleeger, J.W.,** Laboratory studies on core sampling with applications to subtidal meiobenthos collection, *Limnol. Oceanogr.*, 33, 274, 1988.
28. **Vanderborght, J.-P. and Billen, G.,** Vertical distribution of nitrate concentration in interstitial water of marine sediments with nitrification and denitrification, *Limnol. Oceanogr.*, 20, 953, 1975.
29. **Vanderborght, J.-P., Wollast, R., and Billen, G.,** Kinetic models of diagenesis in disturbed sediments. I. Mass transfer properties and silica diagenesis, *Limnol. Oceanogr.*, 22, 787, 1977.
30. **Siever, R.,** A squeezer for extracting interstitial water from modern sediments, *J. Sediment. Petrol.*, 32, 329, 1962.
31. **Kalil, E.K. and Goldhaber, M.,** A sediment squeezer for removal of pore waters without air contact, *J. Sediment. Petrol.*, 43, 553, 1973.
32. **Hartmann, M.,** An apparatus for the recovery of interstitial water from Recent sediments, *Deep Sea Res.*, 12, 225, 1965.
33. **Presley, B.J., Brooks, R.R., and Kappel, H.M.,** A simple squeezer for removal of interstitial water from ocean sediments, *J. Mar. Res.*, 25, 355, 1967.
34. **Kriukov, P.A. and Komorova, N.A.,** On squeezing fluids from clays at very high pressures (Russian), *Dokl. Akad. Nauk SSSR*, 99, 617, 1954.
35. **Manheim, F.T.,** A hydraulic squeezer for obtaining interstitial water from consolidated and unconsolidated sediments, U.S. Geol. Survey Prof. Paper 550-C, 1966, 256.
36. **Brown, F.S., Baedecker, M.J., Nissenbaum, A., and Kaplan, I.R.,** Early diagenesis in a reducing fjord, Saanich Inlet, British Columbia. III. Changes in organic constituents of sediments, *Geochim. Cosmochim. Acta*, 36, 1185, 1972.
37. **Kriukov, P.A.,** Sovremennye issledovanyafyziko-khimicheskikh svoistv pochv, in *Rukovodstvo Polevykh i Laboratornykh Issledovanii Pochv*, Vol. 2, Moscow Izdat. Akad. Nauk, Moscow, U.S.S.R., 1947.
38. **Manheim, F.T.,** Disposable syringe techniques for obtaining small quantities of pore water from unconsolidated sediments, *J. Sediment. Petrol.*, 38, 666, 1972.
39. **Davison, W., Woof, C., and Turner, D.R.,** Handling and measurement techniques for anoxic interstitial waters, *Nature*, 295, 582, 1982.

40. **Howes, B.L., Dacey, J.W.H., and Wakeham, S.G.,** Effects of sampling technique on measurements of porewater constituents in salt marsh sediments, *Limnol. Oceanogr.*, 30, 221, 1985.
41. **Jahnke, R.A.,** A simple, reliable, and inexpensive pore-water sampler, *Limnol. Oceanogr.*, 33, 483, 1988.
42. **Aller, R.C.,** Experimental studies of changes produced by deposit feeders on porewater sediment, and overlying water chemistry, *Am. J. Sci.*, 278, 1185, 1978.
43. **Bender, M., Martin, W., Hess, J., Sayles, F., Ball, L., and Lambert, C.,** A whole-core squeezer for interfacial pore-water sampling, *Limnol. Oceanogr.*, 32, 1214, 1987.
44. **Hartmann, M. and Muller, P.J.,** Trace metals in interstitial waters from central Pacific Ocean sediments, in *The Dynamic Environment of the Ocean Floor*, Fanning, K.A. and Manheim, F.T., Eds., Lexington Books, D.C. Heath, Lexington, MA, 1982, 285.
45. **Reeburgh, W.S.,** An improved interstitial water sampler, *Limnol. Oceanogr.*, 12, 163, 1967.
46. **Robbins, J.A. and Gustinis, J.,** A squeezer for efficient extraction of pore water from small volumes of anoxic sediment, *Limnol. Oceanogr.*, 21, 905, 1976.
47. **Klinkhammer, G.P.,** Early diagenesis in sediments from the eastern equatorial Pacific. II. Pore water metal results, *Earth Planet. Sci. Lett.*, 49, 81, 1980.
48. **Hines, M.E., Knollmeyer, S.L., and Tugel, J.B.,** Sulfate reduction and other sedimentary biogeochemistry in a northern New England salt marsh, *Limnol. Oceanogr.*, 34, 578, 1989.
49. **Scholl, D.W.,** Techniques for removing interstitial water from coarse-grained sediments for chemical analyses, *Sedimentology*, 2, 156, 1963.
50. **Glass, G.E. and Poldoski, J.E.,** Interstitial water components and exchange across the water sediment interface in Lake Superior, *Verh. Int. Verein. Limnol.*, 19, 405, 1975.
51. **Kinniburgh, D.G. and Miles, D.L.,** Extraction and chemical analysis of interstitial water from soils and rocks, *Environ. Sci. Technol.*, 17, 362, 1983.
52. **van Raaphorst, W. and Brinkman, A.G.,** The calculation of transport coefficients of phosphate and calcium fluxes across the sediment-water interface, from experiments with undisturbed sediment cores, *Water Sci. Technol.*, 17, 941, 1984.
53. **Simon, N.S., Kennedy, M.M., and Massoni, C.S.,** Evaluation and use of a diffusion-controlled sampler for determining chemical and dissolved oxygen gradients at the sediment-water interface, *Hydrobiologia*, 126, 135, 1985.
54. **Hesslein, R.H.,** An *in situ* sampler for close interval pore water studies, *Limnol. Oceanogr.*, 21, 912, 1976.
55. **Mayer, L.M.,** Chemical water sampling in lakes and sediments with dialysis bags, *Limnol. Oceanogr.*, 21, 909, 1976.
56. **Sayles, F.L., Wilson, T.R.S., Hume, D.N., and Mangelsdorf, P.C., Jr.,** *In situ* sampler for marine sedimentary pore waters: evidence for potassium depletion and calcium enrichment, *Science*, 181, 154, 1973.
57. **Tessier, A., Carignan, R., Dubreuil, B., and Rapin, F.,** Partitioning of zinc between the water column and the oxic sediments in lakes, *Geochim. Cosmochim. Acta*, 53, 1511, 1989.
58. **Carignan, R. and Flett, R.J.,** Postdepositional mobility of phosphorus in lake sediments, *Limnol. Oceanogr.*, 26, 361, 1981.
59. **Martens, C.S. and Klump, J.V.,** Biogeochemical cycling in an organic-rich coastal marine basin. I. Methane sediment-water exchange process, *Geochim. Cosmochim. Acta*, 44, 471, 1980.
60. **Kipphut, G.W. and Martens, C.S.,** Biogeochemical cycling in an organic-rich coastal marine basin. III. Dissolved gas transport in methane-saturated sediments, *Geochim. Cosmochim. Acta*, 46, 2049, 1982.
61. **Kelly, C.A., Rudd, J.W.M., Furutani, A., and Schindler, D.W.,** Effects of lake acidification on rates of organic matter decomposition in sediments, *Limnol. Oceanogr.*, 29, 687, 1984.
62. **Carignan, R. and Tessier, A.,** Zinc deposition in acid lakes: the role of diffusion, *Science*, 228, 1524, 1985.
63. **Montgomery, J.R., Zimmermann, C.F., and Price, M.T.,** The collection, analysis and variation of nutrients in estuarine pore water, *Estuarine Coastal Mar. Sci.*, 9, 203, 1979.
64. **Montgomery, J.R., Zimmermann, C., Peterson, G., and Price, M.,** Diel variations of dissolved ammonia and phosphate in estuarine sediment pore water, *Fla. Sci.*, 46, 407, 1983.
65. **Carlson, P.R., Yarbro, L.A., Zimmermann, C.F., and Montgomery, J.R.,** Pore water chemistry of an overwash mangrove island, *Fla. Sci.*, 46, 239, 1983.
66. **Montgomery, J.R., Price, M.T., Holt, J., and Zimmermann, C.,** A close-interval sampler for collection of sediment pore waters for nutrient analyses, *Estuaries*, 4, 75, 1981.

67. **Hertkorn-Obst, U., Wendeler, H., Feuerstein, T., and Schmitz, W.,** A device for sampling interstitial water out of river and lake beds, *Environ. Technol. Lett.*, 3, 263, 1982.
68. **Bottomley, E.Z. and Bayly, I.L.,** A sediment porewater sampler used in root zone studies of the submerged macrophyte, *Myriophyllum spicatum*, *Limnol. Oceanogr.*, 29, 671, 1984.
69. **Hopner, T.,** Design and use of a diffusion sampler for interstitial water from fine grained sediments, *Environ. Technol. Lett.*, 2, 187, 1981.
70. **van Eck, G.Th.M. and Smits, J.G.C.,** Calculation of nutrient fluxes across the sediment-water interface in shallow lakes, in *Sediments and Water Interactions*, Sly, P.G., Ed., Springer-Verlag, New York, 1986, 293.
71. **Kepkay, P.E., Cooke, R.C., and Bowen, A.J.,** Molecular diffusion and the sedimentary environment: results from the in situ determination of whole sediment diffusion coefficients, *Geochim. Cosmochim. Acta*, 45, 1401, 1981.
72. **Carignan, R.,** Interstitial water sampling by dialysis: methodological notes, *Limnol. Oceanogr.*, 29, 667, 1984.
73. **Reeburgh, W.S. and Erickson, R.E.,** A "dipstick" sampler for rapid, continuous chemical profiles in sediments, *Limnol. Oceanogr.*, 27, 556, 1982.
74. **Sayles, F.L., Mangelsdorf, P.C., Jr., Wilson, T.R.S., and Hume, D.N.,** A sampler for the *in situ* collection of marine sedimentary pore waters, *Deep Sea Res.*, 23, 259, 1976.
75. **Sayles, F.L.,** The composition and diagenesis of interstitial solutions. I. Fluxes across the seawater-sediment interface in the Atlantic Ocean, *Geochim. Cosmochim. Acta*, 43, 527, 1979.
76. **Murray, J.W. and Grundmanis, V.,** Oxygen consumption in pelagic marine sediments, *Science*, 209, 1527, 1980.
77. **Jahnke, R., Heggie, D., Emerson, S., and Grundmanis, V.,** Pore waters of the central Pacific Ocean: nutrient results, *Earth Planet. Sci. Lett.*, 61, 233, 1982.
78. **Murray, J.W., Grundmanis, V., and Smethie, W.M.,** Interstitial water chemistry in the sediments of Saanich Inlet, *Geochim. Cosmochim. Acta*, 42, 1011, 1978.
79. **van der Loeff, R.M.M.,** Time variation in interstitial nutrient concentrations at an exposed subtidal station in the Dutch Wadden Sea, *Neth. J. Sea Res.*, 14, 123, 1980.
80. **Brinkman, A.G., van Raaphorst, W., and Lijklema, L.,** *In situ* sampling of interstitial water from lake sediments, *Hydrobiologia*, 92, 659, 1982.
81. **Whiticar, M.J.,** Determination of interstitial gases and fluids in sediment collected with an *in situ* sampler, *Anal. Chem.*, 54, 1796, 1982.
82. **Barnes, R.O.,** An *in situ* interstitial water sampler for use in unconsolidated sediments, *Deep Sea Res.*, 20, 1125, 1973.
83. **Crill, P.M. and Martens, C.S.,** Spatial and temporal fluctuations of methane production in anoxic coastal marine sediments, *Limnol. Oceanogr.*, 28, 1117, 1983.
84. **Lidstrom, M.E. and Somers, L.,** Seasonal study of methane oxidation in Lake Washington, *Appl. Environ. Microbiol.*, 47, 1255, 1984.
85. **Andreae, M.O.,** Dimethylsulfide in the water column and the sediment porewaters of the Peru upwelling area, *Limnol. Oceanogr.*, 30, 1208, 1985.
86. **Kuivila, K.M., Murray, J.W., Devol, A.H., Lidstrom, M.E., and Reimers, C.E.,** Methane cycling in the sediments of Lake Washington, *Limnol. Oceanogr.*, 33, 571, 1988.
87. **Conger, P.S.,** Ebullition of Gases from Marsh and Lake Waters, Publ. 56, Chesapeake Biol. Lab., 1943.
88. **Gould, H.R.,** Character of the accumulation sediment, gas content, in Comprehensive Survey of Sedimentation in Lake Mead, 1948-9. U.S. Geol. Survey Prof. Paper 295, 1960, 180.
89. **Reeburgh, W.S.,** Measurements of Gases in Sediments, Ph.D. thesis, Johns Hopkins University, Baltimore, MD, 1967.
90. **Chau, Y.K., Snodgrass, W.J., and Wong, P.T.S.,** A sampler for collecting evolved gases from sediment, *Water Res.*, 11, 807, 1977.
91. **Ohle, W.,** Ebullition of gases from sediment, conditions, and relationship to primary production of lakes, *Verh. Int. Verein. Limnol.*, 20, 957, 1978.
92. **Ohle, W.,** Mineral impact of lakes as background for their biogenic dynamics, *Hydrobiologia*, 72, 51, 1980.
93. **Bartlett, K.B., Crill, P.M., Sebacher, D.I., Harriss, R.C., Wilson, J.O., and Melack, J.M.,** Methane flux from the central Amazon floodplain, *J. Geophys. Res.*, 93, 1571, 1988.
94. **Chanton, J.P., Martens, C.S., and Kelley, C.A.,** Gas transport from methane-saturated, tidal freshwater and wetland sediments, *Limnol. Oceanogr.*, 34, 807, 1989.
95. **Reeburgh, W.S. and Heggie, D.T.,** Microbial methane consumption reactions and their effect on methane distributions in freshwater and marine environments, *Limnol. Oceanogr.*, 22, 1, 1977.

96. **Reeburgh, W.S. and Alperin, M.J.,** Studies on anaerobic methane oxidation, SCOPE/UNEP Sonderband Heft 66, Mitt. Geol.-Palaont. Inst., University of Hamburg, 367, 1988.
97. **Reeburgh, W.S.,** Determination of gases in sediments, *Environ. Sci. Technol.*, 2, 140, 1968.
98. **Reeburgh, W.S.,** Observations of gases in Chesapeake Bay sediments, *Limnol. Oceanogr.*, 14, 1969, 367.
99. **Swinnerton, J.W., Linnenbom, V.J., and Cheek, C.H.,** Determination of dissolved gases in aqueous solutions by gas chromatography, *Anal. Chem.*, 34, 483, 1962.
100. **Swinnerton, J.W., Linnenbom, V.J., and Cheek, C.H.,** Revised sampling procedure for determination of dissolved gases in solution by gas chromatography, *Anal. Chem.*, 34, 1509, 1962.
101. **Linnenbom, V.J., Swinnerton, J.W., and Cheek, C.H.,** Statistical evaluation of gas chromatography for the determination of dissolved gases in sea water, Report 6344, U.S. Naval Research Laboratory, Washington, D.C., 1966.
102. **von Reusmann, G.,** Eine gaschromatographische Methode zur automatischen Bestimmung der im Meerwasser gelosten Gase, *Kiel. Meeresforsch.*, 24, 14, 1968.
103. **Weiss, R.F. and Craig, H.,** Precise shipboard determination of dissolved nitrogen, oxygen, argon and total inorganic carbon by gas chromatography, *Deep Sea Res.*, 20, 291, 1973.
104. **Smith, K.A.,** Gas chromatographic analysis of the soil atmosphere, in *Soil Analysis Instrumental Techniques and Related Procedures*, Smith, K.A., Ed., Marcel Dekker, New York, 1983, 407.
105. **Conrad, R. and Schutz, H.,** Methods of studying methanogenic bacteria and methanogenic activities in aquatic environments, in *Methods in Aquatic Bacteriology*, Austin, B., Ed., John Wiley & Sons, New York, 1988, 301.
106. **Martens, C.S.,** A method for measuring dissolved gases in pore water, *Limnol. Oceanogr.*, 19, 525, 1974.
107. **Martens, C.S., and Berner, R.A.,** Methane production in the interstitial waters of sulfate-depleted marine sediments, *Science*, 185, 1167, 1974.
108. **Martens, C.S. and Berner, R.A.,** Interstitial water chemistry of anoxic Long Island Sound sediments. I. Dissolved gases, *Limnol. Oceanogr.*, 22, 10, 1977.
109. **Mountfort, D.O. and Asher, R.A.,** Role of sulfate reduction versus methanogenesis in terminal carbon flow in polluted intertidal sediment of Waimea Inlet, Nelson, New Zealand, *Appl. Environ. Microbiol.*, 42, 252, 1981.
110. **Lovley, D.R. and Klug, M.J.,** Intermediary metabolism of organic matter in the sediments of a eutrophic lake, *Appl. Environ. Microbiol.*, 43, 552, 1982.
111. **Lovley, D.R. and Goodwin, S.,** Hydrogen concentrations as an indicator of the predominant terminal electron-accepting reactions in aquatic sediments, *Geochim. Cosmochim. Acta*, 52, 2993, 1988.
112. **Naguib, M.,** Kinetics of acetate and methanol conversion into methane in eutrophic sediments and its application to anaerobic systems, *Water Sci. Technol.*, 20, 61, 1988.
113. **Alperin, M.J. and Reeburgh, W.S.,** Geochemical observations supporting anaerobic methane oxidation, in *Microbial Growth on C-1 Compounds*, Crawford, R.L. and Hanson, R.S., Eds., American Society Microbiology, Washington, D.C., 1984, 282.
114. **Adams, D.D., Matisoff, G., and Snodgrass, W.J.,** Flux of reduced chemical constituents (Fe^{2+}, Mn^{2+}, NH_4^+ and CH_4) and sediment oxygen demand in Lake Erie, *Hydrobiologia*, 92, 405, 1982.
115. **Adams, D.D. and Fendinger, N.J.,** Early diagenesis of organic matter in the Recent sediments of Lake Erie and Hamilton Harbour. I. Carbon gas geochemistry, in *Sediments and Water Interactions*, Sly, P.G., Ed., Springer-Verlag, New York, 1986, 305.
116. **Adams, D.D. and van Eck, G.Th.M.,** Biogeochemical cycling of organic carbon in the sediments of the Grote Rug reservoir, *Arch. Hydrobiol. Beih.*, 31, 319, 1988.
117. **Novelli, P.C., Scranton, M.I., and Michener, R.H.,** Hydrogen distributions in marine sediments, *Limnol. Oceanogr.*, 32, 565, 1987.
118. **Berelson, W.M., Hammond, D.E., and Fuller, C.,** Radon-222 as a tracer for mixing in the water column and benthic exchange in the southern California borderland, *Earth Planet. Sci. Lett.*, 61, 41, 1982.
119. **Gruebel, K.A. and Martens, C.S.,** Radon-222 tracing of sediment-water chemical transport in an estuarine sediment, *Limnol. Oceanogr.*, 29, 587, 1984.
120. **Bernard, B.B., Brooks, J.M., and Sackett, W.M.,** Light hydrocarbons in Recent Texas continental shelf and slope sediments, *J. Geophys. Res.*, 83, 4053, 1978.
121. **Rashid, M.A., Vilks, G., and Leonard, J.D.,** Geological environment of a methane-rich Recent sedimentary basin in the Gulf of St. Lawrence, *Chem. Geol.*, 15, 83, 1975.
122. **Seitzinger, S., Nixon, S., Pilson, M.E.Q., and Burke, S.,** Denitrification and N_2O production in near-shore marine sediments, *Geochim. Cosmochim. Acta*, 44, 1853, 1980.

123. **Atkinson, L.P. and Hall, J.R.,** Methane distribution and production in the Georgia salt marsh, *Estuarine Coastal Mar. Sci.*, 4, 677, 1976.
124. **Swinnerton, J.W. and Linnenbom, V.J.,** Determination of the C_1 to C_4 hydrocarbons in seawater by gas chromatography, *J. Chromatogr.*, 5, 570, 1967.
125. **Fendinger, N.J.,** Distribution and Related Fluxes of Dissolved Pore Water Gases (CH_4, N_2, CO_2) in the Sediments of Lake Erie and Two Polluted Harbors, M.Sc. thesis, Wright State University, Dayton, OH., 1981.
126. **Fendinger, N.J. and Adams, D.D.,** Nitrogen gas supersaturation in the Recent sediments of Lake Erie and two polluted harbors, *Water Res.*, 21, 1371, 1986.
127. **McAullife, C.,** GC determination of solutes by multiple phase equilibration, *Chem. Tech.*, 1, 46, 1971.
128. **Bassette, R., Ozeris, S., and Whitnah, C.H.,** Gas chromatographic analysis of head space gas of dilute aqueous solutions, *Anal. Chem.*, 34, 1540, 1962.
129. **Kepner, R.E., Maarse, H., and Strating, J.,** Gas chromatographic head space techniques for the qantitative determination of volatile components in multicomponent aqueous solutions, *Anal. Chem.*, 36, 77, 1964.
130. **Kaplan, W., Valiela, I., and Teal, J.M.,** Denitrification in a salt marsh ecosystem, *Limnol. Oceanogr.*, 24, 726, 1979.
131. **Conrad, R., Phelps, T.J., and Zeikus, J.G.,** Gas metabolism evidence in support of the juxtaposition of hydrogen-producing and methanogenic bacteria in sewage slude and lake sediments, *Appl. Environ. Microbiol.*, 50, 595, 1985.
132. **Naguib, M.,** A rapid method for the quantitative estimation of dissolved methane and its application in ecological research, *Arch. Hydrobiol.*, 82, 66, 1978.
133. **Stainton, M.P.,** A syringe gas-stripping procedure for gas-chromatographic determination of dissolved inorganic and organic carbon in fresh water and carbonates in sediments, *J. Fish. Res. Board Can.*, 30, 1441, 1973.
134. **Williams, R.T. and Crawford, R.L.,** Methane production in Minnesota peatlands, *Appl. Environ. Microbiol.*, 47, 1266, 1984.
135. **Reeburgh, W.S.,** Anaerobic methane oxidation: rate depth distributions in Skan Bay sediments, *Earth Planet. Sci. Lett.*, 47, 345, 1980.
136. **Warford, A.L., Kosiur, D.R., and Doose, P.R.,** Methane production in Santa Barbara Basin sediments, *Geomicrobiol. J.*, 1, 117, 1979.
137. **Martens, C.S. and Goldhaber, M.B.,** Early diagenesis in transitional sedimentary environments of the White Oak River estuary, North Carolina, *Limnol. Oceanogr.*, 23, 428, 1978.
138. **Luther, G.W., Church, T.M., Scudlark, J.R., and Cosman, M.,** Inorganic and organic sulfur cycling in salt-marsh pore waters, *Science*, 232, 746, 1986.
139. **Rudd, J.W.M. and Hamilton, R.D.,** Two samplers for monitoring dissolved gases in lake water and sediments, *Limnol. Oceanogr.*, 20, 902, 1975.
140. **Lerman, A.,** *Geochemical Processes Water and Sediment Environments*, John Wiley & Sons, New York, 1979.
141. **Winfrey, M.R. and Zeikus, J.G.,** Effect of sulfate on carbon and electron flow during microbial methanogenesis in freshwater sediments, *Appl. Environ. Microbiol.*, 33, 275, 1977.
142. **Welch, H.E., Rudd, J.W.M., and Schindler, D.W.,** Methane addition to an arctic lake in winter, *Limnol. Oceanogr.*, 25, 100, 1980.
143. **Kipphut, G.W. and Martens, C.S.,** Biogeochemical cycling in an organic-rich coastal marine basin. III. Dissolved gas transport in methane-saturated sediments, *Geochim. Cosmochim. Acta*, 46, 2049, 1982.
144. **Bartlett, K.B., Harriss, R.C., and Sebacher, D.I.,** Methane flux from coastal salt marshes, *J. Geophys. Res.*, 90, 5710, 1985.
145. **Sebacher, D.I., Harriss, R.C., and Bartlett, K.B.,** Methane emissions to the atmosphere through aquatic plants, *J. Environ. Qual.*, 14, 40, 1985.
146. **Kuivila, K.M., Murray, J.W., and Devol, A.H.,** Methane production, sulfate reduction and competition for substrates in the sediments of Lake Washington, *Geochim. Cosmochim. Acta*, 53, 409, 1989.
147. **Kuivila, K.M., Murray, J.W., and Devol, A.H.,** Methane production in the sulfate-depleted sediments of two marine basins, *Geochim. Cosmochim. Acta*, 54, 403, 1990.
148. **Kuivila, K.M. and Lovley, D.R.,** Dissolved hydrogen concentrations in sulfate reducing and methanogenic sediments, in *Cycling of Reduced Gases in the Hydrosphere*, Adams, D.D., Seitzinger, S.P., and Crill, P.M., Eds., E. Schweizerbartsche Verlagsbuchhandlungen, Stuttgart, Germany, submitted.

149. **Morgan, M.D., Diegmann, and Spratt, H.G., Jr.,** Fate of ropogenic sulfur in a cedar dominated wetland, in *Cycling of Reduced Gases in the Hydrosphere*, Adams, D.D., Seitzinger, S.P., and Crill, P.M., Eds., E. Schweizerbart'sche Verlagsbuchhandlungen, Stuttgart, Germany, submitted.
150. **Short, F.T., Davis, M.W., Gibson, R.A., and Zimmerman, C.F.,** Evidence for phosphorus limitation in carbonate sediments of the seagrass *Syringodium filiforme, Estuarine Coastal Shelf Sci.*, 20, 419, 1985.

Chapter 8
CASE STUDIES

José M. Azcue, Jean R. D. Guimarães, Alena Mudroch, Paul Mudroch, and Olaf Malm

This chapter contains three examples of fieldwork, particularly sediment sampling, carried out under studies of the effects of past and present metal mining on the environment. In the examples, the purpose and objectives of the studies are outlined to show how important it is to have good information of the study area with encountered environmental problems, and define the objectives prior to planning the fieldwork, particularly the selection of locations for sampling stations. Further, logistics of the fieldwork, all equipment necessary to carry out the fieldwork, sample handling, and the costs associated with the fieldwork, are described.

The first example, the fieldwork carried out under a study in southwestern Amazon rivers, Brazil, was prepared by Jean R.D. Guimarães and Olaf Malm. The second example from a study in the Northwest Territories, Canada, was prepared by Paul Mudroch. The third example by José M. Azcue and Alena Mudroch describes the fieldwork under a study in British Columbia, Canada.

CASE STUDY 1: BOTTOM AND SUSPENDED SEDIMENT SAMPLING FOR STUDIES ON THE BEHAVIOR OF MERCURY AND OTHER HEAVY METALS IN SOUTHWESTERN AMAZON RIVERS AND A RESERVOIR, BRAZIL

I. INTRODUCTION

Gold mining in the Amazon region is known to have been occurring since the 17th century, but only assumed its present large-scale dimension in the late 1970s. In the past, gravimetric separation procedures were used to extract large gold particles, but from the 1960s on, the introduction of the mercury-gold amalgam technology also made it possible to extract gold present in low concentrations in sediments and soils. Gold is recovered from the amalgam after burning to volatilize mercury (Hg), an activity that is usually performed in open air. Gold mining operations on rivers are now carried out from boats using motor pumps and divers or powerful mechanical dredges that can mine large quantities of sediments taken at depths of up to 30 m. Due to the poor Hg handling procedures, total Hg emissions to the Amazon environment is estimated to be at least 100 tons/year.[1] The socioeconomic importance of the gold mining in the Amazon is high. It directly involves around one million people and is responsible for most of the Brazilian gold production, estimated by Hasse[2] as 135 tons/year, averaged over the last decade. The simplicity of the mining and extraction techniques gives a high mobility to goldminers ("garimpeiros") and the map of gold-mining areas (and consequent Hg pollution)

1-56670-027-2/94/$0.00+$.50

is continually changing. Some areas, however, have been intensively exploited for the last 15 years. This is the case for the mid and upper reach of the Madeira River, in the northwestern reach of the Amazon basin, where gold mining was officially allowed on a 350-km sector of the river, and presently up to 600 mechanical dredges operate simultaneously. This number was higher by a factor of ten during the peak of mining activity, 5 years ago.

The Radioisotopes Laboratory, Rio de Janeiro, Brazil, has been studying the environmental and toxicological consequences of Hg pollution in many areas of the Amazon since 1986, with particular attention to the Madeira River area. Though very little is known about Hg behavior in tropical aquatic systems, the data gathered so far justifies the current concern about the toxicological consequences of Hg releases: approximately one-third of the carnivorous fish caught downstream of the Madeira River gold-mining area during a survey in the 1987-1990 period had total Hg concentrations in excess of the 0.5 μg/g safety limit.[3] The dwellers of the villages along the river depend heavily on fish as a protein source. They frequently exhibit concentrations of total Hg in hair that could lead to adverse health effects.[4] It has recently been demonstrated that Hg in the fish and hair samples is essentially in the form of methylmercury.[5]

II. OBJECTIVES

Mercury is released from gold-mining fields to the atmosphere or to waterways in the metallic form. However, the toxicological consequences of such emissions for populations not occupationally involved in the gold mining will arise from the ingestion of fish containing methylmercury. Mercury must be in the Hg^{+2} form to be methylated, through biological processes that occur mainly in bottom sediments. The main objectives of the present study were therefore:

- To obtain further data on the distribution of Hg and other heavy metals in bottom and suspended sediments of the Madeira River and tributaries
- To obtain the first data on Hg speciation in sediments and sediment pore water
- To evaluate, through radiochemical techniques, the potential net Hg methylation rates in surface sediments

Biological samples such as fish, phytoplankton, and zooplankton were also collected. In this report, only the aspects related to sediment sampling will be described.

III. MATERIALS AND METHODS

A. STUDY AREA

The Madeira River is the largest tributary of the Amazon River. Its flow rate ranges between 20,000 and 49,000 m^3/s, resulting in water level variations of up to 15 m and forming innumerable seasonal lakes. It originates from the Andes mountains and is therefore the most mineralized river of the Amazon basin. The river was classified as a typical white-water river due to its high conductivity and suspended sediment load.[6] Its vast watershed is mostly covered by tropical forest. The gold-mining area on the Madeira River extends from the Bolivian border at the city of Guajara-Mirim, down to the town of Porto Velho (population approximately 286,400, according to the 1991 census), some 350 km downstream (Figure 1). This is a sector of the river that is not suitable for navigation due to the many rapids and waterfalls. The Madeira tributaries in this area are mainly black-water forest streams, with low conductivity and a pH normally ranging from 4 to 6. The streams are not subjected to gold-mining activities but some, like the Mutum-Paraná, are used for mooring and maintenance of the boats with dredging equipment and present elevated Hg levels. The Jamari River is one of the main Madeira tributaries and can be used as a control in evaluation of Hg contamination since there are no records of past or present gold-mining activities on its drainage basin. This river was impounded in 1989 for hydroelectric generation,

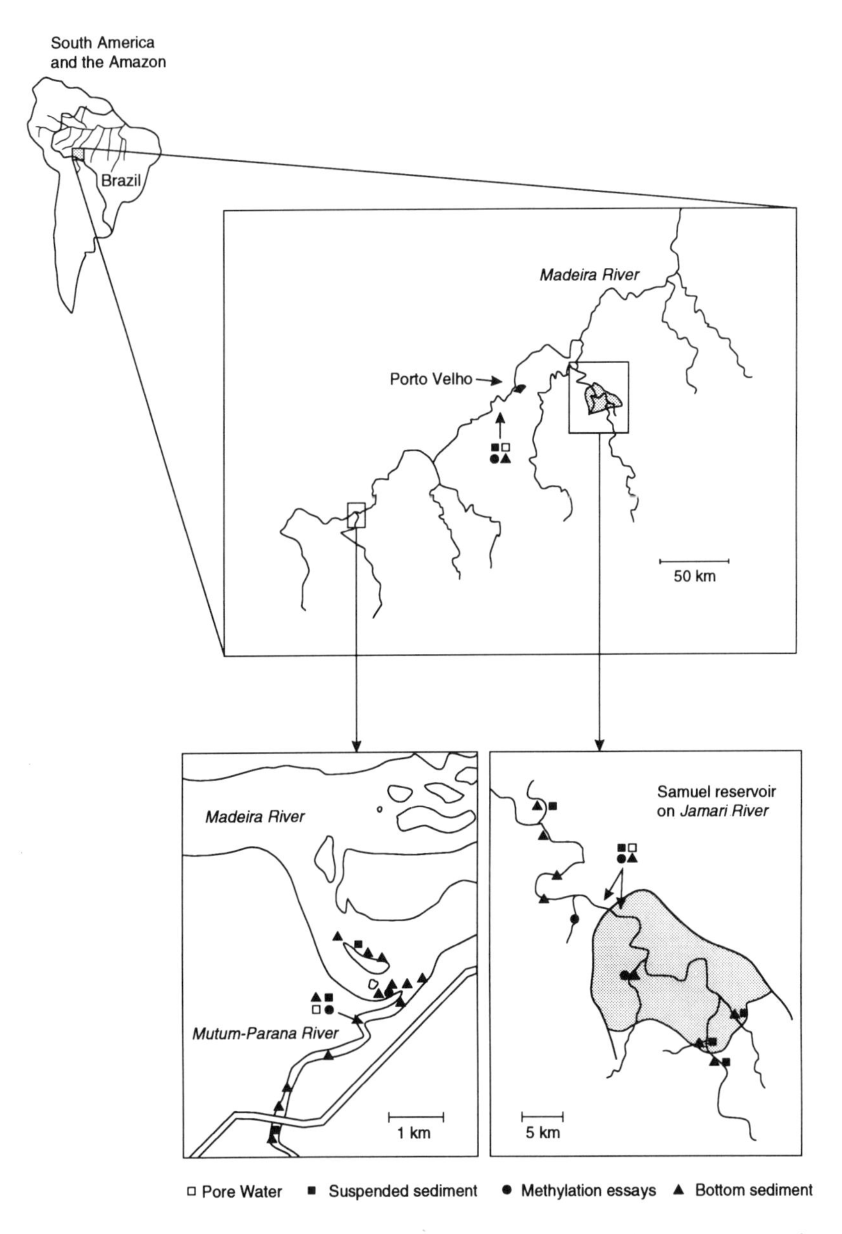

FIGURE 1. Study areas and sampling stations on the Madeira River, Brazil.

forming a 450-km^2 shallow reservoir of flooded forest that can be considered representative of other hydroelectric projects in the Amazon basin, due to its dimension and recent construction. The presence of the Samuel reservoir on the Jamari River was an additional reason to include the latter in this study, since man-made lakes tend to present high Hg levels in fish even when Hg concentrations in water and sediments are relatively very low.

B. SELECTION OF SAMPLING STATIONS

Fine sediments are the best indicators of heavy metals concentration and transport in water systems. Sampling stations were therefore established preferably at depositional areas near the river margins or island banks and as well as upstream rapids and waterfalls where sediments tend to be trapped. Sampling stations on the Madeira River are described in Figure 1. They included the rapids upstream of Porto Velho, the Teotonio waterfalls, and different sampling points near the mouth of the Mutum-Paraná River since this river is, as mentioned before, a significant Hg source to the Madeira River. A total of six sampling stations were established on the Mutum-Paraná to allow comparison with previous detailed surveys of Hg distribution in this river.

The selection of sampling stations on the Jamari River was based on the influence of the Samuel reservoir where oxygen concentrations are low, since the forest was not cleared before flooding, except on a 0.25-km^2 area near the dam. The sampling stations in the reservoir were located to represent the inputs from the different streams that drain into the lake. Downstream sampling stations were chosen at distances of up to 35 km from the dam, which was the longest distance that could be safely reached in a one-day trip and, coincidentally, the point where oxygen concentrations in surface water are back to their normal levels. Sediment pore water and surface sediment samples, the latter for Hg methylation assays, were taken at locations shown in Figure 1.

C. LOGISTICS

The fieldwork described herein was performed during the dry season, in a two-week period in October 1991, involving four research scientists (two M.Sc.s and two Ph.D.s). The work was made possible through the collaboration of different institutions from the State of Rondonia, and arranged well in advance by colleagues from the Rondonia University. Eletronorte Centrais Eletricas, the state energy company that runs the Samuel project, allowed us to use its accommodations both in Porto Velho and near the Samuel dam, the latter including a laboratory with good basic infrastructure, such as large benches, faucets, a fridge, two freezers, and air conditioning. Eletronorte also lent a small truck, a 5-m aluminium boat with a 20-hp engine and a very experienced pilot who was essential to the success of our work on the Samuel reservoir, a labyrinth of semi-submerged forest.

For the sampling work at the Madeira River, the kitchen of the Eletronorte accommodations at Porto Velho was used as the field laboratory. There was no sufficient space available in the fridge and freezer, therefore some samples were accommodated in a freezer at the home of a colleague from the Rondonia University. It was initially planned to establish a field lab in one of the villages on the highway that crosses the Madeira River gold-mining area, to minimize the distances travelled daily to the sampling areas. The high malaria incidence in this region and a quick inspection of the possible accommodations contributed

TABLE 1
List of Equipment Used in Fieldwork

Item	Use
Polyethylene bags, various sizes and thickness	Bottom sediment, soil, and fish samples
250-ml High density polyethylene flasks with large mouth, screw cap	Sediment samples for laboratory Hg methylation assays
100-ml Polyethylene flasks, large mouth, screw cap	Zooplankton and phytoplankton samples
Same as above, narrow mouth, containing 5 ml of 20% $K_2Cr_2O_7$ in concentrated HNO_3	Pore water samples for Hg analysis
300-ml Brown glass flasks, with ground joints	Sample incubation for *in situ* Hg methylation assays
Automatic pipettes, various volumes and respective disposable tips	Preparation of solutions, adding reagents to samples
Nylon fishing net, 2.5-mesh	Fish sampling
4-m Aluminium telescopic stick with plastic cup	Near-shore bottom sediment sampling
Eckman type sampler and 25-m cable	Bottom sediment sampling
Custom-made bottom dredge and 30-m cable	Bottom sediment sampling
Leather gloves	Pulling ropes and cables
Battery operated peristaltic pump with silicon and rubber tubing	Suspended sediment sampling, surface sediment sampling for Hg methylation assays
Stainless steel 47- and 140-mm membrane filter holders	Suspended sediment sampling
0.45-μm Millipore filters, 47- and 100-mm diameter	Sampling of suspended sediment
Perforated Perspex cylinders lined with dialysis film	Sampling pore water for Hg and other heavy metals analysis
PVC tubes with ceramic heads, perforated rubber stoppers and clamped tube	Same as above
Hand-operated vacuum pump	Generating vacuum in the tubes described above
Iceboxes, 50 and 100 l	Sample and gear transportation, shipping of frozen and refrigerated samples
300-l Glass fiber box	Shipping of equipment and samples requiring no refrigeration
Battery-operated temperature and dissolved oxygen meter with 5-m cable	Measuring oxygen, temperature, and pH

TABLE 1 (continued)

Item	Use
Lead shield for 50-ml flask	Hg-203 transport
Geiger-Muller ratemeter	External exposure measurements, spill checking
Dosimetric film badge	Radiation dose measurement
Plastic beakers and volumetric flasks	Preparation of solutions
4 *N* HCl solution	Acidification of samples used in methylation assays
Concentrated HNO_3 solution	Preservation of samples,cleaning glassware and apparatus
20% $K_2Cr_2O_7$ in concentrated HNO_3	Preservation of water samples
PVC tubes with caps and lead weights	Holding glass flasks used for methylation assays
Latex gloves	Protection against acid and/or radioactive solutions
Device for heat-sealing polyethylene bags	Sealing acid and/or radioactive samples
Waterproof marker, pencils, notebooks, etc.	Sample identification, recording notes
Adhesive tape	Sealing polyethylene bags and iceboxes
Maps and reports, aerial photographs	Orientation in the field
Insect repellent	Reduce malaria risk

TABLE 2
Timetable for Field Activities

Day 1: Contacts with local environmental authorities, unpacking and set-up of field lab at Samuel reservoir, meeting with Eletronorte personnel

Day 2: Bottom and suspended sediments sampling at the Jamari River, downstream of the Samuel reservoir

Day 3: Pore water, zoo- and phytotoplankton sampling and methylation assays at the Jamari River, first station downstream of the reservoir (3), fish and bottom invertebrates sampling at the Samuel reservoir (1)

Days 4&5: Bottom and suspended sediments sampling and methylation assays at the Samuel reservoir

Day 6: Lab work, packing of equipment to be used in field and lab work at the Madeira River and Porto Velho

Day 7: Travel to Porto Velho, unpacking, travel to the Mutum-Paraná River to arrange boat rental and check possible accommodations. Set-up of field lab at Porto Velho

TABLE 2 (continued)

Day 8: Bottom and suspended sediments sampling at the Madeira River and at the Mutum-Paraná River mouth

Day 9: Bottom and suspended sediments, sediment pore water and fish sampling, and methylation assays at the lower section of the Mutum-Paraná River (2). Bottom and suspended sediments sampling at the upstream section

Day 10: Bottom and suspended sediments, sediment pore water, zoo- and phytotoplankton and fish sampling, and methylation assays at Teotonio waterfalls on the Madeira River

Day 11: Bottom and suspended sediments sampling at the rapids upstream of Porto Velho. Soil and dust sampling in the Porto Velho urban area, gold dealers district

Day 12: Water and sludge sampling at the Porto Velho water treatment plant. Packing of samples left at the Samuel field lab (2), and packing of remaining samples and equipment (2).

Note: Numbers in brackets represent the number of team members involved in the described activity.

to our decision to keep our base at Porto Velho and travel an extra 300 km daily. The Rondonia State Environmental Agency lent a jeep and a driver. A 3.5-m aluminium boat with a 10-hp engine was rented at one of the floating villages that give support to the gold-mining dredges. A small wooden canoe was towed behind, to allow access to the shallow upper reaches of the Mutum-Paraná River. Table 1 lists the equipment used in the sampling, and Table 2 shows details of the logistics in the sampling trip.

IV. DETAILS OF THE SAMPLING PROCEDURES

A. BOTTOM SEDIMENT

Sampling procedures for bottom sediments varied according to the characteristics of sediments at each sampling station. When water flow was high and/or the bottom was not too irregular, sediments were collected with a specially designed bottom dredge slowly trawled upstream. An Eckman dredge was used in still waters or locations with lots of submerged branches and trunks. Shallow or marginal sites were sampled with a high-density polyethylene cup attached on the extremity of a 2-m aluminium stick whose length could be doubled by screwing on an additional stick. Samples of up to 2 kg, collected for the determination of particle size distribution, and concentration of total Hg and other heavy metals were transferred to half-filled thick polyethylene bags. The bags were closed after removing all air above the sample by tightly twisting their opening and folding the bag into a U shape. The closed bags were secured by adhesive tape, a safe and cheap sealing procedure. The process was repeated so each sample was double-packed. Samples for the determination of Hg chemical forms and availability of Hg and for Hg methylation assays were transferred to large-mouth polyethylene flasks that were overfilled, to avoid unnecessary sample contact with air. When water was shallow and sufficiently clear, the samples for methylation assays were taken by aspiration with a peristaltic pump.

All sample containers were labelled with a waterproof thin-point marker with the date and station name or description. The marking was made on two positions on the sample, to avoid blurring by friction during transport. During the fieldwork, all samples were kept in iceboxes. Upon arrival at the field laboratory, samples for Hg speciation and methylation assays were transferred to a fridge and samples for total Hg analysis to a freezer. When possible, samples in plastic bags requiring freezing were stored in the freezer in open iceboxes, to ensure an efficient utilization of the space in the icebox.

B. SUSPENDED SEDIMENTS

Suspended sediments were sampled at each station by using 47- and 140-mm diameter membranes of 0.45-μm porosity, filtering, respectively, 1 and 5 l of water collected immediately below the surface. Depending on water turbidity, this demanded the use of many filters at each sampling station. The filters were folded in four and kept in labelled individual zip-lock polyethylene bags. The bags were frozen as soon as it was possible.

C. SEDIMENT PORE WATER

Sediment pore water was obtained by introducing a PVC tube with a ceramic head, closed at the other end with a rubber stopper and perforated by a clamped rubber tube, at different depths in the sediment. A hand pump was used to make a vacuum in the tube and pore water was collected after 2 to 5 h, in acid-washed 100-ml polyethylene flasks containing 5 ml of 20% $K_2Cr_2O_7$ in concentrated HNO_3. Pore water samples were stored in the freezer.

D. METHYLATION ASSAYS

Surface sediment slurries were incubated with 45 kBq of Hg-203 in 300-ml dark glass flasks with ground joints. Two samples and an acidified control were stacked in a PVC tube, immersed, and attached with nylon strings at the investigation sites or in their proximities. Incubations were started immediately after sampling or at most 4 h later. Sample incubations were stopped by acidification after 18 to 24 h. Rubber belts were used to secure the glass stoppers and samples and controls were sealed in polyethylene bags and stored in a freezer until Me_2O_3Hg extraction.[7]

The cost related to the fieldwork described in this paper is outlined in Table 3.

V. SUGGESTIONS

- Suspended sediment sampling was time consuming and relatively expensive (see Table 3) but of little value for Hg determination since filter contamination caused erratic results. Continuous flow centrifugation would be more effective for this purpose.
- One of the water samples assayed for Hg methylation was too full and leaked when frozen, though not causing any spill thanks to the sealed polyethylene bags.
- Due to the high water temperature (25 to 30°C), when fishing nets were left in place overnight most fishes were already rotten early in the morning or seriously damaged by carnivores. Nets should therefore not stand for more than 4 h or so.
- Malaria is endemic in most gold-mining areas of the Amazon. Therefore, fieldwork must be organized in a way to avoid outdoor work at dawn or dusk, periods when the mosquitoes are more active. The use of protective clothes and insect repellent during the day is also highly recommended.

TABLE 3
Estimation of Costs Related to Sampling (salaries not included)

Items	% of Total Cost
Air travel, (Rio-Porto Velho-Rio), four persons	47.5
Shipping of samples and equipment as extra luggage	16.8
Meals for four persons, 13 days	11.9
Radioactive tracers	4.9
Boat rental at the Madeira River, two days	4.9
Diesel, gasoline, and oil for borrowed cars and boats	4
Membrane filters	4
Other expendable supplies (chemicals, tape, cables, etc.)	3
Labor costs for car and boat drivers work at night and during weekends	3

Note: Travel expenses were covered by the Environmental Research Group (MFG), Germany.

CASE STUDY 2: SEDIMENT SAMPLING IN THE VICINITY OF DISCOVERY GOLD MINE, NORTHWEST TERRITORIES, CANADA

I. INTRODUCTION

The mining and milling for gold and other elements of economical interest began in the Northwest Territories, Canada, in the early 1900s. By the 1940s, several large mines were in operation without any environmental restrictions.[8,9] Mine tailings and wastewater from the mining and milling operations were discharged to the surrounding environment without treatment.

To determine the environmental impact from this type of mining operation, the Northwest Territories District Office of the Environmental Protection Service, Yellowknife, initiated biological and geochemical studies at selected abandoned mines in 1978 and 1979, and continued with the studies during 1986. The studies conducted in 1978 and 1979 at Discovery Mine at Giauque Lake showed increased concentrations of mercury and other metals above the background concentrations in bottom sediments in the lake.

The 1986 study design was based on the knowledge that bottom sediments accumulate many contaminants entering an aquatic ecosystem. The sediment particles are derived from erosion of the banks, direct contribution from soil creep, from aeolian transport, and, at relatively infrequent intervals, by floods. Suspended particles from these sources are sorted out by currents in the lake, and become deposited on the lake bottom according to their settling velocity. The background concentrations of major and trace elements in sediments correspond to the geochemical composition of soils and bedrock of the lake's drainage basin, and vary over a narrow range. The geochemistry of the bottom sediments can be altered by the introduction of different materials from pollution sources. It was shown that bottom sediments provide an excellent record of input of pollutants into a lake.[10,11] However, for their capability to accumulate and release different contaminants and nutrients, bottom sediments become recognized as an *in situ* pollution source. Different changes of environmental conditions at the sediment-water interface, such as pH, redox potential, mixing and resuspension of sediment particles, etc. may bring about a release of pollutants from the sediment into lake water. Moreover, sediment-associated pollutants can affect the health of benthic invertebrates, and enter the food chain. Consequently, a determination of pollutants in the sediments at several properly selected sites in a lake provides necessary information on the extent of the pollution. Knowledge of basic geochemistry of the sediments is necessary for recognizing the changes in the material deposited on the lake bottom, and for information on natural, i.e., background, concentrations of many elements or compounds, which may be introduced to the lake from anthropogenic sources.

Knowledge of pollutant concentrations and their relationship to different sediment components is the first step for the assessment of the availability and release of pollutants into the lake's ecosystem. One possible mechanism of the release of pollutants from the sediment is the uptake by rooted plants. Several reports demonstrated uptake of elements by plants growing in polluted sediments dredged from lakes and disposed of in marshes or upland.[12,13] Consequently, plants can be used as a good indicator of the availability of pollutants in soils and sediments, and recycling of the pollutants in the aquatic ecosystem.

The objectives of the 1986 study of contamination at Discovery Mine were based on results from previous studies at the site, and on the present knowledge of the behavior of pollutants in aquatic ecosystems as discussed above. The study was carried out to provide information for evaluation of impact of an abandoned gold mine on an aquatic ecosystem in the area.

II. OBJECTIVES

The main objectives of the 1986 study were as follows:

- To determine physical and chemical characteristics of runoff from abandoned mine tailings
- To determine the geochemical characteristics of the tailings that are in contact with surface water bodies
- To determine the extent of leaching of different elements from the tailings, and their spatial distribution in the bottom sediment of lakes and streams that are in contact with the tailings
- To determine the accumulation of these elements by plants colonizing the tailings

To achieve the above objectives, the following sampling strategy was developed, followed by a sampling program carried out in 1986:

1. Collection of runoff samples from the tailings, where available; additional water samples were collected from pits developed in the tailings
2. Collection of samples from land-based and submerged tailings
3. Collection of different plants growing on the land-based and submerged tailings, and on the land in the vicinity of the tailings
4. Collection of sediment cores from lakes adjacent to the abandoned mining site

This report describes only the methods, equipment, and cost for item number 4, i.e., the sampling of lake bottom sediments in the vicinity of the abandoned gold-mining site.

III. MATERIALS AND METHODS

A. STUDY AREA

The mining site is located 83 km north-northeast of Yellowknife, Northwest Territories (NWT), Canada, at the southwestern end of Giauque Lake (Figure 2). Gold and silver were recovered at the mine which operated between 1944 and 1969. The geology of the location is mainly volcanic and sedimentary rocks of the Yellowknife group. The main volcanic mass is comprised of garnetiferous hornblende-feldspar gneiss, which forms a wide belt across the site. North and east of this mass lies a complex of greywacke and argillite, altered to quartz-feldspar-mica schists and nodular schists. Gold-bearing quartz veins occur within meta-greywacke and slate north of and enclosed in the belt of basic volcanics. Metallic minerals constitute less than 1% of the ore. Pyrrhotite is the chief metallic mineral with minor pyrite, galena, arsenopyrite, and chalcopyrite.[14] The mining operations included mercury amalgamation and cyanidation, using zinc dust and lead salt.

Mine tailings, in excess of 1,100,000 tons, were piped to an undyked area where solids were precipitated. Between 1965 and 1968, tailings were discharged directly into Giauque Lake. The tailings material consisted of 75% particles smaller than about 74 μm before the separation of sand from the tailings for backfill. Quartz was the major component of the tailings with less than 1% pyrite and other metallic minerals.[15] The tailings, used for backfilling and construction of an airstrip at the mining site, have been eroded and transported into Round and Giauque Lakes (Figure 2).

Giauque Lake consists of one main depositional basin, from which the water flows relatively slowly in a south-southwesterly direction. Thistlethwaite Lake, located immediately upstream of the study site, was selected as a control site (Figure 3).

B. SELECTION OF SAMPLING STATIONS

Previous studies carried out at the mining site indicated that major and trace elements associated with sediments were transported from the site into Giauque Lake. Therefore, the location of the bottom sediment sampling stations was chosen to obtain general information on the effects of the tailings eroding from the mining site. Three sampling stations were located along a transect extending from the mining site to the deepest part of the lake. One sampling station was located in the center of the eastern arm of Giauque Lake, and one sampling station was located in the northwestern part of Thistlethwaite Lake, which was used as a control site (Figure 3). To investigate the variations in the concentrations of major and trace elements in the sediments at the sampling stations and within Giauque Lake, triplicate samples were collected at the four sampling stations located in the lake.

C. LOGISTICS

A detailed work plan was prepared about one month before the fieldwork. The work plan was finalized after compilation and evaluation of all relevant data from previous studies, and after all members of the field sampling party

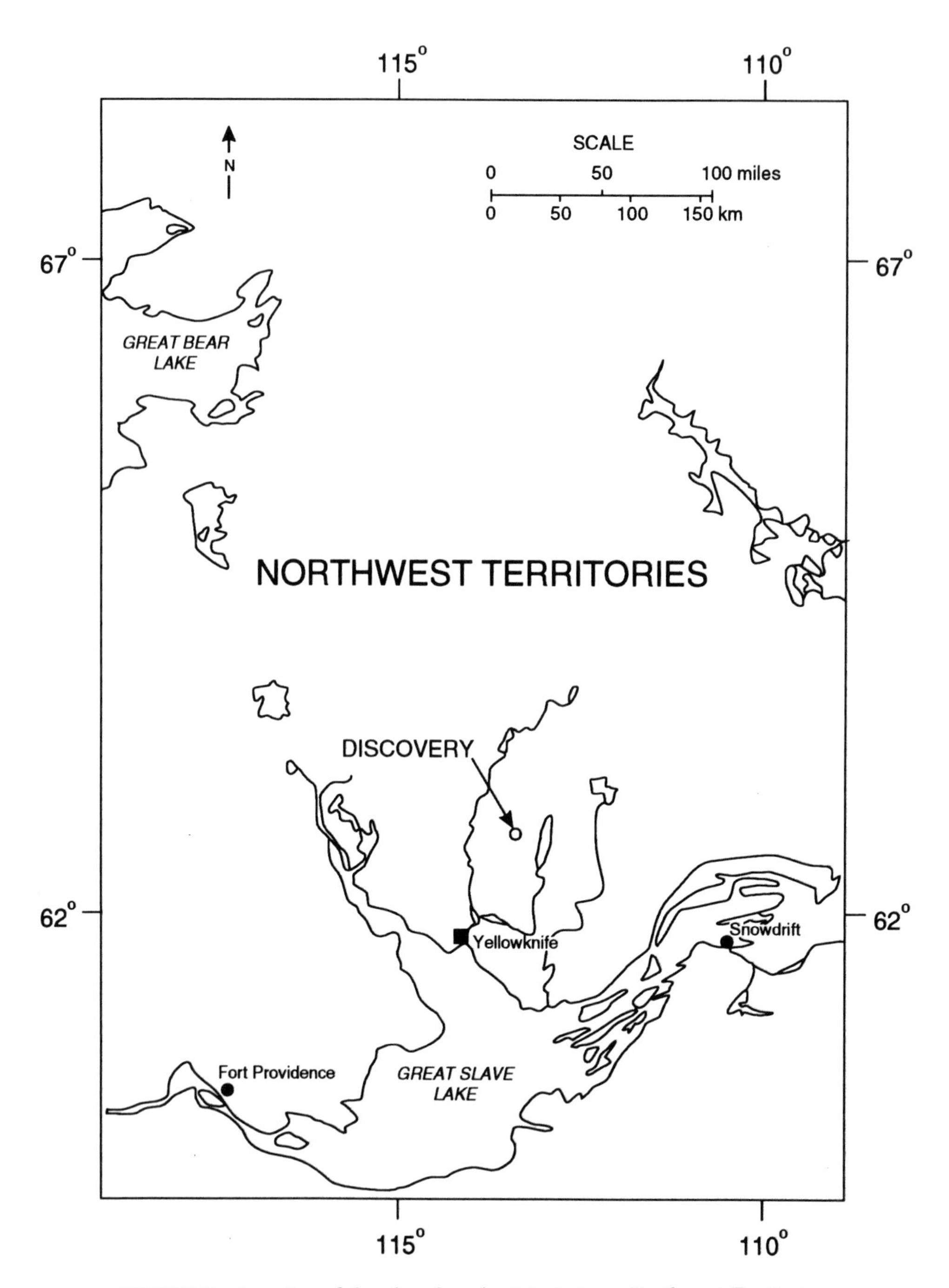

FIGURE 2. Location of the abandoned mining site in Northwest Territories.

finalized the outline of the entire sampling program for the site. Most of the field equipment was assembled in a laboratory located in Yellowknife, NWT. Some of the equipment for sediment coring was shipped from the Canada Centre for Inland Waters in Burlington, Ontario, to the laboratory in Yellowknife.

Sediment sampling was carried out by two people in the first week of September 1986. The equipment used for the sampling is listed in Table 4. The sediment sampling was carried out within one day. However, one-half of a day was spent preparing the equipment prior to the field trip. The subsampling of

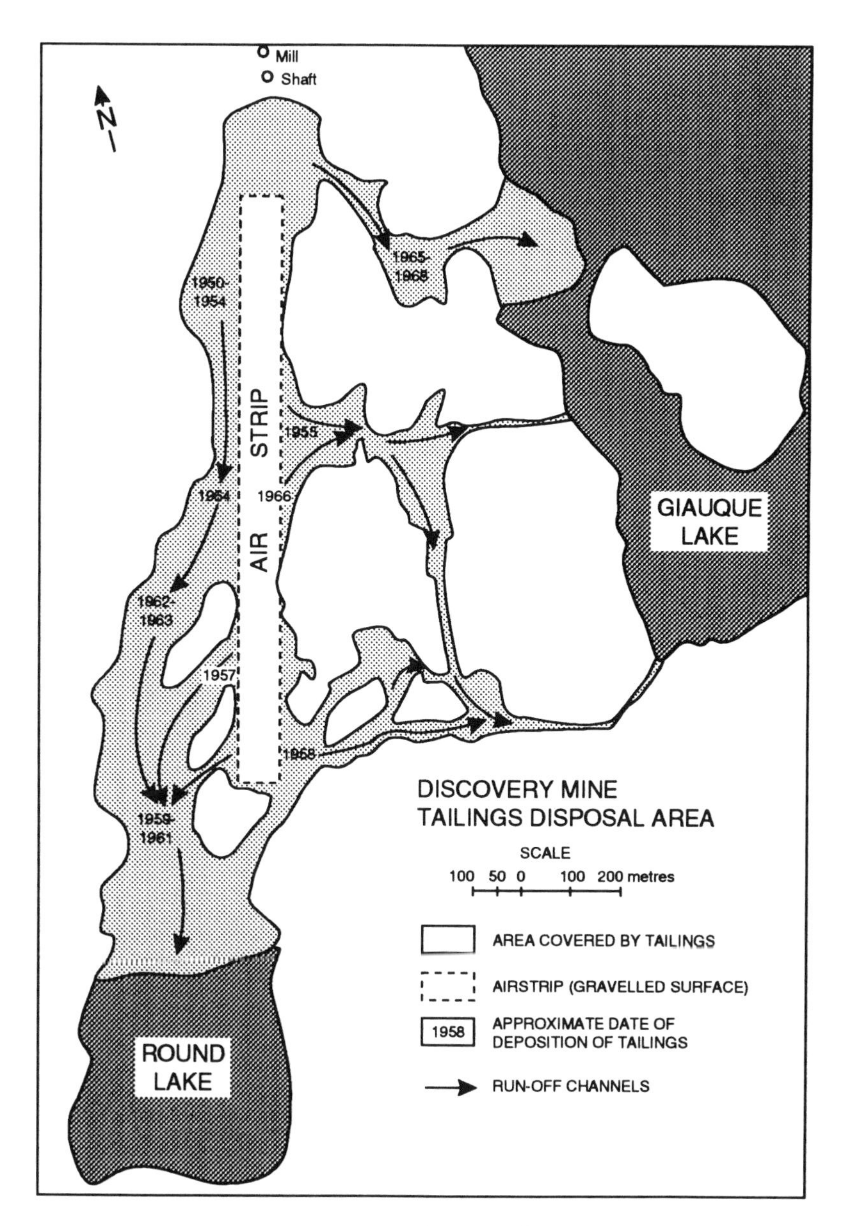

FIGURE 2. (continued).

collected sediments was carried out at the sampling site. A 3-m inflatable Zodiac boat with a 20-hp engine was used for the collection of the sediment samples.

All sampling equipment were loaded into one vehicle on the day prior to departure. In the morning of the departure day, the equipment was unloaded from the vehicle into a Turbo-Beaver float plane. Because of the low ambient temperature, about freezing point, during the sampling day it was not necessary to carry ice into the field to preserve the samples during the transportation from the sampling site to the laboratory in Yellowknife.

IV. DETAILS OF THE SAMPLING PROCEDURES

The sediment sampling party, consisting of two people, was dropped off by plane, along with all necessary equipment, on the shore of Thistlethwaite Lake.

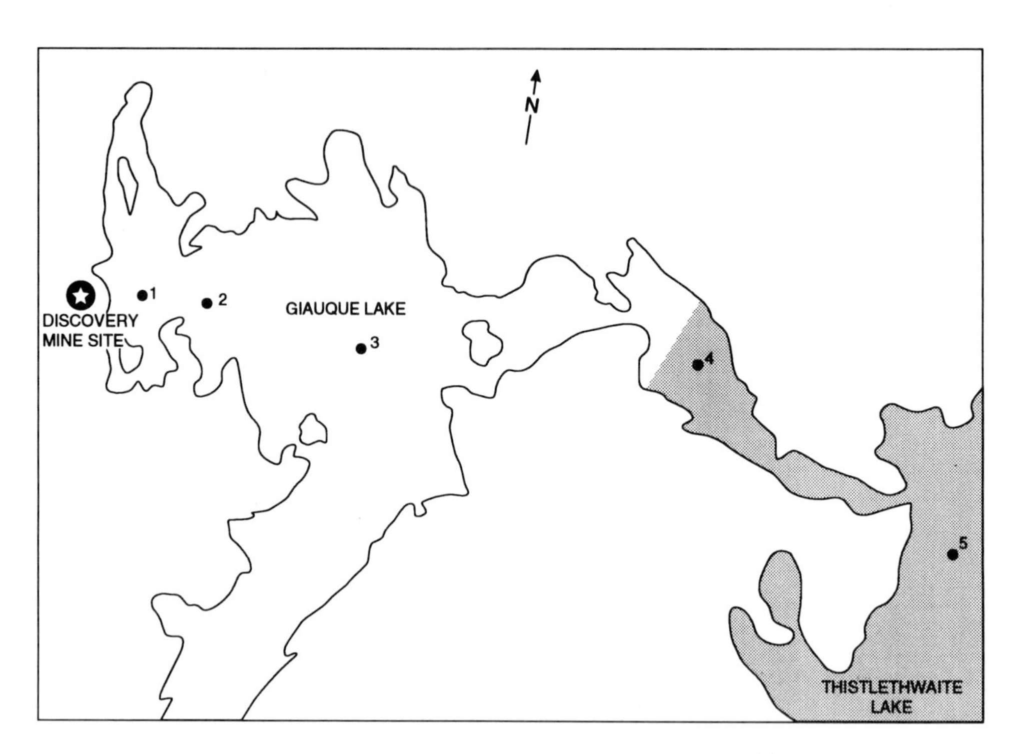

FIGURE 3. Sediment sampling station in Giauque and Thistlethwaite Lakes.

The party commenced the sediment sampling by inflating the boat and assembling a devit and winch, while the plane, with two other members of the sampling party, proceeded to the mining site. Upon finishing the sampling at the sampling station in Thistlethwaite Lake (Figure 3), all equipment had to be disassembled and carried over an approximately 500-m long portage to the western end of Giauque Lake. At this point, all equipment were reassembled and sampling of Giauque Lake was carried out.

A modified Kajak-Brinkhurst corer equipped with a 6.6-cm inside diameter plastic liner was used to obtain sediment cores. The coring device was lowered from the boat using a specially designed devit and winch. The devit was placed on the boat and attached into the slots intended for oars. A small, hand-operated winch was attached to the devit and was used for lowering and raising the corer.

The bottom and top of retrieved sediment cores were capped with plastic caps immediately upon retrieval. Capped cores were labeled and tied to the side of the boat to prevent any disturbance of the sediment in the core tubes by the boat movement. Sediment cores collected at sampling stations 4 and 5 were transported to the shore next to the mining site for further processing prior to the sampling at stations 1 to 3 (Figure 3). On the shore, the cores were extruded and subsampled into 1-cm subsections using a piston extruder, plastic spatula, and a cutting plate. Each subsection was stored in zip-lock bags, pre-labeled with name of the study site, sampling station number, core number, and sediment depth, at an ambient temperature of about 3°C.

The equipment used for the collection of the sediment cores is listed in Table 4. The cost of borrowing a part of the equipment and for the travel for a party of four (i.e., two people carried out the sediment sampling while the other two carried out the land portion sampling of the study), expressed as the percentage of the total estimated cost for the fieldwork is shown in Table 5.

Table 4
Sampling Equipment Used in the Fieldwork

Item	Use
3-m Inflatable Zodiac boat with a 20-hp motor, life jackets, and paddles	Sampling platform
Frame and hand-operated winch	Lowering and raising of coring equipment
Modified Kajak-Brinkhurst corer with spare valve and parts	Sediment coring
15 Pieces of 1-m long plastic liner, 6.6-cm inside diameter	Sediment coring
Plastic caps	Capping collected sediment cores
20-l Plastic garbage can	Washing and transporting smaller equipment
Ruler	Measuring collected cores
Manual extruder	Extruding cores
1-m Long plastic tubing, 1-cm diameter	Draining water from the top of sediment cores
Plastic coolers	Transportation of sediment samples
Plastic bags (zip-lock type), 25 × 25 cm	Storage of subsamples
Plastic spatula and cutting plate	Subsampling of collected cores
Waterproof magic markers, pens and pencils, notebooks	Labeling sample containers, recording field observations
Toolbox	Assembly of coring device, and if necessary, repairs
Maps, charts	Orientation in field
Radio (party-to-party and long range)	Communication with rest of field party and if necessary communication for pick-up by plane
Second coring device	Spare equipment
Mini-Ponar	Spare equipment
Additional clothing	Personal comfort
First-aid kit and immediate survival kit	Personal safety

V. SUGGESTIONS

- In a sampling program at such a remote area as Discovery Mine in the NWT, all members of the sampling party have to be completely safety-conscious at all times. The weather, particularly sudden changes in temperature and wind speed and direction, which can occur in the month of September in the

TABLE 5
Cost Associated with Sediment Sampling at Giauque Lake

Expenses	% of Total Cost
Borrowing of equipment (shipping)	10
Transportation (to and return from site)	70
Shipping of samples for analysis	10
Meals	5
Chemicals	5

Note: The salaries of all four members of the sampling party were covered by Environment Canada. Therefore these are not included in the cost associated with sediment sampling in Giauque Lake.

NWT can play a very important role in the sampling program. Therefore, an alternative plan should be prepared and used upon changes in the weather conditions.

- Due to the very small sampling platform in the Zodiac inflatable boat, the team carrying out the sediment sampling needs to carefully plan and organize the sampling procedures prior to embarking on the field trip. It has to be remembered that time is an important factor in the execution of the sampling plan.
- The sampling party using the equipment listed in Table 4 should be of average to very good physical condition, particularly for the portaging of all equipment from one lake to the other.
- The transportation of sediment cores in an inflatable boat should be carried out with great care, moving the boat at a slow speed to prevent the disturbance of sediment in the core tubes.
- Due to the remoteness of the study area, where no services or more serious repairs to equipment can be undertaken, all equipment should be double-checked before leaving for the sampling trip. When the cost and weight restrictions for the plane transportation allow, spare and duplicate equipment should be carried along to the sampling site.
- Awareness of the presence of wildlife, particularly bears, is necessary and proper precautions should be taken.

CASE STUDY 3: SEDIMENT SAMPLING IN JACK OF CLUBS LAKE IN BRITISH COLUMBIA, CANADA, AFFECTED BY ABANDONED GOLD MINE TAILINGS

I. INTRODUCTION

Gold mining has been an important industry in Canada for more than a century. Since 1921 Canada has occupied second or third place among the world's gold producing nations, after South Africa and the former U.S.S.R. The wastes generated by gold mining contain elements from other minerals associated with the gold and its host bedrock, such as As, Co, Pb, and Zn, and elements and compounds introduced during the extraction of the gold from the ore, such as cyanides, Hg, etc. Recently, the environmental impact of past gold-mining activities have received considerable attention, particularly the effects of abandoned mine tailings and waste rock on aquatic ecosystems.

The Fraser River Basin in British Columbia, Canada, has been an important gold-mining area since the 1800s. Recently, the Fraser River Action Plan was initiated by Canada's Green Plan to identify and control the contaminants entering the Fraser River from industrial and domestic point and non-point sources. One of the objectives of the Fraser River Action Plan is to virtually eliminate the release of persistent toxic substances, as determined under the priority substances list of the Canadian Environmental Protection Act, entering the waters of the river basin by the year 2000.

Contamination of groundwater and transport of airborne contaminated particles from tailings disposed during past gold-mining activities at Wells in the Fraser River Basin have recently been studied.[16-18] Andrews[16] reported concentrations of Hg in lake trout, for human consumption, from Jack of Clubs Lake, at Wells, above the federal guideline (0.5 μg/g Hg). However, limited information is available on the effects of the abandoned gold mine tailings on the aquatic ecosystem of Jack of Clubs Lake and Willow River, a tributary of the Fraser River. Therefore, a study was initiated in 1992 to investigate the effects of the abandoned gold mine tailings on the environment. The study was part of a multidisciplinary investigation of the effects of mine wastes on aquatic ecosystems in Canada carried out by the National Water Research Institute, Burlington, Ontario, Canada.

II. OBJECTIVES

The main objectives of the study were

- To conduct a reconnaissance study to determine the distribution of major and trace elements in different environmental compartments, particularly suspended and bottom sediments and sediment pore water
- To evaluate the effects of the major and trace elements in sediments on benthic community structure in Jack of Clubs Lake
- To evaluate the abandoned mine tailings as source of contaminants to Jack of Clubs Lake
- To examine the transport of major and trace elements from the abandoned tailings into the Fraser River system

The final goal of the investigation was to evaluate the impact of the past gold-mining activities on Jack of Clubs Lake environment and recommend a remedial action to prevent further contamination of the ecosystem at Wells with respect to the objectives of the Fraser River Action Plan.

The sediment sampling of Jack of Clubs Lake was part of the reconnaissance study carried out at Wells in August 1992.[19] Vegetation, mine tailings, water, and groundwater were collected in addition to the sediments in the reconnaissance study. However, this report deals only with methods, equipment, and cost relevant to the sampling of suspended and bottom sediments and sediment pore water.

III. MATERIALS AND METHODS

A. STUDY AREA

Jack of Clubs Lake is located in the Cariboo Range (Figure 4), an important mining area in British Columbia since the 1800s. The town of Wells (present population about 200), at the northeastern shore of Jack of Clubs Lake, was founded in 1932 and built on the tailings from the old Lowhee Mine. During its 33 years of operation at Wells, the Cariboo Gold Quartz Mining Company produced in excess of five million dollars' worth of gold. The tailings from the milling and gold extraction at the Cariboo Gold Quartz Mine were discharged into the northeastern end of Jack of Clubs Lake, changing the original morphometry of the lake. At present, tailings deposit about 4.5-m thick cover approximately 25 hectares of land adjacent to the lake. The tailings are dissected by Highway 26, Lowhee Creek, and Willow River (Figure 4).

Lowhee Creek flows through the abandoned tailings of the Cariboo Gold Quartz Mine before it empties into the northeastern end of Jack of Clubs Lake. During spring runoff the waters of the creek flood an extensive area of the tailings. Most of the sediments that have accumulated near the mouth of the creek are remnants of extensive hydraulic mining activities that occurred during the gold rush.[16] Jack of Clubs Lake is 2.4 km long and 0.5 km wide, with a mean depth of 19 m and a maximum depth of 63 m. Its flushing rate is extremely rapid, averaging 0.08 years.[16] A man-made channel dug through the tailings at the northeastern end of the lake drains into the Willow River, the only outlet of the lake, which flows for 130 km before discharging into the Fraser River. Part of the water flowing through the channel originates as groundwater seepage from the tailings and does not reach the lake.

Bowron Lake, located about 30 km east of Wells, was selected as a control or reference lake for the study. The surface area of the lake is over 121,600 hectares. The lake is inside the Bowron Lake Provincial Park, and there are no reports of past or present mining activities along the lake. The Bowron River drains the northern side of the lake and enters the Fraser River near Prince George. The Cariboo River drains the eastern and southern sides of the lake and flows south to join the Quesnel River, one of the tributaries of the Fraser River.

B. SELECTION OF SAMPLING STATIONS

There were no data available on the geochemical composition and distribution of sediments in Jack of Clubs Lake. Therefore, the bottom sediments sampling stations were selected to obtain general information on the effects of gold mine tailings deposited on the northeastern shore of the lake. Sampling stations were selected randomly along a transect from the southwest to the northeast of the lake (Figure 4). Two additional sampling stations were added at each side of the lake with the assumption that the bottom sediments near the tailings disposal site will be more heterogeneous and contaminated than those at the opposite side of the lake. Because the identical sampling stations were further used for sampling water, small cork buoys, painted with fluorescent paint and labelled with the number of the sampling station, were anchored at each station. Two bottom sediment sampling stations were selected at the deepest area of Bowron Lake, which was considered a control site.

To determine the possible transport of particle-associated contaminants from Jack of Clubs Lake by Willow River into the Fraser River system, suspended sediments were collected at three sampling stations (Figure 4). The

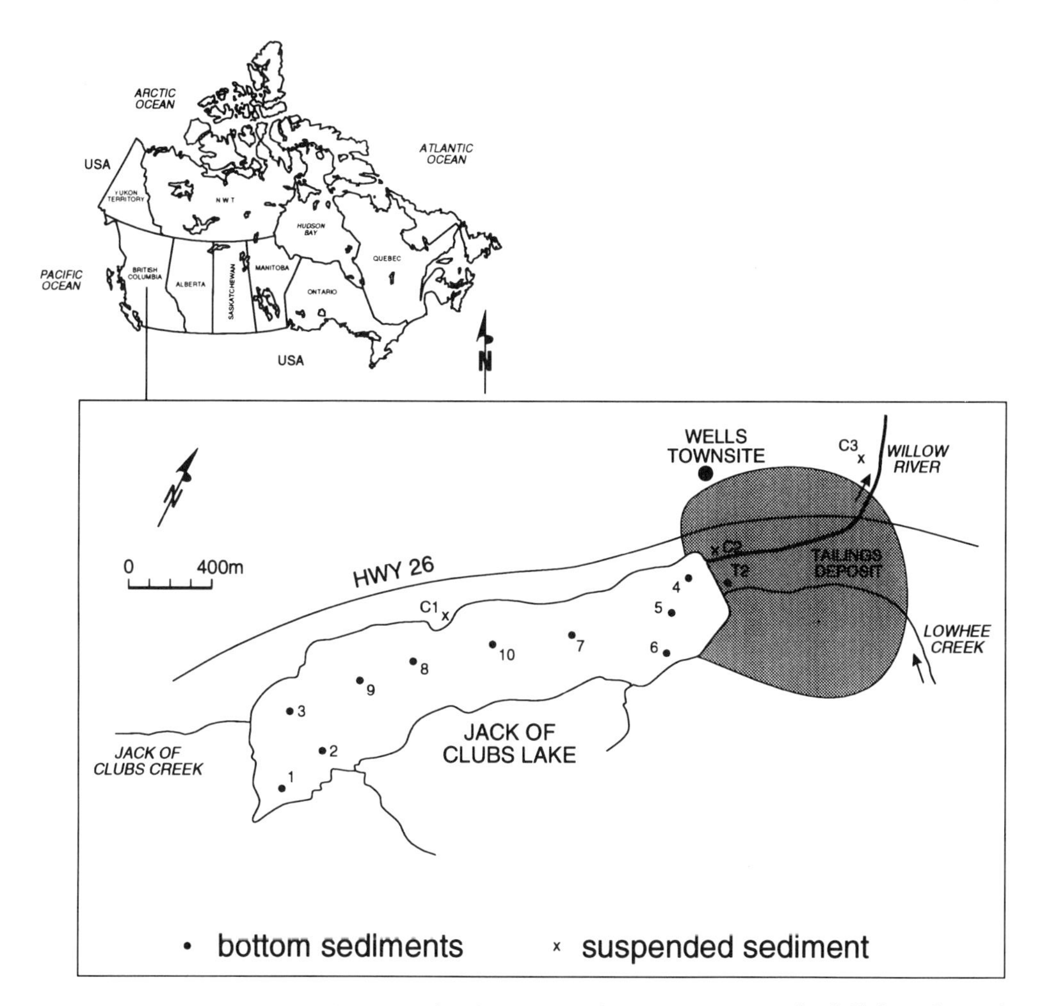

FIGURE 4. Study site and location of sediment sampling stations in Jack of Clubs Lake and Willow River.

location of the stations was selected after collection and visual inspection of the bottom sediments in Jack of Clubs Lake. The first suspended sediment sampling station was in the center of Jack of Clubs Lake where the bottom sediments appeared relatively uncontaminated by the tailings (C1 in Figure 4). The second sampling station was in the middle of the outlet of Willow River from Jack of Clubs Lake (C2 in Figure 4), and the third was in Willow River about 1 km downstream of its outlet from the lake (C3 in Figure 4).

C. LOGISTICS

A detailed work plan was prepared about three months before the fieldwork. The work plan was prepared after meeting and consultation with all scientists involved in the investigation to accommodate their needs for samples, including listing of all sampling equipment, sample containers, sample storage conditions, sample processing in the field, and transport of samples for the laboratory studies at the National Water Research Institute in Burlington, Ontario.

Sediment sampling was carried out during the first week of August 1992 by a field party consisting of two research scientists and one technical assistant. The different samples collected and the equipment used for sampling are listed in Table 6. Table 7 shows the timetable of the field activities. A 3-m aluminum boat with a 9.5-hp engine was used for the collection of the bottom sediment samples in the lake. The boat was lent by the local office of the Ministry of the Environment at Williams Lake, British Columbia, about 100 km southwest of Wells, and was delivered by the personnel from the Williams Lake office to Wells before the arrival of the field party.

All sampling equipment were shipped about 14 days ahead of the sampling trip by air freight to Prince George (population 67,600), British Columbia, about 250 km north of Wells. Arrangements were made to deliver and store all shipped equipment at the warehouse of the British Columbia Ministry of the Environment in Prince George. A car was rented at the airport in Prince George upon arrival of the field party from Toronto, Ontario, to Prince George, British Columbia (about a 3,000-km distance). Further, it was arranged to rent a truck in Prince George on the way from the airport to pick up the equipment stored at the warehouse. The equipment was loaded onto the truck, and the field party with both vehicles continued on the trip to Wells. Accommodation was booked about three months ahead in a motel in Wells near the shore of Jack of Clubs Lake. The accommodation units contained a large kitchen and dinning room with fridge, freezers, stoves, and large tables that were used for some processing, packing, and storage of collected samples. Coolers with ice cubes purchased at the local grocery store or gas station were used for temporary storage of the samples during the sampling on the lake. Upon returning to the motel each evening, the samples were sorted, logged, processed, and stored in the fridges or freezers according to the work plan prepared before the field trip. Arrangements were made for the shipment of all samples to the National Water Research Institute in Burlington, Ontario (about 60 km west of Toronto), one day before departure. Samples requiring storage at 4°C were shipped by refrigerated truck, while larger frozen samples were packed into coolers with dry ice and shipped by air cargo. The small frozen samples were packed into a cooler and transported as extra luggage by the scientists and technician on the return flight from Prince George to Toronto.

IV. DETAILS OF THE SAMPLING PROCEDURES

A. BOTTOM SEDIMENTS

Sediment samples were collected for determination of particle size distribution, total concentrations of major and trace elements, and determination of chemical forms of trace elements and microbial and enzymatic activities.

1. Surficial Sediments (0 to 5 cm)

Sediment samples were collected at ten stations in Jack of Clubs Lake and two stations in Bowron Lake using a mini-Ponar grab sampler. This sampler encompasses an area of about 230 cm^2 and penetrates to an average depth of 5 cm. Upon retrieval of the sampler, one-half of the sediment in the mini-Ponar grab was immediately subsampled into a 1-l glass jar (prewashed with 5% HNO_3 and rinsed with deionized distilled water) using a porcelain spoon. The subsampling was carried out with care and containers filled to the top to avoid unnecessary contact between sediment and air to prevent oxidation of the sediment. The other half of the sediment in the mini-Ponar was placed on a glass tray (40 × 30 × 7 cm) by opening the sampler's jaws over the glass tray. The sediment on the tray was homogenized by mixing with a porcelain spoon and subsampled into prepared containers: plastic bags (for determination of

TABLE 6
Sampling Equipment Used in the Fieldwork

Item	Use
3-m Aluminum boat with 9.5-hp engine, life jackets, a pair of oars, and an anchor	Sampling on the lake
Mini-Ponar grab sampler with spare parts and 70-m long nylon line	Surface sediment sampling
Modified Kajak-Brinkhurst corer with spare valves and parts	Sediment coring
Ten pieces of 1-m long plastic liner, 6.6-cm inside diameter	Sediment coring
Plastic red and yellow core caps	Capping sediment cores
150-l Plastic garbage can	Storing sediment cores on the boat
Porcelain spoon and a glass tray (40 × 30 × 7 cm)	Homogenization of sediments
Ruler or meter stick	Sediment cores logging and subsampling
Piston extruder	Extruding sediment cores
1-m Long plastic tubing, 1.5-cm diameter	Draining water from the top of the sediment cores
Cork buoys painted with fluorescent paint	Marking the sampling stations in the lake
200-m Nylon/cotton line	Setting the buoys in the lake
Kalil and Goldhaber squeeze	Sediment pore water sampling
Nitrogen gas cylinder with regulator	Maintaining oxygen-free atmosphere during the pore water sampling
Plastic glove box	Collection of sediment pore water
Westfalia continuous flow centrifuge with spare parts	Suspended sediments sampling
4-l Plastic buckets with lids, lined with plastic bags	Collection of surface sediments for bioassays
250-Mesh brass sieve	Sediment sieving for determination of benthic community
4% Formaldehyde solution	Preservation of benthic organisms samples
Plastic coolers	Temporary storage of samples requiring refrigeration, and shipping frozen samples with dry ice
Plastic containers with lids (60 × 30 × 50 cm)	Temporary storage of collected samples

TABLE 6 (continued)

Item	Use
Plastic bags (12 × 20 cm)	Sediment samples for determination of particle size distribution
Plastic vials with caps (about 150 ml)	Surface sediments for determination of major and trace elements and subsampled sediment cores sections
Glass jars (250 ml) with Al-foil cover and caps pre-washed with pesticide-grade hexane and dried	Surface sediments for determination of PAH and PCB
1-l Glass jars (pre-washed with 5% HNO_3 and rinsed with DDW)	Surface sediments for speciation of trace elements determination of microbial and enzyme activity
Plastic whirl-pak bags (13 × 23 cm)	Sediments for determination of benthic community structure
Plastic vials (100 ml)	Collecting benthic organisms after sieving
1-l Plastic tubs with lids	Suspended sediments from the centrifuge
Glass vials (25 ml) with lids(pre-washed with 5% HNO_3 and rinsed with DDW and acidified by two drops of ultra-pure concentrated HNO_3)	Pore water samples
0.45-μm Millipore filters	Collection of pore water
Plastic spatula and cutting plate	Subsampling sediment cores
Waterproof magic markers, pens and pencils, notebooks	Labeling sample containers, recording field observations, and sample logging
Toolbox	Adjustment and repairs of sampling equipment
Leather gloves	Pulling nylon rope during retrieval of sediment samplers
Large aluminum boxes with locks (up to 1 × 1 × 1m)	Shipment of equipment
Topographic maps of the area, relevant reports	Orientation in the field, access roads, distances, detailed information on the study site

particle size distribution), plastic vials (for determination of total concentrations of major and trace elements), and glass jars washed with pesticide-grade hexane, dried, and covered with hexane-washed Al-foil (for determination of polycyclic aromatic hydrocarbons - PAH, and polychlorinated biphenyls - PCB). These jars were only half-filled with the sediment to avoid breaking the glass containers upon expansion of the sediment during freezing.

All containers with the samples were labeled with a waterproof magic marker (station number, name of the lake, sampling date). Samples in plastic bags and vials were stored in a large plastic container on the boat. Samples in the glass jars

TABLE 7
Timetable of Field Activities in Sediment Sampling[1]

	Morning	Afternoon
Monday	(2) Unloading sampling equipment (1) Set-up field laboratory in the motel	(2) Establish sampling stations in the lake (1) Preparation for lab work
Tuesday	(3) Stations #1,2 [SB,SS]	(2) Station #3 [SB,SS,SC,PW] (1) Sieving for benthic organisms
Wednesday	(2) Stations #4,5 [SB,SS] (1) SU at station C1	(2) Station #6 [SB,SS,SC,PW] (1) Sieving for benthic organisms
Thursday	(2) Stations #7,8 [SB,SS] (1) SU at station C2	(2) Station #9 [SB,SS,SC,PW] (1) Sieving for benthic organisms
Friday	(2) Station #10 [SB,SS,PW] (1) SS at station C3	(3) Laboratory work and sieving for benthic organisms
Saturday	(3) Control site [SS,SC,PW]	(1) Start packing (2) Laboratory work
Sunday	(3) Packing equipment and samples; departure	

[1] See Figure 4 for location of sediment sampling stations.
Note: Numbers in brackets represent number of people involved in the activity.

Abbreviations: SC = sediment cores; SS = sediment grab; SB = sediment for bioassays; SU = suspended sediment; PW= pore water

were stored in a plastic cooler containing ice. Upon return to the motel, the samples from the cooler, for the determination of chemical species of trace elements and microbial and enzymatic activity, were transferred into a fridge. The samples in the glass jars, for the determination of PAH and PCB, were transferred into a freezer. Samples in the plastic containers which did not require refrigeration or freezing were stored at room temperature in large plastic containers.

2. Sediment Cores

A modified Kajak-Brinkhurst corer equipped with a plastic liner of 6.6-cm inside diameter was used to obtain sediment cores for determination of concentration profiles of major and trace elements in the sediment. Sediment cores were collected at selected stations in Jack of Clubs and Bowron Lakes. The bottom and the top of retrieved sediment cores were capped with plastic caps (yellow and red caps were used for capping the top and the bottom of the core, respectively). Capped cores were stored vertically in a plastic garbage can on the boat to prevent disturbance of the sediment in the core tubes by the movement of the boat. The cores were transported to the shore and divided vertically into 1-cm subsections using a piston extruder, plastic spatula, and a cutting plate. The subsections were stored in labelled plastic vials (name of the lake, sampling station, sediment depth, time of collection) at room temperature. Selected cores were subsampled into 1-cm sections for determination of chemical forms and availability of trace elements. These subsamples were stored in plastic vials in the dark at 4°C.

3. Sediment Samples for Determination of Benthic Invertebrates

Sediments for determination of benthic invertebrate community structure were collected by a method similar to that for collection of sediment cores. Only the top 10 cm of sediments from each core was used for the determination of benthic organisms. Thus, the surface area of the sample was 34.2 cm^2, and the volume about 340 cm^3. Benthic community structure was examined at all ten stations in five replicates. Each replicate was extruded into a plastic whirl-pak bag. Sieving of the sample was conducted in the field using a 250-mesh sieve. A garden hose lent by the owner of the motel in Wells was used for washing the sediments through the sieve. The residue on the sieve was transferred into a labelled plastic vial (name of the lake, sampling station, date of sampling) and preserved in 4% formaldehyde for sorting and identification of the benthic organisms in the laboratory. The samples were stored in the refrigerator.

4. Sediment Samples for Toxicity Testing

Sediment toxicity to benthic organisms was determined at the ten sampling stations in Jack of Clubs Lake (Figure 4). Sediments for toxicity testing were collected by a mini-Ponar grab sampler. Five grab samples were collected at each station. The samples were transferred from the sampler into 4-l plastic buckets with lids, each lined with plastic. Filled buckets were stored on ice and shipped by refrigerated truck to the laboratory at the National Water Research Institute in Burlington for testing. The storage period before testing ranged from 100 to 180 days. Earlier experiments[20] have shown that storage up to 168 days does not effect the test results.

B. SUSPENDED SEDIMENTS

Suspended sediments were collected at three locations in the study area: at the center of Jack of Clubs Lake, 3 m below the water surface; from the outflow of the lake (the Willow River); and from the Willow River at the end of its passage through the town of Wells (Figure 4). To collect the suspended sediments, about 2,000 l of water were pumped into a Westfalia continuous flow centrifuge at a flow rate of 4 l/min. The collected material was washed from the centrifuge bowl with a minimum amount of water from Jack of Clubs Lake into 1-l plastic tubs with lids. The samples were stored in the freezer in the motel room and transported frozen to the laboratory at the National Water Research Institute in Burlington for freeze-drying and analyses.

C. SEDIMENT PORE WATER

Sediment pore water was recovered by squeezing sediments collected by the modified Kajak-Brinkhurst corer in both Jack of Clubs and Bowron Lakes. Vertical sections of the sediment (0 to 5, and 5 to 10 cm) were extruded from the core liner under nitrogen atmosphere maintained within a plastic glove box. Sediment sections were squeezed separately to extract the pore water. The time between the sediment collection and termination of squeezing was less than 5 h. A squeezer assembly, designed by Kalil and Goldhaber[21] was used for

TABLE 8
Cost Associated with Sediment Sampling at Wells, British Columbia

Expenses	% of Total Cost
Shipping of equipment	25
Return flight Toronto - Prince George	24
Rent of car and truck[1]	20
Accommodation and meals	17
Shipping of samples	5.6
Sample containers	4
Temporary storage of samples[2]	4
Chemicals	0.4

[1] Including the cost for gasoline.
[2] Including the cost of ice for coolers and dry ice for shipment of samples.

Note: All sampling equipment and large shipping containers were the property of the National Water Research Institute in Burlington. The salaries of two scientists and one technician involved in the sediment sampling were covered by Canada's Department of the Environment.

the squeezing. In the squeezing process, the pore water was filtered through a 0.45-μm Millipore filter. Pore water samples from individual sediment sections were collected into glass vials pre-washed with 5% HNO_3 and rinsed with deionized distilled water (DDW) and pre-acidified with two drops of ultra-pure concentrated HNO_3. The pore water samples were stored in the refrigerator at 4°C and transported by refrigerated truck to the laboratory at the National Water Research Institute for analyses.

Materials used for bottom and suspended sediments and pore water sampling and sample preparation in the field are listed in Table 6. A timetable of the sediment sampling program is outlined in Table 7. The cost for travel and accommodation for three participants for sediment sampling, shipping of the equipment, etc. expressed as percentage of the total estimated cost for the sediment sampling at Wells, B.C., is shown in Table 8.

V. SUGGESTIONS

The quantity and quality of the bottom and suspended sediments and pore water samples were sufficient to obtain requested information from the reconnaissance study at Wells, B.C. However, the following problem was encountered during the bottom sediment sampling in Jack of Clubs Lake. The nylon/cotton line attached to the buoys marking the sampling stations in the lake shrunk, pulling the painted corks used as the buoys below the water surface. Despite the clear water, it took some time to find the buoys below the water surface for accurate location of individual sampling stations. Therefore, it is necessary to use a non-shrinkable line with the buoys.

REFERENCES

1. **Pfeiffer, W.C. and Lacerda, L.D.,** Mercury inputs to the Amazon basin, *Environ. Technol. Lett.*, 9, 325-330, 1988.
2. **Hasse, R.F.**, A comercialização do mercúrio no Brasil. In *Consequências da Garimpagem no Ambito Social e Ambiental da Amazônia,* Mathis, A. and Rehaag, R., Eds., Proceedings of the Symposium, Belém do Pará, Brazil, 1992.
3. **Malm,O.**, *Contaminação ambiental e humana por mercúrio na região garimpeira de ouro do Rio Madeira.* Ph.D. thesis, Federal University of Rio de Janeiro, 1991.
4. **Malm, O., Pfeiffer, W.C., Souza, C.M.M., and Reuther, R.**, Mercury pollution due to gold mining in the Madeira river basin, *Ambio,* 19,11, 1990.
5. **Malm, O., Branches, F., Akagi, H., Castro, M., Pfeiffer, W.C., Harada, M., Bastos, W.R., and Kato, H.**, Mercury and methylmercury in fish and human hair from the Tapajós river basin, Brasil. Submitted to Sci. Tot. Envir.
6. **Sioli, H.**, *Amazônia: Fundamentos da Ecologia da Maior Região de Florestas Tropicais,* Editora Vozes, Petrópolis, 1985.
7. **Guimarães, J.R.D., Malm, O., and Pfeiffer, W.C.,** A simplified radiochemical technique for measurements of net mercury methylation rates in aquatic systems near goldmining areas, Amazon, Brazil. Submitted to Sci. Tot. Envir.
8. **Lord, C.S.,** Mineral Industry of Mackenzie Northwest Territories, Vol. 5, 64, Geological Survey Canada, 1951.
9. **Environmental Protection Service**, Metal Mining Liquid Effluent Regulations and Guidelines, Environmental Protection Service Regulations, Codes and Protocols Report, EPS-1-WP-77-1, Ottawa, Ontario, 1977.
10. **Durham, R.W. and Oliver, B.G.**, History of Lake Ontario contamination by sediment radiodating and chlorinated hydrocarbon analysis, *J. Great Lakes Res.*, 9, 160, 1983.
11. **Mudroch, A., Joshi, S.R., Sutherland, D., Mudroch, P., and Dickson, K.M.**, Geochemistry of sediments in the Back Bay and Yellowknife Bay of the Great Slave Lake, *Environ. Geol. Water Sci.*, 14, 35, 1989.
12. **Mayers, R.A., McIntosh, A.W., and Anderson, V.L.**, Uptake of cadmium and lead by a rooted aquatic macrophytes (*Elodea Canadensis*), *Ecology*, 58, 1176, 1977.
13. **Lee, C.R., Smart, R.M., Sturgis, T.C., Gordon, R.N., and Landin, M.C.**, Prediction of heavy metal uptake by marsh plants based on chemical extraction of heavy metals from dredged material, Technical Report D-78-6, U.S. Army Engineer Waterways Experiment Station, Vicksburg, Mississippi, 1978.
14. **Mudroch, P.**, Survey of contaminants at abandoned mines in particular Discovery mine, Thompson-Lundmark mine, Camlaren mine, Rayrock mine, Northwest Territories, Unpublished Report, Environmental Protection, Conservation and Protection, Environment Canada, Ocean Dumping Program, Ottawa, Ontario, 1988.
15. **Boojum Research Ltd.**, A synthesis and analysis of information on inactive hardrock mine sites in Yukon and Northwest Territories, Volume II: Northwest Territories, Report prepared for the Indian and Northern Affairs, Canada, Northern Environment Directorate, 1986.
16. **Andrews, K.I.**, *A Preliminary Investigation of Public Health and Environmental Impacts of Abandoned Mine Tailings at Wells, B.C.*, Williams Lake, Ministry of Environment, Province of British Columbia, 1989.
17. **Galbraith D.M.**, *Reclamation Assessment - Wells Tailings*, MEMPR, Resources Management Branch, 1991.
18. **Rescan**, *Tailing Site Investigation and Remediation Planning at Wells, British Columbia,* Rescan Environmental Services, Ltd., Vancouver, B.C., 1990.
19. **Mudroch A., Hall G.E.M., Azcue J., Jackson T.A., Reynoldson T., and Rosa F.**, Preliminary report on the effects of abandoned mine tailings at Wells, B.C., on the aquatic ecosystem of Jack of Clubs Lake, Department of Environment Fraser River Action Plan 1993-07, 1993.
20. **Reynoldson T.B., Thompson S.P., and Bamsey J.L.**, A sediment bioassay using the turbificid oligochaete worm *Tubifex tubifex*, *Environ. Toxicol. Chem.* 10, 1061, 1991.
21. **Kalil E.K. and Goldhaber, M.**, A sediment squeezer for removal of pore water without air contact, *J. Sediment Petrol.*, 43, 553, 1973.

INDEX

D

E

F

M

N

O

P

Q

R

S

T

U

V

W

Z